Paul A. M. Dirac

The Principles of Quantum Mechanics

Bow Wow Press
Los Angeles

Copyright:
Emended Fourth Revised Edition © 2025 Bow Wow Press
The publisher grants noncommercial use without permission. Changes of
the book contents and redistribution of the PDF file are prohibited.

Identifiers:
ISBN: 978-1-950217-13-7 (Paperback)
ISBN: 978-1-950217-14-4 (Hardcover)
Library of Congress Control Number: 2025940378

About the Publisher:
Bow Wow Press is a nonprofit scientific publisher established in 2018.
The publisher will be managed completely by donations. We set book
prices as low as possible, offer free PDF files on the publisher's website,
grant noncommercial use without permission, ...

Preface to the Fourth Edition

The main change from the third edition is that the chapter on quantum electrodynamics has been rewritten. The quantum electrodynamics given in the third edition describes the motion of individual charged particles moving through the electromagnetic field, in close analogy with classical electrodynamics. It is a form of theory in which the number of charged particles is conserved and it cannot be generalized to allow of variation of the number of charged particles.

In present-day high-energy physics the creation and annihilation of charged particles is a frequent occurrence. A quantum electrodynamics which demands conservation of the number of charged particles is therefore out of touch with physical reality. So I have replaced it by a quantum electrodynamics which includes creation and annihilation of electron-positron pairs. This involves abandoning any close analogy with classical electron theory, but provides a closer description of nature. It seems that the classical concept of an electron is no longer a useful model in physics, except possibly for elementary theories that are restricted to low-energy phenomena.

P. A. M. D.

St. John's College, Cambridge
11 May 1957

Note to the Revision of the Fourth Edition

The opportunity has been taken of revising parts of Chapter XII ('Quantum electrodynamics') and of adding two new sections on interpretation and applications.

P. A. M. D.

St. John's College, Cambridge
26 May 1967

From the Preface to the First Edition

The methods of progress in theoretical physics have undergone a vast change during the present century. The classical tradition has been to consider the world to be an association of observable objects (particles, fluids, fields, etc.) moving about according to definite laws of force, so that one could form a mental picture in space and time of the whole scheme. This led to a physics whose aim was to make assumptions about the mechanism and forces connecting these observable objects, to account for their behaviour in the simplest possible way. It has become increasingly evident in recent times, however, that nature works on a different plan. Her fundamental laws do not govern the world as it appears in our mental picture in any very direct way, but instead they control a substratum of which we cannot form a mental picture without introducing irrelevancies. The formulation of these laws requires the use of the mathematics of transformations. The important things in the world appear as the invariants (or more generally the nearly invariants, or quantities with simple transformation properties) of these transformations. The things we are immediately aware of are the relations of these nearly invariants to a certain frame of reference, usually one chosen so as to introduce special simplifying features which are unimportant from the point of view of general theory.

The growth of the use of transformation theory, as applied first to relativity and later to the quantum theory, is the essence of the new method in theoretical physics. Further progress lies in the direction of making our equations invariant under wider and still wider transformations. This state of affairs is very satisfactory from a philosophical point of view, as implying an increasing recognition of the part played by the observer in himself introducing the regularities that appear in his observations, and a lack of arbitrariness in the ways of nature, but it makes things less easy for the learner of physics. The new theories, if one looks apart from their mathematical setting, are built up from physical concepts which cannot be explained in terms of things previously known to the student, which cannot even be explained adequately in words at all. Like the fundamental concepts (e.g. proximity, identity) which every one must learn on his arrival into the world, the newer concepts of physics can be mastered only by long familiarity with their properties and uses.

From the mathematical side the approach to the new theories presents no difficulties, as the mathematics required (at any rate that which is required for the development of physics up to the present) is not essentially different from what has been current for a considerable time. Mathematics is the tool specially suited for dealing with abstract concepts of any kind and there is no limit to its power in this field. For this reason a book on the new physics, if not purely descriptive of experimental work, must be essentially mathematical. All the same the mathematics is only a tool and one should learn to hold the physical ideas in one's mind without reference to the mathematical form. In this book I have tried to keep the physics to the forefront, by beginning with an entirely physical chapter and in the later work examining the physical meaning underlying the formalism wherever possible. The amount of theoretical ground one has to cover before being able to solve problems of real practical value is rather large, but this circumstance is an inevitable consequence of the fundamental part played by transformation theory and is likely to become more pronounced in the theoretical physics of the future.

With regard to the mathematical form in which the theory can be presented, an author must decide at the outset between two methods. There is the symbolic method, which deals directly in an abstract way with the quantities of fundamental importance (the invariants, etc., of the transformations) and there is the method of coordinates or representations, which deals with sets of numbers corresponding to these quantities. The second of these has usually been used for the presentation of quantum mechanics (in fact it has been used practically exclusively with the exception of Weyl's book *Gruppentheorie und Quantenmechanik*). It is known under one or other of the two names 'Wave Mechanics' and 'Matrix Mechanics' according to which physical things receive emphasis in the treatment, the states of a system or its dynamical variables. It has the advantage that the kind of mathematics required is more familiar to the average student, and also it is the historical method.

The symbolic method, however, seems to go more deeply into the nature of things. It enables one to express the physical laws in a neat and concise way, and will probably be increasingly used in the future as it becomes better understood and its own special mathematics gets developed. For this reason I have chosen the symbolic method, introducing the representatives later merely as an aid to practical calculation. This has necessitated a complete break from the historical line of development, but this break is an advantage through enabling the approach to the new ideas to be made as direct as possible.

P. A. M. D.

St. John's College, Cambridge
29 May 1930

Contents

Preface to the Fourth Edition	v
From the Preface to the First Edition	vii

Chapter I. The Principle of Superposition	1
1. The Need for a Quantum Theory	1
2. The Polarization of Photons	4
3. Interference of Photons	7
4. Superposition and Indeterminacy	9
5. Mathematical Formulation of the Principle	13
6. Bra and Ket Vectors	17
Chapter II. Dynamical Variables and Observables	23
7. Linear Operators	23
8. Conjugate Relations	26
9. Eigenvalues and Eigenvectors	29
10. Observables	35
11. Functions of Observables	41
12. The General Physical Interpretation	45
13. Commutability and Compatibility	49
Chapter III. Representations	53
14. Basic Vectors	53
15. The δ Function	58
16. Properties of the Basic Vectors	62
17. The Representation of Linear Operators	67
18. Probability Amplitudes	73
19. Theorems about Functions of Observables	77
20. Developments in Notation	79
Chapter IV. The Quantum Conditions	85
21. Poisson Brackets	85
22. Schrödinger's Representation	90

23.	The Momentum Representation	96
24.	Heisenberg's Principle of Uncertainty	99
25.	Displacement Operators	101
26.	Unitary Transformations	105

Chapter V. The Equations of Motion — 111
27.	Schrödinger's Form for the Equations of Motion	111
28.	Heisenberg's Form for the Equations of Motion	114
29.	Stationary States	119
30.	The Free Particle	121
31.	The Motion of Wave Packets	124
32.	The Action Principle	128
33.	The Gibbs Ensemble	134

Chapter VI. Elementary Applications — 139
34.	The Harmonic Oscillator	139
35.	Angular Momentum	143
36.	Properties of Angular Momentum	148
37.	The Spin of the Electron	153
38.	Motion in a Central Field of Force	156
39.	Energy-levels of the Hydrogen Atom	161
40.	Selection Rules	164
41.	The Zeeman Effect for the Hydrogen Atom	170

Chapter VII. Perturbation Theory — 173
42.	General Remarks	173
43.	The Change in the Energy-levels Caused by a Perturbation	174
44.	The Perturbation Considered as Causing Transitions	178
45.	Application to Radiation	182
46.	Transitions Caused by a Perturbation Independent of the Time	184
47.	The Anomalous Zeeman Effect	187

Chapter VIII. Collision Problems — 193
48.	General Remarks	193
49.	The Scattering Coefficient	196
50.	Solution with the Momentum Representation	201
51.	Dispersive Scattering	207
52.	Resonance Scattering	209
53.	Emission and Absorption	212

Chapter IX.	Systems Containing Several Similar Particles	217
54.	Symmetrical and Antisymmetrical States	217
55.	Permutations as Dynamical Variables	221
56.	Permutations as Constants of the Motion	223
57.	Determination of the Energy-levels	226
58.	Application to Electrons	229
Chapter X.	Theory of Radiation	235
59.	An Assembly of Bosons	235
60.	The Connexion between Bosons and Oscillators	237
61.	Emission and Absorption of Bosons	243
62.	Application to Photons	246
63.	The Interaction Energy between Photons and an Atom	250
64.	Emission, Absorption, and Scattering of Radiation	256
65.	An Assembly of Fermions	260
Chapter XI.	Relativistic Theory of the Electron	265
66.	Relativistic Treatment of a Particle	265
67.	The Wave Equation for the Electron	267
68.	Invariance under a Lorentz Transformation	270
69.	The Motion of a Free Electron	273
70.	Existence of the Spin	276
71.	Transition to Polar Variables	280
72.	The Fine-structure of the Energy-levels of Hydrogen	282
73.	Theory of the Positron	287
Chapter XII.	Quantum Electrodynamics	291
74.	The Electromagnetic Field in the Absence of Matter	291
75.	Relativistic Form of the Quantum Conditions	295
76.	The Dynamical Variables at One Time	299
77.	The Supplementary Conditions	304
78.	Electrons and Positrons by Themselves	308
79.	The Interaction	315
80.	The Physical Variables	320
81.	Interpretation	324
82.	Applications	328
Index		331

CHAPTER I

The Principle of Superposition

1. The Need for a Quantum Theory

Classical mechanics has been developed continuously from the time of Newton and applied to an ever-widening range of dynamical systems, including the electromagnetic field in interaction with matter. The underlying ideas and the laws governing their application form a simple and elegant scheme, which one would be inclined to think could not be seriously modified without having all its attractive features spoilt. Nevertheless it has been found possible to set up a new scheme, called quantum mechanics, which is more suitable for the description of phenomena on the atomic scale and which is in some respects more elegant and satisfying than the classical scheme. This possibility is due to the changes which the new scheme involves being of a very profound character and not clashing with the features of the classical theory that make it so attractive, as a result of which all these features can be incorporated in the new scheme.

The necessity for a departure from classical mechanics is clearly shown by experimental results. In the first place the forces known in classical electrodynamics are inadequate for the explanation of the remarkable stability of atoms and molecules, which is necessary in order that materials may have any definite physical and chemical properties at all. The introduction of new hypothetical forces will not save the situation, since there exist general principles of classical mechanics, holding for all kinds of forces, leading to results in direct disagreement with observation. For example, if an atomic system has its equilibrium disturbed in any way and is then left alone, it will be set in oscillation and the oscillations will get impressed on the surrounding electromagnetic field, so that their frequencies may be observed with a spectroscope. Now whatever the laws of force governing the equilibrium, one would expect to be able to include the various frequencies in a scheme comprising certain fundamental frequencies and their harmonics. This is not observed to be the case. Instead, there is observed a new and unexpected connexion between the frequencies, called Ritz's Combination Law of Spectroscopy, according to which all the frequencies can be expressed as differences between certain terms, the number of terms being much less than the number of frequencies. This law is quite unintelligible from the classical standpoint.

One might try to get over the difficulty without departing from classical mechanics by assuming each of the spectroscopically observed frequencies to be a fundamental frequency with its own degree of freedom, the laws of force being such that the harmonic vibrations do not occur. Such a theory will not do, however, even apart from the fact that it would give no explanation of the Combination Law, since it would immediately bring one into conflict with the experimental evidence on specific heats. Classical statistical mechanics enables one to establish a general connexion between the total number of degrees of freedom of an assembly of vibrating systems and its specific heat. If one assumes all the spectroscopic frequencies of an atom to correspond to different degrees of freedom, one would get a specific heat for any kind of matter very much greater than the observed value. In fact the observed specific heats at ordinary temperatures are given fairly well by a theory that takes into account merely the motion of each atom as a whole and assigns no internal motion to it at all.

This leads us to a new clash between classical mechanics and the results of experiment. There must certainly be some internal motion in an atom to account for its spectrum, but the internal degrees of freedom, for some classically inexplicable reason, do not contribute to the specific heat. A similar clash is found in connexion with the energy of oscillation of the electromagnetic field in a vacuum. Classical mechanics requires the specific heat corresponding to this energy to be infinite, but it is observed to be quite finite. A general conclusion from experimental results is that oscillations of high frequency do not contribute their classical quota to the specific heat.

As another illustration of the failure of classical mechanics we may consider the behaviour of light. We have, on the one hand, the phenomena of interference and diffraction, which can be explained only on the basis of a wave theory; on the other, phenomena such as photo-electric emission and scattering by free electrons, which show that light is composed of small particles. These particles, which are called photons, have each a definite energy and momentum, depending on the frequency of the light, and appear to have just as real an existence as electrons, or any other particles known in physics. A fraction of a photon is never observed.

Experiments have shown that this anomalous behaviour is not peculiar to light, but is quite general. All material particles have wave properties, which can be exhibited under suitable conditions. We have here a very striking and general example of the breakdown of classical mechanics—not merely an inaccuracy in its laws of motion, but *an inadequacy of its concepts to supply us with a description of atomic events.*

The necessity to depart from classical ideas when one wishes to account for the ultimate structure of matter may be seen, not only from experimentally established facts, but also from general philosophical grounds. In a classical explanation of the constitution of matter, one would assume it to be made up of a large number of small constituent parts and one would postulate laws for the behaviour of these parts, from which the laws of the matter in bulk could be deduced. This would not complete the explanation, however, since the question of the structure and stability of the constituent parts is left untouched. To go into this question, it becomes necessary to postulate that each constituent part is itself made up of smaller parts, in terms of which its behaviour is to be explained. There is clearly no end to this procedure, so that one can never arrive at the ultimate structure of matter on these lines. So long as *big* and *small* are merely relative concepts, it is no help to explain the big in terms of the small. It is therefore necessary to modify classical ideas in such a way as to give an absolute meaning to size.

At this stage it becomes important to remember that science is concerned only with observable things and that we can observe an object only by letting it interact with some outside influence. An act of observation is thus necessarily accompanied by some disturbance of the object observed. We may define an object to be big when the disturbance accompanying our observation of it may be neglected, and small when the disturbance cannot be neglected. This definition is in close agreement with the common meanings of big and small.

It is usually assumed that, by being careful, we may cut down the disturbance accompanying our observation to any desired extent. The concepts of big and small are then purely relative and refer to the gentleness of our means of observation as well as to the object being described. In order to give an absolute meaning to size, such as is required for any theory of the ultimate structure of matter, we have to assume that *there is a limit to the fineness of our powers of observation and the smallness of the accompanying disturbance—a limit which is inherent in the nature of things and can never be surpassed by improved technique or increased skill on the part of the observer.* If the object under observation is such that the unavoidable limiting disturbance is negligible, then the object is big in the absolute sense and we may apply classical mechanics to it. If, on the other hand, the limiting disturbance is not negligible, then the object is small in the absolute sense and we require a new theory for dealing with it.

A consequence of the preceding discussion is that we must revise our ideas of causality. Causality applies only to a system which is left undisturbed. If a system is small, we cannot observe it without producing a serious disturbance and hence we cannot expect to find any causal connexion between the results of our observations. Causality will still be assumed to apply to undisturbed systems and the equations which will be set up to describe an undisturbed system will be differential equations expressing a causal connexion between conditions at one time and conditions at a later time. These equations will be in close correspondence with the equations of classical mechanics, but they will be connected only indirectly with the results of observations. There is an unavoidable indeterminacy in the calculation of observational results, the theory enabling us to calculate in general only the probability of our obtaining a particular result when we make an observation.

2. The Polarization of Photons

The discussion in the preceding section about the limit to the gentleness with which observations can be made and the consequent indeterminacy in the results of those observations does not provide any quantitative basis for the building up of quantum mechanics. For this purpose a new set of accurate laws of nature is required. One of the most fundamental and most drastic of these is the *Principle of Superposition of States*. We shall lead up to a general formulation of this principle through a consideration of some special cases, taking first the example provided by the polarization of light.

It is known experimentally that when plane-polarized light is used for ejecting photo-electrons, there is a preferential direction for the electron emission. Thus the polarization properties of light are closely connected with its corpuscular properties and one must ascribe a polarization to the photons. One must consider, for instance, a beam of light plane-polarized in a certain direction as consisting of photons each of which is plane-polarized in that direction and a beam of circularly polarized light as consisting of photons each circularly polarized. Every photon is in a certain *state of polarization*, as we shall say. The problem we must now consider is how to fit in these ideas with the known facts about the resolution of light into polarized components and the recombination of these components.

Let us take a definite case. Suppose we have a beam of light passing through a crystal of tourmaline, which has the property of letting through only light plane-polarized perpendicular to its optic axis. Classical electrodynamics tells us what will happen for any given polarization of the incident beam. If this beam is polarized perpendicular to the optic axis, it will all go through the crystal; if parallel to the axis, none of it will go through; while if polarized at an angle α to the axis, a fraction $\sin^2 \alpha$ will go through. How are we to understand these results on a photon basis?

A beam that is plane-polarized in a certain direction is to be pictured as made up of photons each plane-polarized in that direction. This picture leads to no difficulty in the cases when our incident beam is polarized perpendicular or parallel to the optic axis. We merely have to suppose that each photon polarized perpendicular to the axis passes unhindered and unchanged through the crystal, while each photon polarized parallel to the axis is stopped and absorbed. A difficulty arises, however, in the case of the obliquely polarized incident beam. Each of the incident photons is then obliquely polarized and it is not clear what will happen to such a photon when it reaches the tourmaline.

A question about what will happen to a particular photon under certain conditions is not really very precise. To make it precise one must imagine some experiment performed having a bearing on the question and inquire what will be the result of the experiment. Only questions about the results of experiments have a real significance and it is only such questions that theoretical physics has to consider.

In our present example the obvious experiment is to use an incident beam consisting of only a single photon and to observe what appears on the back side of the crystal. According to quantum mechanics the result of this experiment will be that sometimes one will find a whole photon, of energy equal to the energy of the incident photon, on the back side and other times one will find nothing. When one finds a whole photon, it will be polarized perpendicular to the optic axis. One will never find only a part of a photon on the back side. If one repeats the experiment a large number of times, one will find the photon on the back side in a fraction $\sin^2 \alpha$ of the total number of times. Thus we may say that the photon has a probability $\sin^2 \alpha$ of passing through the tourmaline and appearing on the back side polarized perpendicular to the axis and a probability $\cos^2 \alpha$ of being absorbed. These values for the probabilities lead to the correct classical results for an incident beam containing a large number of photons.

In this way we preserve the individuality of the photon in all cases. We are able to do this, however, only because we abandon the determinacy of the classical theory. The result of an experiment is not determined, as it would be according to classical ideas, by the conditions under the control of the experimenter. The most that can be predicted is a set of possible results, with a probability of occurrence for each.

The foregoing discussion about the result of an experiment with a single obliquely polarized photon incident on a crystal of tourmaline answers all that can legitimately be asked about what happens to an obliquely polarized photon when it reaches the tourmaline. Questions about what decides whether the photon is to go through or not and how it changes its direction of polarization when it does go through cannot be investigated by experiment and should be regarded as outside the domain of science. Nevertheless some further description is necessary in order to correlate the results of this experiment with the results of other experiments that might be performed with photons and to fit them all into a general scheme. Such further description should be regarded, not as an attempt to answer questions outside the domain of science, but as an aid to the formulation of rules for expressing concisely the results of large numbers of experiments.

The further description provided by quantum mechanics runs as follows. It is supposed that a photon polarized obliquely to the optic axis may be regarded as being partly in the state of polarization parallel to the axis and partly in the state of polarization perpendicular to the axis. The state of oblique polarization may be considered as the result of some kind of superposition process applied to the two states of parallel and perpendicular polarization. This implies a certain special kind of relationship between the various states of polarization, a relationship similar to that between polarized beams in classical optics, but which is now to be applied, not to beams, but to the states of polarization of one particular photon. This relationship allows any state of polarization to be resolved into, or expressed as a superposition of, any two mutually perpendicular states of polarization.

When we make the photon meet a tourmaline crystal, we are subjecting it to an observation. We are observing whether it is polarized parallel or perpendicular to the optic axis. The effect of making this observation is to force the photon entirely into the state of parallel or entirely into the state of perpendicular polarization. It has to make a sudden jump from being partly in each of these two states to being entirely in one or other of them. Which of the two states it will jump into cannot be predicted, but is governed only by probability laws. If it jumps into the parallel state it gets absorbed and if it jumps into the perpendicular state it passes through the crystal and appears on the other side preserving this state of polarization.

3. Interference of Photons

In this section we shall deal with another example of superposition. We shall again take photons, but shall be concerned with their position in space and their momentum instead of their polarization. If we are given a beam of roughly monochromatic light, then we know something about the location and momentum of the associated photons. We know that each of them is located somewhere in the region of space through which the beam is passing and has a momentum in the direction of the beam of magnitude given in terms of the frequency of the beam by Einstein's photo-electric law—momentum equals frequency multiplied by a universal constant. When we have such information about the location and momentum of a photon we shall say that it is in a definite *translational state*.

We shall discuss the description which quantum mechanics provides of the interference of photons. Let us take a definite experiment demonstrating interference. Suppose we have a beam of light which is passed through some kind of interferometer, so that it gets split up into two components and the two components are subsequently made to interfere. We may, as in the preceding section, take an incident beam consisting of only a single photon and inquire what will happen to it as it goes through the apparatus. This will present to us the difficulty of the conflict between the wave and corpuscular theories of light in an acute form.

Corresponding to the description that we had in the case of the polarization, we must now describe the photon as going partly into each of the two components into which the incident beam is split. The photon is then, as we may say, in a translational state given by the superposition of the two translational states associated with the two components. We are thus led to a generalization of the term 'translational state' applied to a photon. For a photon to be in a definite translational state it need not be associated with one single beam of light, but may be associated with two or more beams of light which are the components into which one original beam has been split.[1] In the accurate mathematical theory each translational state is associated with one of the wave functions of ordinary wave optics, which wave function may describe either a single beam or two or more beams into which one original beam has been split. Translational states are thus superposable in a similar way to wave functions.

Let us consider now what happens when we determine the energy in one of the components. The result of such a determination must be either the whole photon or nothing at all. Thus the photon must change suddenly from being partly in one beam and partly in the other to being entirely in one of the beams. This sudden change is due to the disturbance in the translational state of the photon which the observation necessarily makes. It is impossible to predict in which of the two beams the photon will be found. Only the probability of either result can be calculated from the previous distribution of the photon over the two beams.

One could carry out the energy measurement without destroying the component beam by, for example, reflecting the beam from a movable mirror and observing the recoil. Our description of the photon allows us to infer that, *after* such an energy measurement, it would not be possible to bring about any interference effects between the two components. So long as the photon is partly in one beam and partly in the other, interference can occur when the two beams are superposed, but this possibility disappears when the photon is forced entirely into one of the beams by an observation. The other beam then no longer enters into the description of the photon, so that it counts as being entirely in the one beam in the ordinary way for any experiment that may subsequently be performed on it.

On these lines quantum mechanics is able to effect a reconciliation of the wave and corpuscular properties of light. The essential point is the association of each of the translational states of a photon with one of the wave functions of ordinary wave optics. The nature of this association cannot be pictured on a basis of classical mechanics, but is something entirely new. It would be quite wrong to picture the photon and its associated wave as interacting in the way in which particles and waves can interact in classical mechanics. The association can be interpreted only statistically, the wave function giving us information about the probability of our finding the photon in any particular place when we make an

---

[1] The circumstance that the superposition idea requires us to generalize our original meaning of translational states, but that no corresponding generalization was needed for the states of polarization of the preceding section, is an accidental one with no underlying theoretical significance.

observation of where it is.

Some time before the discovery of quantum mechanics people realized that the connexion between light waves and photons must be of a statistical character. What they did not clearly realize, however, was that the wave function gives information about the probability of *one* photon being in a particular place and not the probable number of photons in that place. The importance of the distinction can be made clear in the following way. Suppose we have a beam of light consisting of a large number of photons split up into two components of equal intensity. On the assumption that the intensity of a beam is connected with the probable number of photons in it, we should have half the total number of photons going into each component. If the two components are now made to interfere, we should require a photon in one component to be able to interfere with one in the other. Sometimes these two photons would have to annihilate one another and other times they would have to produce four photons. This would contradict the conservation of energy. The new theory, which connects the wave function with probabilities for one photon, gets over the difficulty by making each photon go partly into each of the two components. Each photon then interferes only with itself. Interference between two different photons never occurs.

The association of particles with waves discussed above is not restricted to the case of light, but is, according to modern theory, of universal applicability. All kinds of particles are associated with waves in this way and conversely all wave motion is associated with particles. Thus all particles can be made to exhibit interference effects and all wave motion has its energy in the form of quanta. The reason why these general phenomena are not more obvious is on account of a law of proportionality between the mass or energy of the particles and the frequency of the waves, the coefficient being such that for waves of familiar frequencies the associated quanta are extremely small, while for particles even as light as electrons the associated wave frequency is so high that it is not easy to demonstrate interference.

4. Superposition and Indeterminacy

The reader may possibly feel dissatisfied with the attempt in the two preceding sections to fit in the existence of photons with the classical theory of light. He may argue that a very strange idea has been introduced—the possibility of a photon being partly in each of two states of polarization, or partly in each of two separate beams—but even with the help of this strange idea no satisfying picture of the fundamental single-photon processes has been given. He may say further that this strange idea did not provide any information about experimental results for the experiments discussed, beyond what could have been obtained from an elementary consideration of photons being guided in some vague way by waves. What, then, is the use of the strange idea?

In answer to the first criticism it may be remarked that the main object of physical science is not the provision of pictures, but is the formulation of laws governing phenomena and the application of these laws to the discovery of new phenomena. If a picture exists, so much the better; but whether a picture exists or not is a matter of only secondary importance. In the case of atomic phenomena no picture can be expected to exist in the usual sense of the word 'picture', by which is meant a model functioning essentially on classical lines. One may, however, extend the meaning of the word 'picture' to include any *way of looking at the fundamental laws which makes their self-consistency obvious*. With this extension, one may gradually acquire a picture of atomic phenomena by becoming familiar with the laws of the quantum theory.

With regard to the second criticism, it may be remarked that for many simple experiments with light, an elementary theory of waves and photons connected in a vague statistical way would be adequate to account for the results. In the case of such experiments quantum mechanics has no further information to give. In the great majority of experiments, however, the conditions are too complex for an elementary theory of this kind to be applicable and some more elaborate scheme, such as is provided by quantum mechanics, is then needed. The method of description that quantum mechanics gives in the more complex cases is applicable also to the simple cases and although it is then not really necessary for accounting for the experimental results, its study in these simple cases is perhaps a suitable introduction to its study in the general case.

There remains an overall criticism that one may make to the whole scheme, namely, that in departing from the determinacy of the classical theory a great complication is introduced into the description of Nature, which is a highly undesirable feature. This complication is undeniable, but it is offset by a great simplification, provided by the general *principle of superposition of states*, which we shall now go on to consider. But first it is necessary to make precise the important concept of a 'state' of a general atomic system.

Let us take any atomic system, composed of particles or bodies with specified properties (mass, moment of inertia, etc.) interacting according to specified laws of force. There will be various possible motions of the particles or bodies consistent with the laws of force. Each such motion is called a *state* of the system. According to classical ideas one could specify a state by giving numerical values to all the coordinates and velocities of the various component parts of the system at some instant of time, the whole motion being then completely determined. Now the argument of pp. 3 and 4 shows that we cannot observe a *small* system with that amount of detail which classical theory supposes. The limitation in the power of observation puts a limitation on the number of data that can be assigned to a state. Thus a state of an atomic system must be specified by fewer or more indefinite data than a complete set of numerical values for all the coordinates and velocities at some instant of time. In the case when the system is just a single photon, a state would be completely specified by a given translational state in the sense of § 3 together with a given state of polarization in the sense of § 2.

A state of a system may be defined as an undisturbed motion that is restricted by as many conditions or data as are theoretically possible without mutual interference or contradiction. In practice the conditions could be imposed by a suitable preparation of the system, consisting perhaps in passing it through various kinds of sorting apparatus, such as slits and polarimeters, the system being left undisturbed after the preparation. The word 'state' may be used to mean either the state at one particular time (after the preparation), or the state throughout the whole of time after the preparation. To distinguish these two meanings, the latter will be called a 'state of motion' when there is liable to be ambiguity.

The general principle of superposition of quantum mechanics applies to the states, with either of the above meanings, of any one dynamical system. It requires us to assume that between these states there exist peculiar relationships such that whenever the system is definitely in one state we can consider it as being partly in each of two or more other states. The original state must be regarded as the result of a kind of *superposition* of the two or more new states, in a way that cannot be conceived on classical ideas. Any state may be considered as the result of a superposition of two or more other states, and indeed in an infinite number of ways. Conversely any two or more states may be superposed to give a new state. The procedure of expressing a state as the result of superposition of a number of other states is a mathematical procedure that is always permissible, independent of any reference to physical conditions, like the procedure of resolving a wave into Fourier components. Whether it is useful in any particular case, though, depends on the special physical conditions of the problem under consideration.

In the two preceding sections examples were given of the superposition principle applied to a system consisting of a single photon. § 2 dealt with states differing only with regard to the polarization and § 3 with states differing only with regard to the motion of the photon as a whole.

The nature of the relationships which the superposition principle requires to exist between the states of any system is of a kind that cannot be explained in terms of familiar physical concepts. One cannot in the classical sense picture a system being partly in each of two states and see the equivalence of this to the system being completely in some other state. There is an entirely new idea involved, to which one must get accustomed and in terms of which one must proceed to build up an exact mathematical theory, without having any detailed classical picture.

When a state is formed by the superposition of two other states, it will have properties that are in some vague way intermediate between those of the two original states and that approach more or less closely to those of either of them according to the greater or less 'weight' attached to this state in the superposition process. The new state is completely defined by the two original states when their relative weights in the superposition process are known, together with a certain phase difference, the exact meaning of weights and phases being provided in the general case by the mathematical theory. In the case of the polarization of a photon their meaning is that provided by classical optics, so that, for example, when two perpendicularly plane polarized states are superposed with equal weights, the new state may be circularly polarized in either direction, or linearly polarized at an angle $(1/4)\pi$, or else elliptically polarized, according to the phase difference.

The non-classical nature of the superposition process is brought out clearly if we consider the superposition of two states, A and B, such that there exists an observation which, when made on the system in state A, is certain to lead to one particular result, a say, and when made on the system in state B is certain to lead to some different result, b say. What will be the result of the observation when made on the system in the superposed state? The answer is that the result will be sometimes a and sometimes b, according to a probability law depending on the relative weights of A and B in the superposition process. It will never be different from both a and b. *The intermediate character of the state formed by superposition thus expresses itself through the probability of a particular result for an observation being intermediate between the corresponding probabilities for the original states,*[1] *not through the result itself being intermediate between the corresponding results for the original states.*

In this way we see that such a drastic departure from ordinary ideas as the assumption of superposition relationships between the states is possible only on account of the recognition of the importance of the disturbance accompanying an observation and of the consequent indeterminacy in the result of the observation. When an observation is made on any atomic system that is in a given state, in general the result will not be determinate, i.e., if the experiment is repeated several times under identical conditions several different results may be obtained. It is a law of nature, though, that if the experiment is repeated a large number of times, each particular result will be obtained in a definite fraction of the total number of times, so that there is a definite *probability* of its being obtained. This probability is what the theory sets out to calculate. Only in special cases when the probability for some result is unity is the result of the experiment determinate.

The assumption of superposition relationships between the states leads to a mathematical theory in which the equations that define a state are linear in the unknowns. In consequence of this, people have tried to establish analogies with systems in classical mechanics, such as vibrating strings or membranes, which are governed by linear equations and for which, therefore, a superposition principle holds. Such analogies have led to the name 'Wave Mechanics' being sometimes given to quantum mechanics. It is important to remember, however, that *the superposition that occurs in quantum mechanics is of an essentially different nature from any occurring in the classical theory*, as is shown by the fact that the quantum superposition principle demands indeterminacy in the results of observations in order to be capable of a sensible physical interpretation. The analogies are thus liable to be misleading.

5. Mathematical Formulation of the Principle

A profound change has taken place during the present century in the opinions physicists have held on the mathematical foundations of their subject. Previously they supposed that the principles of Newtonian mechanics would provide the basis for the description of the whole of physi-

---

[1] The probability of a particular result for the state formed by superposition is not always intermediate between those for the original states in the general case when those for the original states are not zero or unity, so there are restrictions on the 'intermediateness' of a state formed by superposition.

cal phenomena and that all the theoretical physicist had to do was suitably to develop and apply these principles. With the recognition that there is no logical reason why Newtonian and other classical principles should be valid outside the domains in which they have been experimentally verified has come the realization that departures from these principles are indeed necessary. Such departures find their expression through the introduction of new mathematical formalisms, new schemes of axioms and rules of manipulation, into the methods of theoretical physics.

Quantum mechanics provides a good example of the new ideas. It requires the states of a dynamical system and the dynamical variables to be interconnected in quite strange ways that are unintelligible from the classical standpoint. The states and dynamical variables have to be represented by mathematical quantities of different natures from those ordinarily used in physics. The new scheme becomes a precise physical theory when all the axioms and rules of manipulation governing the mathematical quantities are specified and when in addition certain laws are laid down connecting physical facts with the mathematical formalism, so that from any given physical conditions equations between the mathematical quantities may be inferred and vice versa. In an application of the theory one would be given certain physical information, which one would proceed to express by equations between the mathematical quantities. One would then deduce new equations with the help of the axioms and rules of manipulation and would conclude by interpreting these new equations as physical conditions. The justification for the whole scheme depends, apart from internal consistency, on the agreement of the final results with experiment.

We shall begin to set up the scheme by dealing with the mathematical relations between the states of a dynamical system at one instant of time, which relations will come from the mathematical formulation of the principle of superposition. The superposition process is a kind of additive process and implies that states can in some way be added to give new states. The states must therefore be connected with mathematical quantities of a kind which can be added together to give other quantities of the same kind. The most obvious of such quantities are vectors. Ordinary vectors, existing in a space of a finite number of dimensions, are not sufficiently general for most of the dynamical systems in quantum mechanics. We have to make a generalization to vectors in a space of an infinite number of dimensions, and the mathematical treatment becomes complicated by questions of convergence. For the present, however, we shall deal merely with some general properties of the vectors, properties

5. MATHEMATICAL FORMULATION OF THE PRINCIPLE

which can be deduced on the basis of a simple scheme of axioms, and questions of convergence and related topics will not be gone into until the need arises.

It is desirable to have a special name for describing the vectors which are connected with the states of a system in quantum mechanics, whether they are in a space of a finite or an infinite number of dimensions. We shall call them *ket vectors*, or simply *kets*, and denote a general one of them by a special symbol $|\rangle$. If we want to specify a particular one of them by a label, A say, we insert it in the middle, thus $|A\rangle$. The suitability of this notation will become clear as the scheme is developed.

Ket vectors may be multiplied by complex numbers and may be added together to give other ket vectors, e.g. from two ket vectors $|A\rangle$ and $|B\rangle$ we can form

$$c_1|A\rangle + c_2|B\rangle = |R\rangle, \tag{1}$$

say, where c_1 and c_2 are any two complex numbers. We may also perform more general linear processes with them, such as adding an infinite sequence of them, and if we have a ket vector $|x\rangle$, depending on and labelled by a parameter x which can take on all values in a certain range, we may integrate it with respect to x, to get another ket vector

$$\int |x\rangle \, dx = |Q\rangle$$

say. A ket vector which is expressible linearly in terms of certain others is said to be *dependent* on them. A set of ket vectors are called *independent* if no one of them is expressible linearly in terms of the others.

We now assume that *each state of a dynamical system at a particular time corresponds to a ket vector, the correspondence being such that if a state results from the superposition of certain other states, its corresponding ket vector is expressible linearly in terms of the corresponding ket vectors of the other states, and conversely*. Thus the state R results from a superposition of the states A and B when the corresponding ket vectors are connected by (1).

The above assumption leads to certain properties of the superposition process, properties which are in fact necessary for the word 'superposition' to be appropriate. When two or more states are superposed, the order in which they occur in the superposition process is unimportant, so the superposition process is symmetrical between the states that are superposed. Again, we see from equation (1) that (excluding the case when the coefficient c_1 or c_2 is zero) if the state R can be formed by superposition of the states A and B, then the state A can be formed by superposition of B and R, and B can be formed by superposition of A and R. The superposition relationship is symmetrical between all three states A, B, and R.

A state which results from the superposition of certain other states will be said to be *dependent* on those states. More generally, a state will be said to be *dependent* on any set of states, finite or infinite in number, if its corresponding ket vector is dependent on the corresponding ket vectors of the set of states. A set of states will be called *independent* if no one of them is dependent on the others.

To proceed with the mathematical formulation of the superposition principle we must introduce a further assumption, namely the assumption that by superposing a state with itself we cannot form any new state, but only the original state over again. If the original state corresponds to the ket vector $|A\rangle$, when it is superposed with itself the resulting state will correspond to

$$c_1|A\rangle + c_2|A\rangle = (c_1 + c_2)|A\rangle,$$

where c_1 and c_2 are numbers. Now we may have $c_1 + c_2 = 0$, in which case the result of the superposition process would be nothing at all, the two components having cancelled each other by an interference effect. Our new assumption requires that, apart from this special case, the resulting state must be the same as the original one, so that $(c_1 + c_2)|A\rangle$ must correspond to the same state that $|A\rangle$ does. Now $c_1 + c_2$ is an arbitrary complex number and hence we can conclude that *if the ket vector corresponding to a state is multiplied by any complex number, not zero, the resulting ket vector will correspond to the same state.* Thus a state is specified by the direction of a ket vector and any length one may assign to the ket vector is irrelevant. All the states of the dynamical system are in one-one correspondence with all the possible directions for a ket vector, no distinction being made between the directions of the ket vectors $|A\rangle$ and $-|A\rangle$.

The assumption just made shows up very clearly the fundamental difference between the superposition of the quantum theory and any kind of classical superposition. In the case of a classical system for which a superposition principle holds, for instance a vibrating membrane, when one superposes a state with itself the result is a *different* state, with a different magnitude of the oscillations. There is no physical characteristic of a quantum state corresponding to the magnitude of the classical oscillations, as distinct from their quality, described by the ratios of the amplitudes at different points of the membrane. Again, while there exists a classical state with zero amplitude of oscillation everywhere, namely the state of rest, there does not exist any corresponding state for a quantum system, the zero ket vector corresponding to no state at all.

Given two states corresponding to the ket vectors $|A\rangle$ and $|B\rangle$, the general state formed by superposing them corresponds to a ket vector $|R\rangle$ which is determined by two complex numbers, namely the coefficients c_1 and c_2 of equation (1). If these two coefficients are multiplied by the same factor (itself a complex number), the ket vector $|R\rangle$ will get multiplied by this factor and the corresponding state will be unaltered. Thus only the ratio of the two coefficients is effective in determining the state R. Hence this state is determined by one complex number, or by two real parameters. Thus from two given states, a twofold infinity of states may be obtained by superposition.

This result is confirmed by the examples discussed in §§ 2 and 3. In the example of § 2 there are just two independent states of polarization for a photon, which may be taken to be the states of plane polarization parallel and perpendicular to some fixed direction, and from the superposition of these two a twofold infinity of states of polarization can be obtained, namely all the states of elliptic polarization, the general one of which requires two parameters to describe it. Again, in the example of § 3, from the superposition of two given translational states for a photon a twofold infinity of translational states may be obtained, the general one of which is described by two parameters, which may be taken to be the ratio of the amplitudes of the two wave functions that are added together and their phase relationship. This confirmation shows the need for allowing complex coefficients in equation (1). If these coefficients were restricted to be real, then, since only their ratio is of importance for determining the direction of the resultant ket vector $|R\rangle$ when $|A\rangle$ and $|B\rangle$ are given, there would be only a simple infinity of states obtainable from the superposition.

6. Bra and Ket Vectors

Whenever we have a set of vectors in any mathematical theory, we can always set up a second set of vectors, which mathematicians call the dual vectors. The procedure will be described for the case when the original vectors are our ket vectors.

Suppose we have a number ϕ which is a function of a ket vector $|A\rangle$, i.e. to each ket vector $|A\rangle$ there corresponds one number ϕ, and suppose further that the function is a linear one, which means that the number corresponding to $|A\rangle + |A'\rangle$ is the sum of the numbers corresponding to $|A\rangle$ and to $|A'\rangle$, and the number corresponding to $c|A\rangle$ is c times the number corresponding to $|A\rangle$, c being any numerical factor. Then the number ϕ corresponding to any $|A\rangle$ may be looked upon as the scalar product of that $|A\rangle$ with some new vector, there being one of these new vectors for each linear function of the ket vectors $|A\rangle$. The justification for this way of looking at ϕ is that, as will be seen later [see equations (5) and (6)], the new vectors may be added together and may be multiplied by numbers to give other vectors of the same kind. The new vectors are, of course, defined only to the extent that their scalar products with the original ket vectors are given numbers, but this is sufficient for one to be able to build up a mathematical theory about them.

We shall call the new vectors *bra vectors*, or simply *bras*, and denote a general one of them by the symbol $\langle |$, the mirror image of the symbol for a ket vector. If we want to specify a particular one of them by a label, B say, we write it in the middle, thus $\langle B|$. The scalar product of a bra vector $\langle B|$ and a ket vector $|A\rangle$ will be written $\langle B|A\rangle$, i.e. as a juxtaposition of the symbols for the bra and ket vectors, that for the bra vector being on the left, and the two vertical lines being contracted to one for brevity.

One may look upon the symbols $\langle$ and $\rangle$ as a distinctive kind of brackets. A scalar product $\langle B|A\rangle$ now appears as a complete bracket expression and a bra vector $\langle B|$ or a ket vector $|A\rangle$ as an incomplete bracket expression. We have the rules that *any complete bracket expression denotes a number and any incomplete bracket expression denotes a vector, of the bra or ket kind according to whether it contains the first or second part of the bracket*s.

The condition that the scalar product of $\langle B|$ and $|A\rangle$ is a linear function of $|A\rangle$ may be expressed symbolically by

$$\langle B|(|A\rangle + |A'\rangle) = \langle B|A\rangle + \langle B|A'\rangle, \qquad (2)$$

$$\langle B|(c|A\rangle) = c\langle B|A\rangle, \qquad (3)$$

c being any number.

A bra vector is considered to be completely defined when its scalar product with every ket vector is given, so that if a bra vector has its scalar product with every ket vector vanishing, the bra vector itself must be considered as vanishing. In symbols, if

$$\langle P|A\rangle = 0, \quad \text{all } |A\rangle, \qquad (4)$$

then

$$\langle P| = 0. \qquad (4)$$

The sum of two bra vectors $\langle B|$ and $\langle B'|$ is defined by the condition that its scalar product with any ket vector $|A\rangle$ is the sum of the scalar products of $\langle B|$ and $\langle B'|$ with $|A\rangle$,

$$(\langle B| + \langle B'|)|A\rangle = \langle B|A\rangle + \langle B'|A\rangle, \tag{5}$$

and the product of a bra vector $\langle B|$ and a number c is defined by the condition that its scalar product with any ket vector $|A\rangle$ is c times the scalar product of $\langle B|$ with $|A\rangle$,

$$(c\langle B|)|A\rangle = c\langle B|A\rangle. \tag{6}$$

Equations (2) and (5) show that products of bra and ket vectors satisfy the distributive axiom of multiplication, and equations (3) and (6) show that multiplication by numerical factors satisfies the usual algebraic axioms.

The bra vectors, as they have been here introduced, are quite a different kind of vector from the kets, and so far there is no connexion between them except for the existence of a scalar product of a bra and a ket. We now make the assumption that *there is a one-one correspondence between the bras and the kets, such that the bra corresponding to $|A\rangle + |A'\rangle$ is the sum of the bras corresponding to $|A\rangle$ and to $|A'\rangle$, and the bra corresponding to $c|A\rangle$ is $\bar{c}$ times the bra corresponding to $|A\rangle$, $\bar{c}$ being the conjugate complex number to c.* We shall use the same label to specify a ket and the corresponding bra. Thus the bra corresponding to $|A\rangle$ will be written $\langle A|$.

The relationship between a ket vector and the corresponding bra makes it reasonable to call one of them the conjugate imaginary of the other. Our bra and ket vectors are complex quantities, since they can be multiplied by complex numbers and are then of the same nature as before, but they are complex quantities of a special kind which cannot be split up into real and pure imaginary parts. The usual method of getting the real part of a complex quantity, by taking half the sum of the quantity itself and its conjugate, cannot be applied since a bra and a ket vector are of different natures and cannot be added together. To call attention to this distinction, we shall use the words 'conjugate complex' to refer to numbers and other complex quantities which can be split up into real and pure imaginary parts, and the words 'conjugate imaginary' for bra and ket vectors, which cannot. With the former kind of quantity, we shall use the notation of putting a bar over one of them to get the conjugate complex one.

On account of the one-one correspondence between bra vectors and ket vectors, *any state of our dynamical system at a particular time may be specified by the direction of a bra vector just as well as by the direction of a ket vector.* In fact the whole theory will be symmetrical in its essentials between bras and kets.

Given any two ket vectors $|A\rangle$ and $|B\rangle$, we can construct from them a number $\langle B|A\rangle$ by taking the scalar product of the first with the conjugate imaginary of the second. This number depends linearly on $|A\rangle$ and *antilinearly* on $|B\rangle$, the antilinear dependence meaning that the number formed from $|B\rangle + |B'\rangle$ is the sum of the numbers formed from $|B\rangle$ and from $|B'\rangle$, and the number formed from $c|B\rangle$ is $\bar{c}$ times the number formed from $|B\rangle$. There is a second way in which we can construct a number which depends linearly on $|A\rangle$ and antilinearly on $|B\rangle$, namely by forming the scalar product of $|B\rangle$ with the conjugate imaginary of $|A\rangle$ and taking the conjugate complex of this scalar product. *We assume that these two numbers are always equal,* i.e.

$$\langle B|A\rangle = \overline{\langle A|B\rangle}. \tag{7}$$

Putting $|B\rangle = |A\rangle$ here, we find that the number $\langle A|A\rangle$ must be real. We make the further assumption

$$\langle A|A\rangle > 0, \tag{8}$$

except when $|A\rangle = 0$.

In ordinary space, from any two vectors one can construct a number—their scalar product—which is a real number and is symmetrical between them. In the space of bra vectors or the space of ket vectors, from any two vectors one can again construct a number—the scalar product of one with the conjugate imaginary of the other—but this number is complex and goes over into the conjugate complex number when the two vectors are interchanged. There is thus a kind of perpendicularity in these spaces, which is a generalization of the perpendicularity in ordinary space. We shall call a bra and a ket vector *orthogonal* if their scalar product is zero, and two bras or two kets will be called orthogonal if the scalar product of one with the conjugate imaginary of the other is zero. Further, we shall say that two states of our dynamical system are orthogonal if the vectors corresponding to these states are orthogonal.

The *length* of a bra vector $\langle A|$ or of the conjugate imaginary ket vector $|A\rangle$ is defined as the square root of the positive number $\langle A|A\rangle$. When we are given a state and wish to set up a bra or ket vector to correspond to it, only the direction of the vector is given and the vector itself is undetermined to the extent of an arbitrary numerical factor. It is often convenient to choose this numerical factor so that the vector is of length unity. This procedure is called *normalization* and the vector so chosen is said to be *normalized*. The vector is not completely determined even then, since one can still multiply it by any number of modulus unity, i.e. any number $e^{i\gamma}$ where γ is real, without changing its length. We shall call such a number a *phase factor*.

The foregoing assumptions give the complete scheme of relations between the states of a dynamical system at a particular time. The relations appear in mathematical form, but they imply physical conditions, which will lead to results expressible in terms of observations when the theory is developed further. For instance, if two states are orthogonal, it means at present simply a certain equation in our formalism, but this equation implies a definite physical relationship between the states, which further developments of the theory will enable us to interpret in terms of observational results (see the bottom of p. 35).

CHAPTER II

Dynamical Variables and Observables

7. Linear Operators

In the preceding section we considered a number which is a linear function of a ket vector, and this led to the concept of a bra vector. We shall now consider a ket vector which is a linear function of a ket vector, and this will lead to the concept of a linear operator.

Suppose we have a ket $|F\rangle$ which is a function of a ket $|A\rangle$, i.e. to each ket $|A\rangle$ there corresponds one ket $|F\rangle$, and suppose further that the function is a linear one, which means that the $|F\rangle$ corresponding to $|A\rangle + |A'\rangle$ is the sum of the $|F\rangle$'s corresponding to $|A\rangle$ and to $|A'\rangle$, and the $|F\rangle$ corresponding to $c|A\rangle$ is c times the $|F\rangle$ corresponding to $|A\rangle$, c being any numerical factor. Under these conditions, we may look upon the passage from $|A\rangle$ to $|F\rangle$ as the application of a *linear operator* to $|A\rangle$. Introducing the symbol α for the linear operator, we may write

$$|F\rangle = \alpha|A\rangle,$$

in which the result of α operating on $|A\rangle$ is written like a product of α with $|A\rangle$. We make the rule that in such products *the ket vector must always be put on the right of the linear operator*. The above conditions of linearity may now be expressed by the equations

$$\begin{aligned}\alpha(|A\rangle + |A'\rangle) &= \alpha|A\rangle + \alpha|A'\rangle,\\ \alpha(c|A\rangle) &= c\alpha|A\rangle.\end{aligned} \quad (1)$$

A linear operator is considered to be completely defined when the result of its application to every ket vector is given. Thus a linear operator is to be considered zero if the result of its application to every ket vanishes, and two linear operators are to be considered equal if they produce the same result when applied to every ket.

Linear operators can be added together, the sum of two linear operators being defined to be that linear operator which, operating on any ket, produces the sum of what the two linear operators separately would produce. Thus $\alpha + \beta$ is defined by

$$(\alpha + \beta)|A\rangle = \alpha|A\rangle + \beta|A\rangle \tag{2}$$

for any $|A\rangle$. Equation (2) and the first of equations (1) show that products of linear operators with ket vectors satisfy the distributive axiom of multiplication.

Linear operators can also be multiplied together, the product of two linear operators being defined as that linear operator, the application of which to any ket produces the same result as the application of the two linear operators successively. Thus the product of $\alpha\beta$ is defined as the linear operator which, operating on any ket $|A\rangle$, changes it into that ket which one would get by operating first on $|A\rangle$ with β, and then on the result of the first operation with α. In symbols

$$(\alpha\beta)|A\rangle = \alpha(\beta|A\rangle).$$

This definition appears as the associative axiom of multiplication for the triple product of α, β, and $|A\rangle$, and allows us to write this triple product as $\alpha\beta|A\rangle$ without brackets. However, this triple product is in general not the same as what we should get if we operated on $|A\rangle$ first with α and then with β, i.e. in general $\alpha\beta|A\rangle$ differs from $\beta\alpha|A\rangle$, so that in general $\alpha\beta$ must differ from $\beta\alpha$. *The commutative axiom of multiplication does not hold for linear operators.* It may happen as a special case that two linear operators ξ and η are such that $\xi\eta$ and $\eta\xi$ are equal. In this case we say that ξ *commutes with* η, or that ξ and η *commute*.

By repeated applications of the above processes of adding and multiplying linear operators, one can form sums and products of more than two of them, and one can proceed to build up an algebra with them. In this algebra the commutative axiom of multiplication does not hold, and also the product of two linear operators may vanish without either factor vanishing. But all the other axioms of ordinary algebra, including the associative and distributive axioms of multiplication, are valid, as may easily be verified.

If we take a number k and multiply it into ket vectors, it appears as a linear operator operating on ket vectors, the conditions (1) being fulfilled with k substituted for α. A number is thus a special case of a linear operator. It has the property that it commutes with all linear operators and this property distinguishes it from a general linear operator.

So far we have considered linear operators operating only on ket vectors. We can give a meaning to their operating also on bra vectors, in the following way. Take the scalar product of any bra $\langle B|$ with the ket $\alpha|A\rangle$.

This scalar product is a number which depends linearly on $|A\rangle$ and therefore, from the definition of bras, it may be considered as the scalar product of $|A\rangle$ with some bra. The bra thus defined depends linearly on $\langle B|$, so we may look upon it as the result of some linear operator applied to $\langle B|$. This linear operator is uniquely determined by the original linear operator α and may reasonably be called the same linear operator operating on a bra. In this way our linear operators are made capable of operating on bra vectors.

A suitable notation to use for the resulting bra when α operates on the bra $\langle B|$ is $\langle B|\alpha$, as in this notation the equation which defines $\langle B|\alpha$ is

$$(\langle B|\alpha)|A\rangle = \langle B|(\alpha|A\rangle) \qquad (3)$$

for any $|A\rangle$, which simply expresses the associative axiom of multiplication for the triple product of $\langle B|$, α, and $|A\rangle$. We therefore make the general rule that in a product of a bra and a linear operator, the bra must always be put on the left. We can now write the triple product of $\langle B|$, α, and $|A\rangle$ simply as $\langle B|\alpha|A\rangle$ without brackets. It may easily be verified that the distributive axiom of multiplication holds for products of bras and linear operators just as well as for products of linear operators and kets.

There is one further kind of product which has a meaning in our scheme, namely the product of a ket vector and a bra vector with the ket on the left, such as $|A\rangle\langle B|$. To examine this product, let us multiply it into an arbitrary ket $|P\rangle$, putting the ket on the right, and assume the associative axiom of multiplication. The product is then $|A\rangle\langle B|P\rangle$, which is another ket, namely $|A\rangle$ multiplied by the number $\langle B|P\rangle$, and this ket depends linearly on the ket $|P\rangle$. Thus $|A\rangle\langle B|$ appears as a linear operator that can operate on kets. It can also operate on bras, its product with a bra $\langle Q|$ on the left being $\langle Q|A\rangle\langle B|$, which is the number $\langle Q|A\rangle$ times the bra $\langle B|$. The product $|A\rangle\langle B|$ is to be sharply distinguished from the product $\langle B|A\rangle$ of the same factors in the reverse order, the latter product being, of course, a number.

We now have a complete algebraic scheme involving three kinds of quantities, bra vectors, ket vectors, and linear operators. They can be multiplied together in the various ways discussed above, and the associative and distributive axioms of multiplication always hold, but the commutative axiom of multiplication does not hold. In this general scheme we still have the rules of notation of the preceding section, that any complete bracket expression, containing $\langle$ on the left and $\rangle$ on the right, denotes a number, while any incomplete bracket expression, containing only $\langle$ or $\rangle$, denotes a vector.

With regard to the physical significance of the scheme, we have already assumed that the bra vectors and ket vectors, or rather the directions of these vectors, correspond to the states of a dynamical system at a particular time. We now make the further assumption that *the linear operators correspond to the dynamical variables at that time*. By dynamical variables are meant quantities such as the coordinates and the components of velocity, momentum and angular momentum of particles, and functions of these quantities—in fact the variables in terms of which classical mechanics is built up. The new assumption requires that these quantities shall occur also in quantum mechanics, but with the striking difference that *they are now subject to an algebra in which the commutative axiom of multiplication does not hold.*

This different algebra for the dynamical variables is one of the most important ways in which quantum mechanics differs from classical mechanics. We shall see later on that, in spite of this fundamental difference, the dynamical variables of quantum mechanics still have many properties in common with their classical counterparts and it will be possible to build up a theory of them closely analogous to the classical theory and forming a beautiful generalization of it.

It is convenient to use the same letter to denote a dynamical variable and the corresponding linear operator. In fact, we may consider a dynamical variable and the corresponding linear operator to be both the same thing, without getting into confusion.

8. Conjugate Relations

Our linear operators are complex quantities, since one can multiply them by complex numbers and get other quantities of the same nature. Hence they must correspond in general to complex dynamical variables, i.e. to complex functions of the coordinates, velocities, etc. We need some further development of the theory to see what kind of linear operator corresponds to a real dynamical variable.

Consider the ket which is the conjugate imaginary of $\langle P|\alpha$. This ket depends antilinearly on $\langle P|$ and thus depends linearly on $|P\rangle$. It may therefore be considered as the result of some linear operator operating on $|P\rangle$. This linear operator is called the *adjoint* of α and we shall denote it by $\bar{\alpha}$. With this notation, the conjugate imaginary of $\langle P|\alpha$ is $\bar{\alpha}|P\rangle$.

8. CONJUGATE RELATIONS

In formula (7) of Chapter I put $\langle P|\alpha$ for $\langle A|$ and its conjugate imaginary $\bar{\alpha}|P\rangle$ for $|A\rangle$. The result is

$$\langle B|\bar{\alpha}|P\rangle = \overline{\langle P|\alpha|B\rangle}. \tag{4}$$

This is a general formula holding for any ket vectors $|B\rangle$, $|P\rangle$ and any linear operator α, and it expresses one of the most frequently used properties of the adjoint.

Putting $\bar{\alpha}$ for α in (4), we get

$$\langle B|\bar{\bar{\alpha}}|P\rangle = \overline{\langle P|\bar{\alpha}|B\rangle} = \langle B|\alpha|P\rangle,$$

by using (4) again with $|P\rangle$ and $|B\rangle$ interchanged. This holds for any ket $|P\rangle$, so we can infer from (4) of Chapter I,

$$\langle B|\bar{\bar{\alpha}} = \langle B|\alpha,$$

and since this holds for any bra vector $\langle B|$, we can infer

$$\bar{\bar{\alpha}} = \alpha.$$

Thus *the adjoint of the adjoint of a linear operator is the original linear operator*. This property of the adjoint makes it like the conjugate complex of a number, and it is easily verified that in the special case when the linear operator is a number, the adjoint linear operator is the conjugate complex number. Thus it is reasonable to assume that *the adjoint of a linear operator corresponds to the conjugate complex of a dynamical variable*. With this physical significance for the adjoint of a linear operator, we may call the adjoint alternatively the *conjugate complex linear operator*, which conforms with our notation $\bar{\alpha}$.

A linear operator may equal its adjoint, and is then called *self-adjoint*. It corresponds to a real dynamical variable, so it may be called alternatively a *real linear operator*. Any linear operator may be split up into a real part and a pure imaginary part. For this reason the words 'conjugate complex' are applicable to linear operators and not the words 'conjugate imaginary'.

The conjugate complex of the sum of two linear operators is obviously the sum of their conjugate complexes. To get the conjugate complex of the product of two linear operators α and β, we apply formula (7) of Chapter I with

$$\langle A| = \langle P|\alpha, \quad \langle B| = \langle Q|\bar{\beta},$$

so that

$$|A\rangle = \bar{\alpha}|P\rangle, \quad |B\rangle = \beta|Q\rangle.$$

The result is
$$\langle Q|\overline{\bar\beta\bar\alpha}|P\rangle = \overline{\langle P|\alpha\beta|Q\rangle} = \langle Q|\overline{\alpha\beta}|P\rangle$$
from (4). Since this holds for any $|P\rangle$ and $\langle Q|$, we can infer that
$$\bar\beta\bar\alpha = \overline{\alpha\beta}. \tag{5}$$
Thus *the conjugate complex of the product of two linear operators equals the product of the conjugate complexes of the factors in the reverse order.*

As simple examples of this result, it should be noted that, if ξ and η are real, in general $\xi\eta$ is not real. This is an important difference from classical mechanics. However, $\xi\eta + \eta\xi$ is real, and so is $i(\xi\eta - \eta\xi)$. Only when ξ and η commute is $\xi\eta$ itself also real. Further, if ξ is real, then so is ξ^2 and, more generally, ξ^n with n any positive integer.

We may get the conjugate complex of the product of three linear operators by successive applications of the rule (5) for the conjugate complex of the product of two of them. We have
$$\overline{\alpha\beta\gamma} = \overline{\alpha(\beta\gamma)} = \overline{\beta\gamma}\bar\alpha = \bar\gamma\bar\beta\bar\alpha, \tag{6}$$
so the conjugate complex of the product of three linear operators equals the product of the conjugate complexes of the factors in the reverse order. The rule may easily be extended to the product of any number of linear operators.

In the preceding section we saw that the product $|A\rangle\langle B|$ is a linear operator. We may get its conjugate complex by referring directly to the definition of the adjoint. Multiplying $|A\rangle\langle B|$ into a general bra $\langle P|$ we get $\langle P|A\rangle\langle B|$, whose conjugate imaginary ket is
$$\overline{\langle P|A\rangle}|B\rangle = \langle A|P\rangle|B\rangle = |B\rangle\langle A|P\rangle.$$
Hence
$$\overline{|A\rangle\langle B|} = |B\rangle\langle A|. \tag{7}$$

We now have several rules concerning conjugate complexes and conjugate imaginaries of products, namely equation (7) of Chapter I, equations (4), (5), (6), (7) of this chapter, and the rule that the conjugate imaginary of $\langle P|\alpha$ is $\bar\alpha|P\rangle$. These rules can all be summed up in a single comprehensive rule, *the conjugate complex or conjugate imaginary of any product of bra vectors, ket vectors, and linear operators is obtained by taking the conjugate complex or conjugate imaginary of each factor and reversing the order of all the factors.* The rule is easily verified to hold quite generally, also for the cases not explicitly given above.

THEOREM. *If ξ is a real linear operator and*

$$\xi^m|P\rangle = 0 \tag{8}$$

for a particular ket $|P\rangle$, m being a positive integer, then

$$\xi|P\rangle = 0.$$

To prove the theorem, take first the case when $m = 2$. Equation (8) then gives

$$\langle P|\xi^2|P\rangle = 0,$$

showing that the ket $\xi|P\rangle$ multiplied by the conjugate imaginary bra $\langle P|\xi$ is zero. From the assumption (8) of Chapter I with $\xi|P\rangle$ for $|A\rangle$, we see that $\xi|P\rangle$ must be zero. Thus the theorem is proved for $m = 2$.

Now take $m > 2$ and put

$$\xi^{m-2}|P\rangle = |Q\rangle.$$

Equation (8) now gives

$$\xi^2|Q\rangle = 0.$$

Applying the theorem for $m = 2$, we get

$$\xi|Q\rangle = 0$$

or

$$\xi^{m-1}|P\rangle = 0. \tag{9}$$

By repeating the process by which equation (9) is obtained from (8), we obtain successively

$$\xi^{m-2}|P\rangle = 0, \quad \xi^{m-3}|P\rangle = 0, \quad \ldots, \quad \xi^2|P\rangle = 0, \quad \xi|P\rangle = 0,$$

and so the theorem is proved generally.

9. Eigenvalues and Eigenvectors

We must make a further development of the theory of linear operators, consisting in studying the equation

$$\alpha|P\rangle = a|P\rangle, \tag{10}$$

where α is a linear operator and a is a number. This equation usually presents itself in the form that α is a known linear operator and the number a and the ket $|P\rangle$ are unknowns, which we have to try to choose so as to satisfy (10), ignoring the trivial solution $|P\rangle = 0$.

Equation (10) means that the linear operator α applied to the ket $|P\rangle$ just multiplies this ket by a numerical factor without changing its direction, or else multiplies it by the factor zero, so that it ceases to have a direction. This same α applied to other kets will, of course, in general change both their lengths and their directions. It should be noticed that only the direction of $|P\rangle$ is of importance in equation (10). If one multiplies $|P\rangle$ by any number not zero, it will not affect the question of whether (10) is satisfied or not.

Together with equation (10), we should consider also the conjugate imaginary form of equation

$$\langle Q|\alpha = b\langle Q|, \tag{11}$$

where b is a number. Here the unknowns are the number b and the non-zero bra $\langle Q|$. Equations (10) and (11) are of such fundamental importance in the theory that it is desirable to have some special words to describe the relationships between the quantities involved. If (10) is satisfied, we shall call a an *eigenvalue*[1] of the linear operator α, or of the corresponding dynamical variable, and we shall call $|P\rangle$ an *eigenket* of the linear operator or dynamical variable. Further, we shall say that the eigenket $|P\rangle$ *belongs to* the eigenvalue a. Similarly, if (11) is satisfied, we shall call b an eigenvalue of α and $\langle Q|$ an eigenbra belonging to this eigenvalue. The words eigenvalue, eigenket, eigenbra have a meaning, of course, *only with reference to a linear operator or dynamical variable*.

Using this terminology, we can assert that, if an eigenket of α is multiplied by any number not zero, the resulting ket is also an eigenket and belongs to the same eigenvalue as the original one. It is possible to have two or more independent eigenkets of a linear operator belonging to the same eigenvalue of that linear operator, e.g. equation (10) may have several solutions, $|P1\rangle, |P2\rangle, \ldots$ say, all holding for the same value of a, with the various eigenkets $|P1\rangle, |P2\rangle, \ldots$ independent. In this case it is evident that any linear combination of the eigenkets is another eigenket belonging to the same eigenvalue of the linear operator, e.g.

$$c_1|P1\rangle + c_2|P2\rangle + \cdots$$

is another solution of (10), where $c_1, c_2, \ldots$ are any numbers.

In the special case when the linear operator α of equations (10) and (11) is a number, k say, it is obvious that any ket $|P\rangle$ and bra $\langle Q|$ will satisfy these equations provided a and b equal k. Thus a number considered as a linear operator has just one eigenvalue, and any ket is an eigenket and any bra is an eigenbra, belonging to this eigenvalue.

---

[1] The word 'proper' is sometimes used instead of 'eigen', but this is not satisfactory as the words 'proper' and 'improper' are often used with other meanings. For example, in §§ 15 and 46 the words 'improper function' and 'proper-energy' are used.

9. EIGENVALUES AND EIGENVECTORS

The theory of eigenvalues and eigenvectors of a linear operator α which is not real is not of much use for quantum mechanics. We shall therefore confine ourselves to real linear operators for the further development of the theory. Putting for α the real linear operator ξ, we have instead of equations (10) and (11)

$$\xi|P\rangle = a|P\rangle, \qquad (12)$$

$$\langle Q|\xi = b\langle Q|. \qquad (13)$$

Three important results can now be readily deduced.

(i) *The eigenvalues are all real numbers.* To prove that a satisfying (12) is real, we multiply (12) by the bra $\langle P|$ on the left, obtaining

$$\langle P|\xi|P\rangle = a\langle P|P\rangle.$$

Now from equation (4) with $\langle B|$ replaced by $\langle P|$ and α replaced by the real linear operator ξ, we see that the number $\langle P|\xi|P\rangle$ must be real, and from (8) of § 6, $\langle P|P\rangle$ must be real and not zero. Hence a is real. Similarly, by multiplying (13) by $|Q\rangle$ on the right, we can prove that b is real.

Suppose we have a solution of (12) and we form the conjugate imaginary equation, which will read

$$\langle P|\xi = a\langle P|$$

in view of the reality of ξ and a. This conjugate imaginary equation now provides a solution of (13), with $\langle Q| = \langle P|$ and $b = a$. Thus we can infer

(ii) *The eigenvalues associated with eigenkets are the same as the eigenvalues associated with eigenbras.*

(iii) *The conjugate imaginary of any eigenket is an eigenbra belonging to the same eigenvalue, and conversely.* This last result makes it reasonable to call the state corresponding to any eigenket or to the conjugate imaginary eigenbra an *eigenstate* of the real dynamical variable ξ.

Eigenvalues and eigenvectors of various real dynamical variables are used very extensively in quantum mechanics, so it is desirable to have some systematic notation for labelling them. The following is suitable for most purposes. If ξ is a real dynamical variable, we call its eigenvalues ξ', ξ'', ξ^r, etc. Thus we have a letter by itself denoting a *real dynamical variable* or a *real linear operator*, and the same letter with primes or an index attached denoting a *number*, namely an eigenvalue of what the letter by itself denotes. An eigenvector may now be labelled by the eigenvalue to which it belongs. Thus $|\xi'\rangle$ denotes an eigenket belonging to the eigenvalue ξ' of the dynamical variable ξ. If in a piece of work we deal with more than one eigenket belonging to the same eigenvalue of a dynamical variable, we may distinguish them one from another by means of a further label, or possibly of more than one further labels. Thus, if we are dealing with two eigenkets belonging to the same eigenvalue of ξ', we may call them $|\xi'1\rangle$ and $|\xi'2\rangle$.

THEOREM. *Two eigenvectors of a real dynamical variable belonging to different eigenvalues are orthogonal.*

To prove the theorem, let $|\xi'\rangle$ and $|\xi''\rangle$ be two eigenkets of the real dynamical variable ξ, belonging to the eigenvalues ξ' and ξ'' respectively. Then we have the equations

$$\xi|\xi'\rangle = \xi'|\xi'\rangle, \tag{14}$$

$$\xi|\xi''\rangle = \xi''|\xi''\rangle. \tag{15}$$

Taking the conjugate imaginary of (14), we get

$$\langle\xi'|\xi = \xi'\langle\xi'|.$$

Multiplying this by $|\xi''\rangle$ on the right gives

$$\langle\xi'|\xi|\xi''\rangle = \xi'\langle\xi'|\xi''\rangle$$

and multiplying (15) by $\langle\xi'|$ on the left gives

$$\langle\xi'|\xi|\xi''\rangle = \xi''\langle\xi'|\xi''\rangle.$$

Hence, subtracting,

$$(\xi' - \xi'')\langle\xi'|\xi''\rangle = 0, \tag{16}$$

showing that, if $\xi' \neq \xi''$, $\langle\xi'|\xi''\rangle = 0$ and the two eigenvectors $|\xi'\rangle$ and $|\xi''\rangle$ are orthogonal. This theorem will be referred to as the *orthogonality theorem*.

We have been discussing properties of the eigenvalues and eigenvectors of a real linear operator, but have not yet considered the question of whether, for a given real linear operator, any eigenvalues and eigenvectors exist, and if so, how to find them. This question is in general very difficult to answer. There is one useful special case, however, which is quite tractable, namely when the real linear operator, ξ say, satisfies an algebraic equation

$$\phi(\xi) \equiv \xi^n + a_1\xi^{n-1} + a_2\xi^{n-2} + \cdots + a_n = 0, \tag{17}$$

the coefficients a being numbers. This equation means, of course, that the linear operator $\phi(\xi)$ produces the result zero when applied to any ket vector or to any bra vector.

9. EIGENVALUES AND EIGENVECTORS

Let (17) be the simplest algebraic equation that ξ satisfies. Then it will be shown that

(α) The number of eigenvalues of ξ is n.
(β) There are so many eigenkets of ξ that any ket whatever can be expressed as a sum of such eigenkets.

The algebraic form $\phi(\xi)$ can be factorized into n linear factors, the result being

$$\phi(\xi) \equiv (\xi - c_1)(\xi - c_2) \cdots (\xi - c_n) \tag{18}$$

say, the c's being numbers, not assumed to be all different. This factorization can be performed with ξ a linear operator just as well as with ξ an ordinary algebraic variable, since there is nothing occurring in (18) that does not commute with ξ. Let the quotient when $\phi(\xi)$ is divided by $(\xi - c_r)$ be $\chi_r(\xi)$, so that

$$\phi(\xi) \equiv (\xi - c_r)\chi_r(\xi) \quad (r = 1, 2, \ldots, n).$$

Then, for any ket $|P\rangle$,

$$(\xi - c_r)\chi_r(\xi)|P\rangle = \phi(\xi)|P\rangle = 0. \tag{19}$$

Now $\chi_r(\xi)|P\rangle$ cannot vanish for every ket $|P\rangle$, as otherwise $\chi_r(\xi)$ itself would vanish and we should have ξ satisfying an algebraic equation of degree $n-1$, which would contradict the assumption that (17) is the simplest equation that ξ satisfies. If we choose $|P\rangle$ so that $\chi_r(\xi)|P\rangle$ does not vanish, then equation (19) shows that $\chi_r(\xi)|P\rangle$ is an eigenket of ξ, belonging to the eigenvalue c_r. The argument holds for each value of r from 1 to n, and hence each of the c's is an eigenvalue of ξ. No other number can be an eigenvalue of ξ, since if ξ' is any eigenvalue, belonging to an eigenket $|\xi'\rangle$,

$$\xi|\xi'\rangle = \xi'|\xi'\rangle$$

and we can deduce

$$\phi(\xi)|\xi'\rangle = \phi(\xi')|\xi'\rangle,$$

and since the left-hand side vanishes we must have $\phi(\xi') = 0$.

To complete the proof of (α) we must verify that the c's are all different. Suppose the c's are not all different and c_s occurs m times say, with $m > 1$. Then $\phi(\xi)$ is of the form

$$\phi(\xi) \equiv (\xi - c_s)^m \theta(\xi),$$

with $\theta(\xi)$ a rational integral function of ξ. Equation (17) now gives us

$$(\xi - c_s)^m \theta(\xi)|A\rangle = 0 \tag{20}$$

for any ket $|A\rangle$. Since c_s is an eigenvalue of ξ it must be real, so that $\xi - c_s$ is a real linear operator. Equation (20) is now of the same form as equation (8) with $\xi - c_s$ for ξ and $\theta(\xi)|A\rangle$ for $|P\rangle$. From the theorem connected with equation (8) we can infer that

$$(\xi - c_s)\theta(\xi)|A\rangle = 0.$$

Since the ket $|A\rangle$ is arbitrary,

$$(\xi - c_s)\theta(\xi) = 0,$$

which contradicts the assumption that (17) is the simplest equation that ξ satisfies. Hence the c's are all different and (α) is proved.

Let $\chi_r(c_r)$ be the number obtained when c_r is substituted for ξ in the algebraic expression $\chi_r(\xi)$. Since the c's are all different, $\chi_r(c_r)$ cannot vanish. Consider now the expression

$$\sum_r \frac{\chi_r(\xi)}{\chi_r(c_r)} - 1. \tag{21}$$

If c_s is substituted for ξ here, every term in the sum vanishes except the one for which $r = s$, since $\chi_r(\xi)$ contains $(\xi - c_s)$ as a factor when $r \neq s$, and the term for which $r = s$ is unity, so the whole expression vanishes. Thus the expression (21) vanishes when ξ is put equal to any of the n numbers $c_1, c_2, \ldots, c_n$. Since, however, the expression is only of degree $n - 1$ in ξ, it must vanish identically. If we now apply the linear operator (21) to an arbitrary ket $|P\rangle$ and equate the result to zero, we get

$$|P\rangle = \sum_r \frac{1}{\chi_r c_r} \chi_r(\xi)|P\rangle. \tag{22}$$

Each term in the sum on the right here is, according to (19), an eigenket of ξ, if it does not vanish. Equation (22) thus expresses the arbitrary ket $|P\rangle$ as a sum of eigenkets of ξ, and thus (β) is proved.

As a simple example we may consider a real linear operator σ that satisfies the equation

$$\sigma^2 = 1. \tag{23}$$

Then σ has the two eigenvalues 1 and -1. Any ket $|P\rangle$ can be expressed as

$$|P\rangle = \frac{1}{2}(1 + \sigma)|P\rangle + \frac{1}{2}(1 - \sigma)|P\rangle.$$

It is easily verified that the two terms on the right here are eigenkets of σ, belonging to the eigenvalues 1 and -1 respectively, when they do not vanish.

10. Observables

We have made a number of assumptions about the way in which states and dynamical variables are to be represented mathematically in the theory. These assumptions are not, by themselves, laws of nature, but become laws of nature when we make some further assumptions that provide a physical interpretation of the theory. Such further assumptions must take the form of establishing connexions between the results of observations, on one hand, and the equations of the mathematical formalism on the other.

When we make an observation we measure some dynamical variable. It is obvious physically that the result of such a measurement must always be a real number, so we should expect that any dynamical variable that we can measure must be a real dynamical variable. One might think one could measure a complex dynamical variable by measuring separately its real and pure imaginary parts. But this would involve two measurements or two observations, which would be all right in classical mechanics, but would not do in quantum mechanics, where two observations in general interfere with one another—it is not in general permissible to consider that two observations can be made exactly simultaneously, and if they are made in quick succession the first will usually disturb the state of the system and introduce an indeterminacy that will affect the second. We therefore have to restrict the dynamical variables that we can measure to be real, the condition for this in quantum mechanics being as given in § 8. Not every real dynamical variable can be measured, however. A further restriction is needed, as we shall see later.

We now make some assumptions for the physical interpretation of the theory. *If the dynamical system is in an eigenstate of a real dynamical variable ξ, belonging to the eigenvalue ξ', then a measurement of ξ will certainly give as result the number ξ'.* Conversely, *if the system is in a state such that a measurement of a real dynamical variable ξ is certain to give one particular result* (instead of giving one or other of several possible results according to a probability law, as is in general the case), *then the state is an eigenstate of ξ and the result of the measurement is the eigenvalue of ξ to which this eigenstate belongs.* These assumptions are reasonable on account of the eigenvalues of real linear operators being always real numbers.

Some of the immediate consequences of the assumptions will be noted. If we have two or more eigenstates of a real dynamical variable ξ belonging to the same eigenvalue ξ', then any state formed by superposition of them will also be an eigenstate of ξ belonging to the eigenvalue ξ'. We can infer that if we have two or more states for which a measurement of ξ is certain to give the result ξ', then for any state formed by superposition of them a measurement of ξ will still be certain to give the result ξ'. This gives us some insight into the physical significance of superposition of states. Again, two eigenstates of ξ belonging to different eigenvalues are orthogonal. We can infer that two states for which a measurement of ξ is certain to give two different results are orthogonal. This gives us some insight into the physical significance of orthogonal states.

When we measure a real dynamical variable ξ, the disturbance involved in the act of measurement causes a jump in the state of the dynamical system. From physical continuity, if we make a second measurement of the same dynamical variable ξ immediately after the first, the result of the second measurement must be the same as that of the first. Thus after the first measurement has been made, there is no indeterminacy in the result of the second. Hence, after the first measurement has been made, the system is in an eigenstate of the dynamical variable ξ, the eigenvalue it belongs to being equal to the result of the first measurement. This conclusion must still hold if the second measurement is not actually made. In this way we see that a measurement always causes the system to jump into an eigenstate of the dynamical variable that is being measured, the eigenvalue this eigenstate belongs to being equal to the result of the measurement.

We can infer that, with the dynamical system in any state, *any result of a measurement of a real dynamical variable is one of its eigenvalues.* Conversely, *every eigenvalue is a possible result of a measurement of the dynamical variable for some state of the system*, since it is certainly the result if the state is an eigenstate belonging to this eigenvalue. This gives us the physical significance of eigenvalues. The set of eigenvalues of a real dynamical variable are just the possible results of measurements of that dynamical variable and the calculation of eigenvalues is for this reason an important problem.

Another assumption we make connected with the physical interpretation of the theory is that, *if a certain real dynamical variable ξ is measured with the system in a particular state, the states into which the system may jump on account of the measurement are such that the original state is dependent on them*. Now these states into which the system may jump are all eigenstates of ξ, and hence the original state is dependent on eigenstates of ξ. But the original state may be *any* state, so we can conclude that any state is dependent on eigenstates of ξ. If we define a *complete* *set* of states to be a set such that any state is dependent on them, then our conclusion can be formulated—the eigenstates of ξ form a complete set.

Not every real dynamical variable has sufficient eigenstates to form a complete set. Those whose eigenstates do not form complete sets are not quantities that can be measured. We obtain in this way a further condition that a dynamical variable has to satisfy in order that it shall be susceptible to measurement, in addition to the condition that it shall be real. We call a real dynamical variable whose eigenstates form a complete set an *observable*. Thus any quantity that can be measured is an observable.

The question now presents itself—Can every observable be measured? The answer theoretically is yes. In practice it may be very awkward, or perhaps even beyond the ingenuity of the experimenter, to devise an apparatus which could measure some particular observable, but the theory always allows one to imagine that the measurement can be made.

Let us examine mathematically the condition for a real dynamical variable ξ to be an observable. Its eigenvalues may consist of a (finite or infinite) discrete set of numbers, or alternatively, they may consist of all numbers in a certain range, such as all numbers lying between a and b. In the former case, the condition that any state is dependent on eigenstates of ξ is that any ket can be expressed as a sum of eigenkets of ξ. In the latter case the condition needs modification, since one may have an integral instead of a sum, i.e. a ket $|P\rangle$ may be expressible as an integral of eigenkets of ξ,

$$|P\rangle = \int |\xi'\rangle \, d\xi', \qquad (24)$$

$|\xi'\rangle$ being an eigenket of ξ belonging to the eigenvalue ξ' and the range of integration being the range of eigenvalues, as such a ket is dependent on eigenkets of ξ. Not every ket dependent on eigenkets of ξ can be expressed in the form of the right-hand side of (24), since one of the eigenkets itself cannot, and more generally any sum of eigenkets cannot. The condition for the eigenstates of ξ to form a complete set must thus be formulated, that any ket $|P\rangle$ can be expressed as an integral plus a sum of eigenkets of ξ, i.e.

$$|P\rangle = \int |\xi'c\rangle \, d\xi' + \sum_r |\xi^r d\rangle, \qquad (25)$$

where the $|\xi'c\rangle$, $|\xi^r d\rangle$ are all eigenkets of ξ, the labels c and d being inserted to distinguish them when the eigenvalues ξ' and ξ^r are equal, and where the integral is taken over the whole range of eigenvalues and the sum is taken over any selection of them. If this condition is satisfied in the case when the eigenvalues of ξ consist of a range of numbers, then ξ is an observable.

There is a more general case that sometimes occurs, namely the eigenvalues of ξ may consist of a range of numbers together with a discrete set of numbers lying outside the range. In this case the condition that ξ shall be an observable is still that any ket shall be expressible in the form of the right-hand side of (25), but the sum over r is now a sum over the discrete set of eigenvalues as well as a selection of those in the range.

It is often very difficult to decide mathematically whether a particular real dynamical variable satisfies the condition for being an observable or not, because the whole problem of finding eigenvalues and eigenvectors is in general very difficult. However, we may have good reason on experimental grounds for believing that the dynamical variable can be measured and then we may reasonably assume that it is an observable even though the mathematical proof is missing. This is a thing we shall frequently do during the course of development of the theory, e.g. we shall assume the energy of any dynamical system to be always an observable, even though it is beyond the power of present-day mathematical analysis to prove it so except in simple cases.

In the special case when the real dynamical variable is a number, every state is an eigenstate and the dynamical variable is obviously an observable. Any measurement of it always gives the same result, so it is just a physical constant, like the charge on an electron. A physical constant in quantum mechanics may thus be looked upon either as an observable with a single eigenvalue or as a mere number appearing in the equations, the two points of view being equivalent.

If the real dynamical variable satisfies an algebraic equation, then the result (β) of the preceding section shows that the dynamical variable is an observable. Such an observable has a finite number of eigenvalues. Conversely, any observable with a finite number of eigenvalues satisfies an algebraic equation, since if the observable ξ has as its eigenvalues $\xi', \xi'', \ldots, \xi^n$, then

$$(\xi - \xi')(\xi - \xi'') \ldots (\xi - \xi^n)|P\rangle = 0$$

holds for $|P\rangle$ any eigenket of ξ, and thus it holds for any $|P\rangle$ whatever, because any ket can be expressed as a sum of eigenkets of ξ on account of ξ being an observable. Hence

$$(\xi - \xi')(\xi - \xi'') \ldots (\xi - \xi^n) = 0. \tag{26}$$

10. OBSERVABLES

As an example we may consider the linear operator $|A\rangle\langle A|$, where $|A\rangle$ is a normalized ket. This linear operator is real according to (7), and its square is

$$(|A\rangle\langle A|)^2 = |A\rangle\langle A|A\rangle\langle A| = |A\rangle\langle A| \qquad (27)$$

since $\langle A|A\rangle = 1$. Thus its square equals itself and so it satisfies an algebraic equation and is an observable. Its eigenvalues are 1 and 0, with $|A\rangle$ as the eigenket belonging to the eigenvalue 1 and all kets orthogonal to $|A\rangle$ as eigenkets belonging to the eigenvalue 0. A measurement of the observable thus certainly gives the result 1 if the dynamical system is in the state corresponding to $|A\rangle$ and the result 0 if the system is in any orthogonal state, so the observable may be described as the quantity which determines whether the system is in the state $|A\rangle$ or not.

Before concluding this section we should examine the conditions for an integral such as occurs in (24) to be significant. Suppose $|X\rangle$ and $|Y\rangle$ are two kets which can be expressed as integrals of eigenkets of the observable ξ,

$$|X\rangle = \int |\xi'x\rangle\, d\xi', \quad |Y\rangle = \int |\xi''y\rangle\, d\xi'',$$

x and y being used as labels to distinguish the two integrands. Then we have, taking the conjugate imaginary of the first equation and multiplying by the second

$$\langle X|Y\rangle = \iint \langle \xi'x|\xi''y\rangle\, d\xi'\, d\xi''. \qquad (28)$$

Consider now the single integral

$$\int \langle \xi'x|\xi''y\rangle\, d\xi''. \qquad (29)$$

From the orthogonality theorem, the integrand here must vanish over the whole range of integration except the one point $\xi'' = \xi'$. If the integrand is finite at this point, the integral (29) vanishes, and if this holds for all ξ', we get from (28) that $\langle X|Y\rangle$ vanishes. Now in general $\langle X|Y\rangle$ does not vanish, so in general $\langle \xi'x|\xi'y\rangle$ must be infinitely great in such a way as to make (29) non-vanishing and finite. The form of infinity required for this will be discussed in § 15.

In our work up to the present it has been implied that our bra and ket vectors are of finite length and their scalar products are finite. We see now the need for relaxing this condition when we are dealing with eigenvectors of an observable whose eigenvalues form a range. If we did not relax it, the phenomenon of ranges of eigenvalues could not occur and our theory would be too weak for most practical problems.

Taking $|Y\rangle = |X\rangle$ above, we get the result that in general $\langle \xi'x|\xi'x\rangle$ is infinitely great. We shall assume that if $|\xi'x\rangle \neq 0$

$$\int \langle \xi'x|\xi''x\rangle\, d\xi'' > 0, \tag{30}$$

as the axiom corresponding to (8) of § 6 for vectors of infinite length.

The space of bra or ket vectors when the vectors are restricted to be of finite length and to have finite scalar products is called by mathematicians a *Hilbert space*. The bra and ket vectors that we now use form a more general space than a Hilbert space.

We can now see that the expansion of a ket $|P\rangle$ in the form of the right-hand side of (25) is unique, provided there are not two or more terms in the sum referring to the same eigenvalue. To prove this result, let us suppose that two different expansions of $|P\rangle$ are possible. Then by subtracting one from the other, we get an equation of the form

$$0 = \int |\xi'a\rangle\, d\xi' + \sum_s |\xi^s b\rangle, \tag{31}$$

a and b being used as new labels for the eigenvectors, and the sum over s including all terms left after the subtraction of one sum from the other. If there is a term in the sum in (31) referring to an eigenvalue ξ' not in the range, we get, by multiplying (31) on the left by $\langle \xi'b|$ and using the orthogonality theorem,

$$0 = \langle \xi'b|\xi'b\rangle,$$

which contradicts (8) of § 6. Again, if the integrand in (31) does not vanish for some eigenvalue ξ'' not equal to any ξ^s occurring in the sum, we get, by multiplying (31) on the left by $\langle \xi''a|$ and using the orthogonality theorem,

$$0 = \int \langle \xi''a|\xi'a\rangle\, d\xi',$$

which contradicts (30). Finally, if there is a term in the sum in (31) referring to an eigenvalue ξ' in the range, we get, multiplying (31) on the left by $\langle \xi'b|$,

$$0 = \int \langle \xi'b|\xi'a\rangle\, d\xi' + \langle \xi'b|\xi'b\rangle \tag{32}$$

and multiplying (31) on the left by $|\xi'a\rangle$

$$0 = \int \langle \xi'a|\xi'a\rangle\, d\xi' + \langle \xi'a|\xi'b\rangle. \tag{33}$$

Now the integral in (33) is finite, so $\langle \xi'a|\xi'b\rangle$ is finite and $\langle \xi'b|\xi'a\rangle$ is finite. The integral in (32) must then be zero, so $\langle \xi'b|\xi'b\rangle$ is zero and we again have a contradiction. Thus every term in (31) must vanish and the expansion of a ket $|P\rangle$ in the form of the right-hand side of (25) must be unique.

11. Functions of Observables

Let ξ be an observable. We can multiply it by any real number k and get another observable $k\xi$. In order that our theory may be self-consistent it is necessary that, when the system is in a state such that a measurement of the observable ξ certainly gives the result ξ', a measurement of the observable $k\xi$ shall certainly give the result $k\xi'$. It is easily verified that this condition is fulfilled. The ket corresponding to a state for which a measurement of ξ certainly gives the result ξ' is an eigenket of ξ, $|\xi'\rangle$ say, satisfying

$$\xi|\xi'\rangle = \xi'|\xi'\rangle.$$

This equation leads to

$$k\xi|\xi'\rangle = k\xi'|\xi'\rangle,$$

showing that $|\xi'\rangle$ is an eigenket of $k\xi$ belonging to the eigenvalue $k\xi'$, and thus that a measurement of $k\xi$ will certainly give the result $k\xi'$.

More generally, we may take any real function of ξ, $f(\xi)$ say, and consider it as a new observable which is automatically measured whenever ξ is measured, since an experimental determination of the value of ξ also provides the value of $f(\xi)$. We need not restrict $f(\xi)$ to be real, and then its real and pure imaginary parts are two observables which are automatically measured when ξ is measured. For the theory to be consistent it is necessary that, when the system is in a state such that a measurement of ξ certainly gives the result ξ', a measurement of the real and pure imaginary parts of $f(\xi)$ shall certainly give for results the real and pure imaginary parts of $f(\xi')$. In the case when $f(\xi)$ is expressible as a power series

$$f(\xi) = c_0 + c_1\xi + c_2\xi^2 + \cdots,$$

the c's being numbers, this condition can again be verified by elementary algebra. In the case of more general functions f it may not be possible to verify the condition. The condition may then be used to define $f(\xi)$, which we have not yet defined mathematically. In this way we can get a more general definition of a function of an observable than is provided by power series.

We define $f(\xi)$ in general to be that linear operator which satisfies

$$f(\xi)|\xi'\rangle = f(\xi')|\xi'\rangle \tag{34}$$

for every eigenket $|\xi'\rangle$ of ξ, $f(\xi')$ being a number for each eigenvalue ξ'. It is easily seen that this definition is self-consistent when applied to eigenkets $|\xi'\rangle$ that are not independent. If we have an eigenket $|\xi'A\rangle$ dependent on other eigenkets of ξ, these other eigenkets must all belong to the same eigenvalue ξ', otherwise we should have an equation of the type (31), which we have seen is impossible. On multiplying the equation which expresses $|\xi'A\rangle$ linearly in terms of the other eigenkets of ξ by $f(\xi)$ on the left, we merely multiply each term in it by the number $f(\xi')$, so we obviously get a consistent equation. Further, equation (34) is sufficient to define the linear operator $f(\xi)$ completely, since to get the result of $f(\xi)$ multiplied into an arbitrary ket $|P\rangle$, we have only to expand $|P\rangle$ in the form of the right-hand side of (25) and take

$$f(\xi)|P\rangle = \int f(\xi')|\xi'c\rangle\, d\xi' + \sum_r f(\xi^r)|\xi^r d\rangle. \tag{35}$$

The conjugate complex $\overline{f(\xi)}$ of $f(\xi)$ is defined by the conjugate imaginary equation to (34), namely

$$\langle\xi'|\overline{f(\xi)} = \overline{f(\xi')}\langle\xi'|,$$

holding for any eigenbra $\langle\xi'|$, $\overline{f(\xi')}$ being the conjugate complex function to $f(\xi')$. Let us replace ξ' here by ξ'' and multiply the equation on the right by the arbitrary ket $|P\rangle$. Then we get, using the expansion (25) for $|P\rangle$,

$$\langle\xi''|\overline{f(\xi)}|P\rangle = \overline{f(\xi'')}\langle\xi''|P\rangle$$
$$= \int \overline{f(\xi'')}\langle\xi''|\xi'c\rangle\, d\xi' + \sum_r \overline{f(\xi'')}\langle\xi''|\xi^r d\rangle$$
$$= \int \overline{f(\xi'')}\langle\xi''|\xi'c\rangle\, d\xi' + \overline{f(\xi'')}\langle\xi''|\xi''d\rangle \tag{36}$$

with the help of the orthogonality theorem, $\langle\xi''|\xi''d\rangle$ being understood to be zero if ξ'' is not one of the eigenvalues to which the terms in the sum in (25) refer. Again, putting the conjugate complex function $\overline{f(\xi')}$ for $f(\xi')$ in (35) and multiplying on the left by $\langle\xi''|$, we get

$$\langle\xi''|\overline{f(\xi)}|P\rangle = \int \overline{f(\xi')}\langle\xi''|\xi'c\rangle\, d\xi' + \overline{f(\xi'')}\langle\xi''|\xi''d\rangle.$$

11. FUNCTIONS OF OBSERVABLES

The right-hand side here equals that of (36) since the integrands vanish for $\xi' \neq \xi''$, and hence

$$\langle \xi'' | \overline{f(\xi)} | P \rangle = \langle \xi'' | \bar{f}(\xi) | P \rangle.$$

This holds for $\langle \xi'' |$ any eigenbra and $|P\rangle$ any ket, so

$$\overline{f(\xi)} = \bar{f}(\xi). \tag{37}$$

Thus *the conjugate complex of the linear operator $f(\xi)$ is the conjugate complex function $\bar{f}$ of ξ*.

It follows as a corollary that if $f(\xi')$ is a real function of ξ', $f(\xi)$ is a real linear operator. $f(\xi)$ is then also an observable, since its eigenstates form a complete set, every eigenstate of ξ being also an eigenstate of $f(\xi)$.

With the above definition *we are able to give a meaning to any function f of an observable, provided only that the domain of existence of the function of a real variable $f(x)$ includes all the eigenvalues of the observable*. If the domain of existence contains other points besides these eigenvalues, then the values of $f(x)$ for these other points will not affect the function of the observable. The function need not be analytic or continuous. The eigenvalues of a function f of an observable are just the function f of the eigenvalues of the observable.

It is important to observe that the possibility of defining a function f of an observable requires the existence of a unique number $f(x)$ for each value of x which is an eigenvalue of the observable. Thus the function $f(x)$ must be single-valued. This may be illustrated by considering the question: When we have an observable $f(A)$ which is a real function of the observable A, is the observable A a function of the observable $f(A)$? The answer to this is yes, if different eigenvalues A' of A always lead to different values of $f(A')$. If, however, there exist two different eigenvalues of A, A' and A'' say, such that $f(A') = f(A'')$, then, corresponding to the eigenvalue $f(A')$ of the observable $f(A)$, there will not be a unique eigenvalue of the observable A and the latter will not be a function of the observable $f(A)$.

It may easily be verified mathematically, from the definition, that the sum or product of two functions of an observable is a function of that observable and that a function of a function of an observable is a function of that observable. Also it is easily seen that the whole theory of functions of an observable is symmetrical between bras and kets and that we could equally well work from the equation

$$\langle \xi' | f(\xi) = f(\xi') \langle \xi' | \tag{38}$$

instead of from (34).

We shall conclude this section with a discussion of two examples which are of great practical importance, namely the reciprocal and the square root. The reciprocal of an observable exists if the observable does not have the eigenvalue zero. If the observable α does not have the eigenvalue zero, the reciprocal observable, which we call α^{-1} or $1/\alpha$, will satisfy

$$\alpha^{-1}|\alpha'\rangle = \alpha'^{-1}|\alpha'\rangle, \qquad (39)$$

where $|\alpha'\rangle$ is an eigenket of α belonging to the eigenvalue α'. Hence

$$\alpha\alpha^{-1}|\alpha'\rangle = \alpha\alpha'^{-1}|\alpha'\rangle = |\alpha'\rangle.$$

Since this holds for any eigenket $|\alpha'\rangle$, we must have

$$\alpha\alpha^{-1} = 1. \qquad (40)$$

Similarly,

$$\alpha^{-1}\alpha = 1. \qquad (41)$$

Either of these equations is sufficient to determine α^{-1} completely, provided α does not have the eigenvalue zero. To prove this in the case of (40), let x be any linear operator satisfying the equation

$$\alpha x = 1$$

and multiply both sides on the left by the α^{-1} defined by (39). The result is

$$\alpha^{-1}\alpha x = \alpha^{-1}$$

and hence from (41)

$$x = \alpha^{-1}.$$

Equations (40) and (41) can be used to define the reciprocal, when it exists, of a general linear operator α, which need not even be real. One of these equations by itself is then not necessarily sufficient. If any two linear operators α and β have reciprocals, their product $\alpha\beta$ has the reciprocal

$$(\alpha\beta)^{-1} = \beta^{-1}\alpha^{-1}, \qquad (42)$$

obtained by taking the reciprocal of each factor and reversing their order. We verify (42) by noting that its right-hand side gives unity when multiplied by $\alpha\beta$, either on the right or on the left. This reciprocal law for products can be immediately extended to more than two factors, i.e.,

$$(\alpha\beta\gamma\cdots)^{-1} = \cdots\gamma^{-1}\beta^{-1}\alpha^{-1}.$$

The square root of an observable α always exists, and is real if α has no negative eigenvalues. We write it $\sqrt{\alpha}$ or $\alpha^{1/2}$. It satisfies

$$\sqrt{\alpha}|\alpha'\rangle = \pm\sqrt{\alpha'}|\alpha'\rangle, \qquad (43)$$

$|\alpha'\rangle$ being an eigenket of α belonging to the eigenvalue α'. Hence

$$\sqrt{\alpha}\sqrt{\alpha}|\alpha'\rangle = \sqrt{\alpha'}\sqrt{\alpha'}|\alpha'\rangle = \alpha'|\alpha'\rangle = \alpha|\alpha'\rangle,$$

and since this holds for any eigenket $|\alpha'\rangle$ we must have

$$\sqrt{\alpha}\sqrt{\alpha} = \alpha. \qquad (44)$$

On account of the ambiguity of sign in (43) there will be several square roots. To fix one of them we must specify a particular sign in (43) for each eigenvalue. This sign may vary irregularly from one eigenvalue to the next and equation (43) will always define a linear operator $\sqrt{\alpha}$ satisfying (44) and forming a square-root function of α. If there is an eigenvalue of α with two or more independent eigenkets belonging to it, then we must, according to our definition of a function, have the same sign in (43) for each of these eigenkets. If we took different signs, however, equation (44) would still hold, and hence equation (44) by itself is not sufficient to define $\sqrt{\alpha}$, except in the special case when there is only one independent eigenket of α belonging to any eigenvalue.

The number of different square roots of an observable is 2^n, where n is the total number of eigenvalues not zero. In practice the square root function is used only for observables without negative eigenvalues and the particular square root that is useful is the one for which the positive sign is always taken in (43). This one will be called the *positive square root*.

12. The General Physical Interpretation

The assumptions that we made at the beginning of § 10 to get a physical interpretation of the mathematical theory are of a rather special kind, since they can be used only in connexion with eigenstates. We need some more general assumption which will enable us to extract physical information from the mathematics even when we are not dealing with eigenstates.

In classical mechanics an observable always, as we say, 'has a value' for any particular state of the system. What is there in quantum mechanics corresponding to this? If we take any observable ξ and any two states x and y, corresponding to the vectors $\langle x|$ and $|y\rangle$, then we can form the number $\langle x|\xi|y\rangle$. This number is not very closely analogous to the value which an observable can 'have' in the classical theory, for three reasons, namely, (i) it refers to *two* states of the system, while the classical value always refers to *one*, (ii) it is in general not a real number, and (iii) it is not uniquely determined by the observable and the states, since the vectors $\langle x|$ and $|y\rangle$ contain arbitrary numerical factors. Even if we impose on $\langle x|$ and $|y\rangle$ the condition that they shall be normalized, there will still be an undetermined factor of modulus unity in $\langle x|\xi|y\rangle$. These three reasons cease to apply, however, if we take the two states to be identical and $|y\rangle$ to be the conjugate imaginary vector to $\langle x|$. The number that we then get, namely $\langle x|\xi|x\rangle$, is necessarily real, and also it is uniquely determined when $\langle x|$ is normalized, since if we multiply $\langle x|$ by the numerical factor e^{ic}, c being some real number, we must multiply $|x\rangle$ by e^{-ic} and $\langle x|\xi|x\rangle$ will be unaltered.

One might thus be inclined to make the tentative assumption that the observable ξ 'has the value' $\langle x|\xi|x\rangle$ for the state x, in a sense analogous to the classical sense. This would not be satisfactory, though, for the following reason. Let us take a second observable η, which would have by the above assumption the value $\langle x|\eta|x\rangle$ for this same state. We should then expect, from classical analogy, that for this state the sum of the two observables would have a value equal to the sum of the values of the two observables separately and the product of the two observables would have a value equal to the product of the values of the two observables separately. Actually, the tentative assumption would give for the sum of the two observables the value $\langle x|\xi + \eta|x\rangle$, which is, in fact, equal to the sum of $\langle x|\xi|x\rangle$ and $\langle x|\eta|x\rangle$, but for the product it would give the value $\langle x|\xi\eta|x\rangle$ or $\langle x|\eta\xi|x\rangle$, neither of which is connected in any simple way with $\langle x|\xi|x\rangle$ and $\langle x|\eta|x\rangle$.

However, since things go wrong only with the product and not with the sum, it would be reasonable to call $\langle x|\xi|x\rangle$ the *average* value of the observable ξ for the state x. This is because the average of the sum of two quantities must equal the sum of their averages, but the average of their product need not equal the product of their averages. We therefore make the general assumption that *if the measurement of the observable ξ for the system in the state corresponding to $|x\rangle$ is made a large number of times, the average of all the results obtained will be $\langle x|\xi|x\rangle$, provided $|x\rangle$ is normalized*. If $|x\rangle$ is not normalized, as is necessarily the case if the state x is an eigenstate of some observable belonging to an eigenvalue in a range, the assumption becomes that the average result of a measurement of ξ is proportional to $\langle x|\xi|x\rangle$. This general assumption provides a basis

12. THE GENERAL PHYSICAL INTERPRETATION

for a general physical interpretation of the theory.

The expression that an observable 'has a particular value' for a particular state is permissible in quantum mechanics in the special case when a measurement of the observable is certain to lead to the particular value, so that the state is an eigenstate of the observable. It may easily be verified from the algebra that, with this restricted meaning for an observable 'having a value', if two observables have values for a particular state, then for this state the sum of the two observables (if this sum is an observable[1]) has a value equal to the sum of the values of the two observables separately and the product of the two observables (if this product is an observable[2]) has a value equal to the product of the values of the two observables separately.

In the general case we cannot speak of an observable having a value for a particular state, but we can speak of its having an average value for the state. We can go further and speak of the probability of its having any specified value for the state, meaning the probability of this specified value being obtained when one makes a measurement of the observable. This probability can be obtained from the general assumption in the following way.

Let the observable be ξ and let the state correspond to the normalized ket $|x\rangle$. Then the general assumption tells us, not only that the average value of ξ is $\langle x|\xi|x\rangle$, but also that the average value of any function of ξ, $f(\xi)$ say, is $\langle x|f(\xi)|x\rangle$. Take $f(\xi)$ to be that function of ξ which is equal to unity when $\xi = a$, a being some real number, and zero otherwise. This function of ξ has a meaning according to our general theory of functions of an observable, and it may be denoted by $\delta_{\xi a}$ in conformity with the general notation of the symbol δ with two suffixes given on p. 62 (equation (17)). The average value of this function of ξ is just the probability, P_a say, of ξ having the value a. Thus

$$P_a = \langle x|\delta_{\xi a}|x\rangle. \tag{45}$$

If a is not an eigenvalue of ξ, $\delta_{\xi a}$ multiplied into any eigenket of ξ is zero, and hence $\delta_{\xi a} = 0$ and $P_a = 0$. This agrees with a conclusion of § 10, that any result of a measurement of an observable must be one of its eigenvalues.

---

[1] This is not obviously so, since the sum may not have sufficient eigenstates to form a complete set, in which case the sum, considered as a single quantity, would not be measurable.

[2] Here the reality condition may fail, as well as the condition for the eigenstates to form a complete set.

If the possible results of a measurement of ξ form a range of numbers, the probability of ξ having exactly a particular value will be zero in most physical problems. The quantity of physical importance is then the probability of ξ having a value within a small range, say from a to $a+da$. This probability, which we may call $P(a)\,da$, is equal to the average value of that function of ξ which is equal to unity for ξ lying within the range a to $a+da$ and zero otherwise. This function of ξ has a meaning according to our general theory of functions of an observable. Denoting it by $\chi(\xi)$, we have

$$P(a)\,da = \langle x|\chi(\xi)|x\rangle. \qquad (46)$$

If the range a to $a + da$ does not include any eigenvalues of ξ, we have as above $\chi(\xi) = 0$ and $P(a) = 0$. If $|x\rangle$ is not normalized, the right-hand sides of (45) and (46) will still be proportional to the probability of ξ having the value a and lying within the range a to $a + da$ respectively.

The assumption of § 10, that a measurement of ξ is certain to give the result ξ' if the system is in an eigenstate of ξ belonging to the eigenvalue ξ', is consistent with the general assumption for physical interpretation and can in fact be deduced from it. Working from the general assumption we see that, if $|\xi'\rangle$ is an eigenket of ξ belonging to the eigenvalue ξ', then, in the case of discrete eigenvalues of ξ,

$$\delta_{\xi a}|\xi'\rangle = 0 \quad \text{unless} \quad a = \xi',$$

and in the case of a range of eigenvalues of ξ

$$\chi(\xi)|\xi'\rangle = 0 \quad \text{unless the range } a \text{ to } a + da \text{ includes } \xi'.$$

In either case, for the state corresponding to $|\xi'\rangle$, the probability of ξ having any value other than ξ' is zero.

An eigenstate of ξ belonging to an eigenvalue ξ' lying in a range is a state which cannot strictly be realized in practice, since it would need an infinite amount of precision to get ξ to equal exactly ξ'. The most that could be attained in practice would be to get ξ to lie within a narrow range about the value ξ'. The system would then be in a state approximating to an eigenstate of ξ. Thus an eigenstate belonging to an eigenvalue in a range is a mathematical idealization of what can be attained in practice. All the same such eigenstates play a very useful role in the theory and one could not very well do without them. Science contains many examples of theoretical concepts which are limits of things met with in practice and are useful for the precise formulation of laws of nature, although they are not realizable experimentally, and this is just one more of them. It may be that the infinite length of the ket vectors corresponding to these eigenstates is connected with their unrealizability, and that all realizable states correspond to ket vectors that can be normalized and that form a Hilbert space.

13. Commutability and Compatibility

A state may be simultaneously an eigenstate of two observables. If the state corresponds to the ket vector $|A\rangle$ and the observables are ξ and η, we should then have the equations

$$\xi|A\rangle = \xi'|A\rangle,$$
$$\eta|A\rangle = \eta'|A\rangle,$$

where ξ' and η' are eigenvalues of ξ and η respectively. We can now deduce

$$\xi\eta|A\rangle = \xi\eta'|A\rangle = \xi'\eta'|A\rangle = \xi'\eta|A\rangle = \eta\xi'|A\rangle = \eta\xi|A\rangle,$$

or

$$(\xi\eta - \eta\xi)|A\rangle = 0.$$

This suggests that the chances for the existence of a simultaneous eigenstate are most favourable if $\xi\eta - \eta\xi = 0$ and the two observables commute. If they do not commute a simultaneous eigenstate is not impossible, but is rather exceptional. On the other hand, *if they do commute there exist so many simultaneous eigenstates that they form a complete set*, as will now be proved.

Let ξ and η be two commuting observables. Take an eigenket of η, $|\eta'\rangle$ say, belonging to the eigenvalue η', and expand it in terms of eigenkets of ξ in the form of the right-hand side of (25), thus

$$|\eta'\rangle = \int |\xi'\eta'c\rangle\, d\xi' + \sum_r |\xi^r\eta'd\rangle. \tag{47}$$

The eigenkets of ξ on the right-hand side here have η' inserted in them as an extra label, in order to remind us that they come from the expansion of a special ket vector, namely $|\eta'\rangle$, and not a general one as in equation (25). We can now show that each of these eigenkets of ξ is also an eigenket of η belonging to the eigenvalue η'. We have

$$0 = (\eta - \eta')|\eta'\rangle = \int (\eta - \eta')|\xi'\eta'c\rangle\, d\xi' + \sum_r (\eta - \eta')|\xi^r\eta'd\rangle. \tag{48}$$

Now the ket $(\eta - \eta')|\xi^r\eta'd\rangle$ satisfies

$$\xi(\eta-\eta')|\xi^r\eta'd\rangle = (\eta-\eta')\xi|\xi^r\eta'd\rangle = (\eta-\eta')\xi^r|\xi^r\eta'd\rangle = \xi^r(\eta-\eta')|\xi^r\eta'd\rangle,$$

showing that it is an eigenket of ξ belonging to the eigenvalue ξ^r, and similarly the ket $(\eta-\eta')|\xi'\eta'c\rangle$ is an eigenket of ξ belonging to the eigenvalue ξ'. Equation (48) thus gives an integral plus a sum of eigenkets of ξ equal to zero, which, as we have seen with equation (31), is impossible unless the integrand and every term in the sum vanishes. Hence

$$(\eta - \eta')|\xi'\eta'c\rangle = 0, \quad (\eta - \eta')|\xi^r\eta'd\rangle = 0,$$

so that all the kets appearing on the right-hand side of (47) are eigenkets of η as well as of ξ. Equation (47) now gives $|\eta'\rangle$ expanded in terms of simultaneous eigenkets of ξ and η. Since any ket can be expanded in terms of eigenkets $|\eta'\rangle$ of η, it follows that any ket can be expanded in terms of simultaneous eigenkets of ξ and η, and thus the simultaneous eigenstates form a complete set.

The above simultaneous eigenkets of ξ and η, $|\xi'\eta'c\rangle$ and $|\xi^r\eta'd\rangle$, are labelled by the eigenvalues ξ' and η', or ξ^r and η', to which they belong, together with the labels c and d which may also be necessary. The procedure of using eigenvalues as labels for simultaneous eigenvectors will be generally followed in the future, just as it has been followed in the past for eigenvectors of single observables.

The converse to the above theorem says that, *if ξ and η are two observables such that their simultaneous eigenstates form a complete set, then ξ* and η *commute*. To prove this, we note that, if $|\xi'\eta'\rangle$ is a simultaneous eigenket belonging to the eigenvalues ξ' and η',

$$(\xi\eta - \eta\xi)|\xi'\eta'\rangle = (\xi'\eta' - \eta'\xi')|\xi'\eta'\rangle = 0. \tag{49}$$

Since the simultaneous eigenstates form a complete set, an arbitrary ket $|P\rangle$ can be expanded in terms of simultaneous eigenkets $|\xi'\eta'\rangle$, for each of which (49) holds, and hence

$$(\xi\eta - \eta\xi)|P\rangle = 0$$

and so

$$\xi\eta - \eta\xi = 0.$$

The idea of simultaneous eigenstates may be extended to more than two observables and the above theorem and its converse still hold, i.e. if any set of observables commute, each with all the others, their simultaneous eigenstates form a complete set, and conversely. The same arguments used for the proof with two observables are adequate for the general case; e.g., if we have three commuting observables ξ, η, ζ, we can expand any simultaneous eigenket of ξ and η in terms of eigenkets of ζ and then show that each of these eigenkets of ζ is also an eigenket of ξ and of η. Thus the simultaneous eigenket of ξ and η is expanded in terms of simultaneous eigenkets of ξ, η, and ζ, and since any ket can be expanded in terms of simultaneous eigenkets of ξ and η, it can also be expanded in terms of simultaneous eigenkets of ξ, η, and ζ.

The orthogonality theorem applied to simultaneous eigenkets tells us that two simultaneous eigenvectors of a set of commuting observables are orthogonal if the sets of eigenvalues to which they belong differ in any way.

Owing to the simultaneous eigenstates of two or more commuting observables forming a complete set, we can set up a theory of functions of two or more commuting observables on the same lines as the theory of functions of a single observable given in § 11. If $\xi, \eta, \zeta, \ldots$ are commuting observables, we define a general function f of them to be that linear operator $f(\xi, \eta, \zeta, \ldots)$ which satisfies

$$f(\xi, \eta, \zeta, \ldots) |\xi' \eta' \zeta' \cdots \rangle = f(\xi', \eta', \zeta', \ldots) |\xi' \eta' \zeta' \cdots \rangle, \qquad (50)$$

where $|\xi' \eta' \zeta' \cdots \rangle$ is any simultaneous eigenket of $\xi, \eta, \zeta, \ldots$ belonging to the eigenvalues $\xi', \eta', \zeta', \ldots$. Here f is any function such that $f(a, b, c, \ldots)$ is defined for all values of $a, b, c, \ldots$ which are eigenvalues of $\xi, \eta, \zeta, \ldots$ respectively. As with a function of a single observable defined by (34), we can show that $f(\xi, \eta, \zeta, \ldots)$ is completely determined by (50), that

$$\overline{f(\xi, \eta, \zeta, \ldots)} = \overline{f}(\xi, \eta, \zeta, \ldots),$$

corresponding to (37), and that if $f(a, b, c, \ldots)$ is a real function, $f(\xi, \eta, \zeta, \ldots)$ is real and is an observable.

We can now proceed to generalize the results (45) and (46). Given a set of commuting observables $\xi, \eta, \zeta, \ldots$, we may form that function of them which is equal to unity when $\xi = a, \eta = b, \zeta = c, \ldots$, $a, b, c, \ldots$ being real numbers, and is equal to zero when any of these conditions is not fulfilled. This function may be written $\delta_{\xi a} \delta_{\eta b} \delta_{\zeta c} \cdots$, and is in fact just the product in any order of the factors $\delta_{\xi a}, \delta_{\eta b}, \delta_{\zeta c}, \ldots$ defined as functions of single observables, as may be seen by substituting this product for $f(\xi, \eta, \zeta, \ldots)$ in the left-hand side of (50). The average value of this function for any state is the probability, $P_{abc\ldots}$ say, of $\xi, \eta, \zeta, \ldots$ having the values $a, b, c, \ldots$ respectively for that state. Thus if the state corresponds to the normalized ket vector $|x\rangle$, we get from our general assumption for physical interpretation

$$P_{abc\ldots} = \langle x | \delta_{\xi a} \delta_{\eta b} \delta_{\zeta c} \cdots | x \rangle. \qquad (51)$$

$P_{abc...}$ is zero unless each of the numbers $a, b, c, \ldots$ is an eigenvalue of the corresponding observable. If any of the numbers $a, b, c, \ldots$ is an eigenvalue in a range of eigenvalues of the corresponding observable, $P_{abc...}$ will usually again be zero, but in this case we ought to replace the requirement that this observable shall have exactly one value by the requirement that it shall have a value lying within a small range, which involves replacing one of the δ factors in (51) by a factor like the $\chi(\xi)$ of equation (46). On carrying out such a replacement for each of the observables $\xi, \eta, \zeta, \ldots$, whose corresponding numerical value $a, b, c, \ldots$ lies in a range of eigenvalues, we shall get a probability which does not in general vanish.

If certain observables commute, there exist states for which they all have particular values, in the sense explained at the top of p. 47, namely the simultaneous eigenstates. Thus *one can give a meaning to several commuting observables having values at the same time*. Further, we see from (51) that for any state *one can give a meaning to the probability of particular results being obtained for simultaneous measurements of several commuting observables*. This conclusion is an important new development. In general one cannot make an observation on a system in a definite state without disturbing that state and spoiling it for the purposes of a second observation. One cannot then give any meaning to the two observations being made simultaneously. The above conclusion tells us, though, that in the special case when the two observables commute, the observations are to be considered as non-interfering or *compatible*, in such a way that one *can* give a meaning to the two observations being made simultaneously and can discuss the probability of any particular results being obtained. The two observations may, in fact, be considered as a single observation of a more complicated type, the result of which is expressible by two numbers instead of a single number. *From the point of view of general theory, any two or more commuting observables may be counted as a single observable, the result of a measurement of which consists of two or more numbers*. The states for which this measurement is certain to lead to one particular result are the simultaneous eigenstates.

CHAPTER III

Representations

14. Basic Vectors

In the preceding chapters we set up an algebraic scheme involving certain abstract quantities of three kinds, namely bra vectors, ket vectors, and linear operators, and we expressed some of the fundamental laws of quantum mechanics in terms of them. It would be possible to continue to develop the theory in terms of these abstract quantities and to use them for applications to particular problems. However, for some purposes it is more convenient to replace the abstract quantities by sets of numbers with analogous mathematical properties and to work in terms of these sets of numbers. The procedure is similar to using coordinates in geometry, and has the advantage of giving one greater mathematical power for the solving of particular problems.

The way in which the abstract quantities are to be replaced by numbers is not unique, there being many possible ways corresponding to the many systems of coordinates one can have in geometry. Each of these ways is called a *representation* and the set of numbers that replace an abstract quantity is called the *representative* of that abstract quantity in the representation. Thus the representative of an abstract quantity corresponds to the coordinates of a geometrical object. When one has a particular problem to work out in quantum mechanics, one can minimize the labour by using a representation in which the representatives of the more important abstract quantities occurring in that problem are as simple as possible.

To set up a representation in a general way, we take a complete set of bra vectors, i.e. a set such that any bra can be expressed linearly in terms of them (as a sum or an integral or possibly an integral plus a sum). These bras we call the *basic bras* of the representation. They are sufficient, as we shall see, to fix the representation completely.

Take any ket $|a\rangle$ and form its scalar product with each of the basic bras. The numbers so obtained constitute the representative of $|a\rangle$. They are sufficient to determine the ket $|a\rangle$ completely, since if there is a second ket, $|a_1\rangle$ say, for which these numbers are the same, the difference $|a\rangle - |a_1\rangle$ will have its scalar product with any basic bra vanishing, and hence its scalar product with any bra whatever will vanish and $|a\rangle - |a_1\rangle$ itself will vanish.

We may suppose the basic bras to be labelled by one or more parameters, $\lambda_1, \lambda_2, \ldots, \lambda_u$, each of which may take on certain numerical values. The basic bras will then be written $\langle \lambda_1 \lambda_2 \cdots \lambda_u |$ and the representative of $|a\rangle$ will be written $\langle \lambda_1 \lambda_2 \cdots \lambda_u | a \rangle$. This representative will now consist of a set of numbers, one for each set of values that $\lambda_1, \lambda_2, \ldots, \lambda_u$ may have in their respective domains. Such a set of numbers just forms a *function* of the variables $\lambda_1, \lambda_2, \ldots, \lambda_u$. Thus the representative of a ket may be looked upon either as a set of numbers or as a function of the variables used to label the basic bras.

If the number of independent states of our dynamical system is finite, equal to n say, it is sufficient to take n basic bras, which may be labelled by a single parameter λ taking on the values $1, 2, \ldots, n$. The representative of any ket $|a\rangle$ now consists of the set of n numbers $\langle 1|a\rangle, \langle 2|a\rangle, \ldots, \langle n|a\rangle$, which are precisely the coordinates of the vector $|a\rangle$ referred to a system of coordinates in the usual way. The idea of the representative of a ket vector is just a generalization of the idea of the coordinates of an ordinary vector and reduces to the latter when the number of dimensions of the space of the ket vectors is finite.

In a general representation there is no need for the basic bras to be all independent. In most representations used in practice, however, they are all independent, and also satisfy the more stringent condition that any two of them are orthogonal. The representation is then called an *orthogonal representation*.

Take an orthogonal representation with basic bras $\langle \lambda_1 \lambda_2 \cdots \lambda_u |$, labelled by parameters $\lambda_1, \lambda_2, \ldots, \lambda_u$ whose domains are all real. Take a ket $|a\rangle$ and form its representative $\langle \lambda_1 \lambda_2 \cdots \lambda_u | a \rangle$. Now form the numbers $\lambda_1 \langle \lambda_1 \lambda_2 \cdots \lambda_u | a \rangle$ and consider them as the representative of a new ket $|b\rangle$. This is permissible since the numbers forming the representative of a ket are independent, on account of the basic bras being independent. The ket $|b\rangle$ is defined by the equation

$$\langle \lambda_1 \lambda_2 \cdots \lambda_u | b \rangle = \lambda_1 \langle \lambda_1 \lambda_2 \cdots \lambda_u | a \rangle.$$

The ket $|b\rangle$ is evidently a linear function of the ket $|a\rangle$, so it may be considered as the result of a linear operator applied to $|a\rangle$. Calling this linear operator L_1, we have

$$|b\rangle = L_1 |a\rangle$$

and hence
$$\langle \lambda_1 \lambda_2 \cdots \lambda_u | L_1 | a \rangle = \lambda_1 \langle \lambda_1 \lambda_2 \cdots \lambda_u | a \rangle.$$
This equation holds for any ket $|a\rangle$, so we get
$$\langle \lambda_1 \lambda_2 \cdots \lambda_u | L_1 = \lambda_1 \langle \lambda_1 \lambda_2 \cdots \lambda_u |. \tag{1}$$

Equation (1) may be looked upon as the definition of the linear operator L_1. It shows that *each basic bra is an eigenbra of L_1, the value of the parameter λ_1 being the eigenvalue belonging to it.*

From the condition that the basic bras are orthogonal we can deduce that L_1 is real and is an observable. Let $\lambda_1', \lambda_2', \ldots, \lambda_u'$ and $\lambda_1'', \lambda_2'', \ldots, \lambda_u''$ be two sets of values for the parameters $\lambda_1, \lambda_2, \ldots, \lambda_u$. We have, putting λ''s for the λ's in (1) and multiplying on the right by $|\lambda_1'' \lambda_2'' \cdots \lambda_u''\rangle$, the conjugate imaginary of the basic bra $\langle \lambda_1'' \lambda_2'' \cdots \lambda_u''|$,
$$\langle \lambda_1' \lambda_2' \cdots \lambda_u' | L_1 | \lambda_1'' \lambda_2'' \cdots \lambda_u'' \rangle = \lambda_1' \langle \lambda_1' \lambda_2' \cdots \lambda_u' | \lambda_1'' \lambda_2'' \cdots \lambda_u'' \rangle.$$
Interchanging λ''s and λ'''s,
$$\langle \lambda_1'' \lambda_2'' \cdots \lambda_u'' | L_1 | \lambda_1' \lambda_2' \cdots \lambda_u' \rangle = \lambda_1'' \langle \lambda_1'' \lambda_2'' \cdots \lambda_u'' | \lambda_1' \lambda_2' \cdots \lambda_u' \rangle.$$
On account of the basic bras being orthogonal, the right-hand sides here vanish unless $\lambda_r'' = \lambda_r'$ for all r from 1 to u, in which case the right-hand sides are equal, and they are also real, λ_1' being real. Thus, whether the λ'''s are equal to the λ''s or not,
$$\langle \lambda_1' \lambda_2' \cdots \lambda_u' | L_1 | \lambda_1'' \lambda_2'' \cdots \lambda_u'' \rangle = \overline{\langle \lambda_1'' \lambda_2'' \cdots \lambda_u'' | L_1 | \lambda_1' \lambda_2' \cdots \lambda_u' \rangle}$$
$$= \langle \lambda_1' \lambda_2' \cdots \lambda_u' | \overline{L}_1 | \lambda_1'' \lambda_2'' \cdots \lambda_u'' \rangle$$
from equation (4) of § 8. Since the $\langle \lambda_1' \lambda_2' \cdots \lambda_u' |$'s form a complete set of bras and the $|\lambda_1'' \lambda_2'' \cdots \lambda_u''\rangle$'s form a complete set of kets, we can infer that $L_1 = \overline{L}_1$. The further condition required for L_1 to be an observable, namely that its eigenstates shall form a complete set, is obviously satisfied since it has as eigenbras the basic bras, which form a complete set.

We can similarly introduce linear operators $L_2, L_3, \ldots, L_u$ by multiplying $\langle \lambda_1 \lambda_2 \cdots \lambda_u | a \rangle$ by the factors $\lambda_2, \lambda_3, \ldots, \lambda_u$ in turn and considering the resulting sets of numbers as representatives of kets. Each of these L's can be shown in the same way to have the basic bras as eigenbras and to be real and an observable. The basic bras are simultaneous eigenbras of all the L's. Since these simultaneous eigenbras form a complete set, it follows from a theorem of § 13 that any two of the L's commute.

It will now be shown that, *if $\xi_1, \xi_2, \ldots, \xi_u$ are any set of commuting observables, we can set up an orthogonal representation in which the basic bras are simultaneous eigenbras of $\xi_1, \xi_2, \ldots, \xi_u$.* Let us suppose first that there is only one independent simultaneous eigenbra of $\xi_1, \xi_2, \ldots, \xi_u$ belonging to any set of eigenvalues $\xi'_1, \xi'_2, \ldots, \xi'_u$. Then we may take these simultaneous eigenbras, with arbitrary numerical coefficients, as our basic bras. They are all orthogonal on account of the orthogonality theorem (any two of them will have at least one eigenvalue different, which is sufficient to make them orthogonal) and there are sufficient of them to form a complete set, from a result of § 13. They may conveniently be labelled by the eigenvalues $\xi'_1, \xi'_2, \ldots, \xi'_u$ to which they belong, so that one of them is written $\langle \xi'_1 \xi'_2 \cdots \xi'_u |$.

Passing now to the general case when there are several independent simultaneous eigenbras of $\xi_1, \xi_2, \ldots, \xi_u$ belonging to some sets of eigenvalues, we must pick out from all the simultaneous eigenbras belonging to a set of eigenvalues $\xi'_1, \xi'_2, \ldots, \xi'_u$ a complete subset, the members of which are all orthogonal to one another. (The condition of completeness here means that any simultaneous eigenbra belonging to the eigenvalues $\xi'_1, \xi'_2, \ldots, \xi'_u$ can be expressed linearly in terms of the members of the subset.) We must do this for each set of eigenvalues $\xi'_1, \xi'_2, \ldots, \xi'_u$ and then put all the members of all the subsets together and take them as the basic bras of the representation. These bras are all orthogonal, two of them being orthogonal from the orthogonality theorem if they belong to different sets of eigenvalues and from the special way in which they were chosen if they belong to the same set of eigenvalues, and they form altogether a complete set of bras, as any bra can be expressed linearly in terms of simultaneous eigenbras and each simultaneous eigenbra can then be expressed linearly in terms of the members of a subset. There are infinitely many ways of choosing the subsets, and each way provides one orthogonal representation.

For labelling the basic bras in this general case, we may use the eigenvalues $\xi'_1, \xi'_2, \ldots, \xi'_u$ to which they belong, together with certain additional real variables $\lambda_1, \lambda_2, \ldots, \lambda_v$ say, which must be introduced to distinguish basic vectors belonging to the same set of eigenvalues from one another. A basic bra is then written $\langle \xi'_1 \xi'_2 \cdots \xi'_u \lambda_1 \lambda_2 \cdots \lambda_v |$. Corresponding to the variables $\lambda_1, \lambda_2, \ldots, \lambda_v$ we can define linear operators $L_1, L_2, \ldots, L_v$ by equations like (1) and can show that these linear operators have the basic bras as eigenbras, and that they are real and observables, and that they commute with one another and with the ξ's. The basic bras are now simultaneous eigenbras of all the commuting observables

$\xi_1, \xi_2, \ldots, \xi_u, L_1, L_2, \ldots, L_v$.

Let us define a *complete set of commuting observables* to be a set of observables which all commute with one another and for which there is only one simultaneous eigenstate belonging to any set of eigenvalues. Then the observables $\xi_1, \xi_2, \ldots, \xi_u, L_1, L_2, \ldots, L_v$ form a complete set of commuting observables, there being only one independent simultaneous eigenbra belonging to the eigenvalues $\xi'_1, \xi'_2, \ldots, \xi'_u, \lambda_1, \lambda_2, \ldots, \lambda_v$, namely the corresponding basic bra. Similarly the observables $L_1, L_2, \ldots, L_u$ defined by equation (1) and the following work form a complete set of commuting observables. With the help of this definition the main results of the present section can be concisely formulated thus:

(i) The basic bras of an orthogonal representation are simultaneous eigenbras of a complete set of commuting observables.

(ii) Given a complete set of commuting observables, we can set up an orthogonal representation in which the basic bras are simultaneous eigenbras of this complete set.

(iii) Any set of commuting observables can be made into a complete commuting set by adding certain observables to it.

(iv) A convenient way of labelling the basic bras of an orthogonal representation is by means of the eigenvalues of the complete set of commuting observables of which the basic bras are simultaneous eigenbras.

The conjugate imaginaries of the basic bras of a representation we call the *basic kets* of the representation. Thus, if the basic bras are denoted by $\langle \lambda_1 \lambda_2 \cdots \lambda_u |$, the basic kets will be denoted by $| \lambda_1 \lambda_2 \cdots \lambda_u \rangle$. The representative of a bra $\langle b |$ is given by its scalar product with each of the basic kets, i.e. by $\langle b | \lambda_1 \lambda_2 \cdots \lambda_u \rangle$. It may, like the representative of a ket, be looked upon either as a set of numbers or as a function of the variables $\lambda_1, \lambda_2, \ldots, \lambda_u$. We have

$$\langle b | \lambda_1 \lambda_2 \cdots \lambda_u \rangle = \overline{\langle \lambda_1 \lambda_2 \cdots \lambda_u | b \rangle},$$

showing that *the representative of a bra is the conjugate complex of the representative of the conjugate imaginary ket*. In an orthogonal representation, where the basic bras are simultaneous eigenbras of a complete set of commuting observables, $\xi_1, \xi_2, \ldots, \xi_u$ say, the basic kets will be simultaneous eigenkets of $\xi_1, \xi_2, \ldots, \xi_u$.

We have not yet considered the lengths of the basic vectors. With an orthogonal representation, the natural thing to do is to normalize the basic vectors, rather than leave their lengths arbitrary, and so introduce a further stage of simplification into the representation. However, it is possible to normalize them only if the parameters which label them all take on discrete values. If any of these parameters are continuous variables that can take on all values in a range, the basic vectors are eigenvectors of some observable belonging to eigenvalues in a range and are of infinite length, from the discussion in § 10 (see p. 39 and top of p. 40). Some other procedure is then needed to fix the numerical factors by which the basic vectors may be multiplied. To get a convenient method of handling this question a new mathematical notation is required, which will be given in the next section.

15. The δ Function

Our work in § 10 led us to consider quantities involving a certain kind of infinity. To get a precise notation for dealing with these infinities, we introduce a quantity $\delta(x)$ depending on a parameter x satisfying the conditions

$$\int_{-\infty}^{\infty} \delta(x)\,dx = 1, \quad \delta(x) = 0 \text{ for } x \neq 0. \tag{2}$$

To get a picture of $\delta(x)$, take a function of the real variable x which vanishes everywhere except inside a small domain, of length ϵ say, surrounding the origin $x = 0$, and which is so large inside this domain that its integral over this domain is unity. The exact shape of the function inside this domain does not matter, provided there are no unnecessarily wild variations (for example provided the function is always of order ϵ^{-1}). Then in the limit $\epsilon \to 0$ this function will go over into $\delta(x)$.

$\delta(x)$ is not a function of x according to the usual mathematical definition of a function, which requires a function to have a definite value for each point in its domain, but is something more general, which we may call an 'improper function' to show up its difference from a function defined by the usual definition. Thus $\delta(x)$ is not a quantity which can be generally used in mathematical analysis like an ordinary function, but its use must be confined to certain simple types of expression for which it is obvious that no inconsistency can arise.

The most important property of $\delta(x)$ is exemplified by the following equation,

$$\int_{-\infty}^{\infty} f(x)\,\delta(x)\,dx = f(0), \tag{3}$$

where $f(x)$ is any continuous function of x. We can easily see the validity of this equation from the above picture of $\delta(x)$. The left-hand side of (3) can depend only on the values of $f(x)$ very close to the origin, so that we may replace $f(x)$ by its value at the origin, $f(0)$, without essential error. Equation (3) then follows from the first of equations (2). By making a change of origin in (3), we can deduce the formula

$$\int_{-\infty}^{\infty} f(x)\,\delta(x-a)\,dx = f(a), \tag{4}$$

where a is any real number. Thus *the process of multiplying a function of x by $\delta(x-a)$ and integrating over all x is equivalent to the process of substituting a for x*. This general result holds also if the function of x is not a numerical one, but is a vector or linear operator depending on x.

The range of integration in (3) and (4) need not be from $-\infty$ to ∞, but may be over any domain surrounding the critical point at which the δ function does not vanish. In future the limits of integration will usually be omitted in such equations, it being understood that the domain of integration is a suitable one.

Equations (3) and (4) show that, although an improper function does not itself have a well-defined value, when it occurs as a factor in an integrand the integral has a well-defined value. In quantum theory, whenever an improper function appears, it will be something which is to be used ultimately in an integrand. Therefore it should be possible to rewrite the theory in a form in which the improper functions appear all through only in integrands. One could then eliminate the improper functions altogether. The use of improper functions thus does not involve any lack of rigour in the theory, but is merely a convenient notation, enabling us to express in a concise form certain relations which we could, if necessary, rewrite in a form not involving improper functions, but only in a cumbersome way which would tend to obscure the argument.

An alternative way of defining the δ function is as the differential coefficient $\epsilon'(x)$ of the function $\epsilon(x)$ given by

$$\begin{aligned} \epsilon(x) &= 0 \quad (x<0) \\ &= 1 \quad (x>0). \end{aligned} \tag{5}$$

We may verify that this is equivalent to the previous definition by substituting $\epsilon'(x)$ for $\delta(x)$ in the left-hand side of (3) and integrating by parts. We find, for g_1 and g_2 two positive numbers,

$$\int_{-g_2}^{g_1} f(x)\,\epsilon'(x)\,dx = [f(x)\,\epsilon(x)]_{-g_2}^{g_1} - \int_{-g_2}^{g_1} f'(x)\,\epsilon(x)\,dx$$

$$= f(g_1) - \int_0^{g_1} f'(x)\,dx = f(0),$$

in agreement with (3). The δ function appears whenever one differentiates a discontinuous function.

There are a number of elementary equations which one can write down about δ functions. These equations are essentially rules of manipulation for algebraic work involving δ functions. The meaning of any of these equations is that its two sides give equivalent results as factors in an integrand.

Examples of such equations are

$$\delta(-x) = \delta(x), \tag{6}$$

$$x\,\delta(x) = 0, \tag{7}$$

$$\delta(ax) = a^{-1}\delta(x) \quad (a > 0), \tag{8}$$

$$\delta(x^2 - a^2) = \frac{1}{2}a^{-1}[\delta(x-a) + \delta(x+a)] \quad (a > 0), \tag{9}$$

$$\int \delta(a-x)\,dx\,\delta(x-b) = \delta(a-b), \tag{10}$$

$$f(x)\,\delta(x-a) = f(a)\,\delta(x-a). \tag{11}$$

Equation (6), which merely states that $\delta(x)$ is an even function of its variable x is trivial. To verify (7) take any continuous function of x, $f(x)$. Then

$$\int f(x)x\,\delta(x)\,dx = 0,$$

from (3). Thus $x\,\delta(x)$ as a factor in an integrand is equivalent to zero, which is just the meaning of (7). (8) and (9) may be verified by similar elementary arguments. To verify (10) take any continuous function of a, $f(a)$. Then

$$\int f(a)\,da \int \delta(a-x)\,dx\,\delta(x-b)$$

$$= \int \delta(x-b)\,dx \int f(a)\,da\,\delta(a-x)$$

$$= \int \delta(x-b)\,dx\,f(x) = \int f(a)\,da\,\delta(a-b).$$

Thus the two sides of (10) are equivalent as factors in an integrand with a as variable of integration. It may be shown in the same way that they are equivalent also as factors in an integrand with b as variable of integration, so that equation (10) is justified from either of these points of view.

Equation (11) is also easily justified, with the help of (4), from two points of view.

Equation (10) would be given by an application of (4) with $f(x) = \delta(x - b)$. We have here an illustration of the fact that we may often use an improper function as though it were an ordinary continuous function, without getting a wrong result.

Equation (7) shows that, whenever one divides both sides of an equation by a variable x which can take on the value zero, one should add on to one side an arbitrary multiple of $\delta(x)$, i.e. from an equation

$$A = B \tag{12}$$

one cannot infer

$$\frac{A}{x} = \frac{B}{x},$$

but only

$$\frac{A}{x} = \frac{B}{x} + c\,\delta(x), \tag{13}$$

where c is unknown.

As an illustration of work with the δ function, we may consider the differentiation of $\log x$. The usual formula

$$\frac{d}{dx}\log x = \frac{1}{x} \tag{14}$$

requires examination for the neighbourhood of $x = 0$. In order to make the reciprocal function $1/x$ well defined in the neighbourhood of $x = 0$ (in the sense of an improper function) we must impose on it an extra condition, such as that its integral from $-\epsilon$ to ϵ vanishes. With this extra condition, the integral of the right-hand side of (14) from $-\epsilon$ to ϵ vanishes, while that of the left-hand side of (14) equals $\log(-1)$, so that (14) is not a correct equation. To correct it, we must remember that, taking principal values, $\log x$ has a pure imaginary term $i\pi$ for negative values of x. As x passes through the value zero this pure imaginary term vanishes discontinuously. The differentiation of this pure imaginary term gives us the result $-i\pi\,\delta(x)$, so that (14) should read

$$\frac{d}{dx}\log x = \frac{1}{x} - i\pi\,\delta(x). \tag{15}$$

The particular combination of reciprocal function and δ function appearing in (15) plays an important part in the quantum theory of collision processes (see § 50).

16. Properties of the Basic Vectors

Using the notation of the δ function, we can proceed with the theory of representations. Let us suppose first that we have a single observable ξ forming by itself a complete commuting set, the condition for this being that there is only one eigenstate of ξ belonging to any eigenvalue ξ', and let us set up an orthogonal representation in which the basic vectors are eigenvectors of ξ and are written $\langle\xi'|, |\xi'\rangle$.

In the case when the eigenvalues of ξ are discrete, we can normalize the basic vectors, and we then have

$$\langle\xi'|\xi''\rangle = 0 \quad (\xi' \neq \xi''),$$
$$\langle\xi'|\xi'\rangle = 1.$$

These equations can be combined into the single equation

$$\langle\xi'|\xi''\rangle = \delta_{\xi'\xi''}, \tag{16}$$

where the symbol δ with two suffixes, which we shall often use in the future, has the meaning

$$\begin{aligned}\delta_{rs} &= 0 \quad \text{when} \quad r \neq s \\ &= 1 \quad \text{when} \quad r = s.\end{aligned} \tag{17}$$

In the case when the eigenvalues of ξ are continuous we cannot normalize the basic vectors. If we now consider the quantity $\langle\xi'|\xi''\rangle$ with ξ' fixed and ξ'' varying, we see from the work connected with expression (29) of § 10 that this quantity vanishes for $\xi'' \neq \xi'$ and that its integral over a range of ξ'' extending through the value ξ' is finite, equal to c say. Thus

$$\langle\xi'|\xi''\rangle = c\,\delta(\xi' - \xi'').$$

From (30) of § 10, c is a positive number. It may vary with ξ', so we should write it $c(\xi')$ or c' for brevity, and thus we have

$$\langle\xi'|\xi''\rangle = c'\,\delta(\xi' - \xi''). \tag{18}$$

Alternatively, we have

$$\langle\xi'|\xi''\rangle = c''\,\delta(\xi' - \xi''), \tag{19}$$

where c'' is short for $c(\xi'')$, the right-hand sides of (18) and (19) being equal on account of (11).

Let us pass to another representation whose basic vectors are eigenvectors of ξ, the new basic vectors being numerical multiples of the previous ones. Calling the new basic vectors $\langle\xi'^*|, |\xi'^*\rangle$, with the additional

label * to distinguish them from the previous ones, we have
$$\langle \xi'^* | = k' \langle \xi' |, \quad |\xi'^* \rangle = \overline{k'} |\xi' \rangle,$$
where k' is short for $k(\xi')$ and is a number depending on ξ'. We get
$$\langle \xi'^* | \xi''^* \rangle = k' \overline{k''} \langle \xi' | \xi'' \rangle = k' \overline{k''} c' \, \delta(\xi' - \xi'')$$
with the help of (18). This may be written
$$\langle \xi'^* | \xi''^* \rangle = k' \overline{k'} c' \, \delta(\xi' - \xi'')$$
from (11). By choosing k' so that its modulus is $c'^{-1/2}$, which is possible since c' is positive, we arrange to have
$$\langle \xi'^* | \xi''^* \rangle = \delta(\xi' - \xi''). \tag{20}$$
The lengths of the new basic vectors are now fixed so as to make the representation as simple as possible. The way these lengths were fixed is in some respects analogous to the normalizing of the basic vectors in the case of discrete ξ', equation (20) being of the form of (16) with the δ function $\delta(\xi' - \xi'')$ replacing the δ symbol $\delta_{\xi' \xi''}$ of equation (16). We shall continue to work with the new representation and shall drop the * labels in it to save writing. Thus (20) will now be written
$$\langle \xi' | \xi'' \rangle = \delta(\xi' - \xi''). \tag{21}$$

We can develop the theory on closely parallel lines for the discrete and continuous cases. For the discrete case we have, using (16),
$$\sum_{\xi'} |\xi'\rangle \langle \xi' | \xi'' \rangle = \sum_{\xi'} |\xi'\rangle \delta_{\xi' \xi''} = |\xi''\rangle,$$
the sum being taken over all eigenvalues. This equation holds for any basic ket $|\xi''\rangle$ and hence, since the basic kets form a complete set,
$$\sum_{\xi'} |\xi'\rangle \langle \xi' | = 1. \tag{22}$$
This is a useful equation expressing an important property of the basic vectors, namely, *if $|\xi'\rangle$ is multiplied on the right by $\langle \xi' |$ the resulting linear operator, summed for all ξ', equals the unit operator*. Equations (16) and (22) give the fundamental properties of the basic vectors for the discrete case.

Similarly, for the continuous case we have, using (21),
$$\int |\xi'\rangle d\xi' \langle \xi' | \xi'' \rangle = \int |\xi'\rangle d\xi' \, \delta(\xi' - \xi'') = |\xi''\rangle \tag{23}$$

from (4) applied with a ket vector for $f(x)$, the range of integration being the range of eigenvalues. This holds for any basic ket $|\xi''\rangle$ and hence

$$\int |\xi'\rangle \, d\xi' \, \langle\xi'| = 1. \tag{24}$$

This is of the same form as (22) with an integral replacing the sum. Equations (21) and (24) give the fundamental properties of the basic vectors for the continuous case.

Equations (22) and (24) enable one to expand any bra or ket in terms of the basic vectors. For example, we get for the ket $|P\rangle$ in the discrete case, by multiplying (22) on the right by $|P\rangle$,

$$|P\rangle = \sum_{\xi'} |\xi'\rangle\langle\xi'|P\rangle, \tag{25}$$

which gives $|P\rangle$ expanded in terms of the $|\xi'\rangle$'s and shows that the coefficients in the expansion are $\langle\xi'|P\rangle$, which are just the numbers forming the representative of $|P\rangle$. Similarly, in the continuous case,

$$|P\rangle = \int |\xi'\rangle \, d\xi' \, \langle\xi'|P\rangle, \tag{26}$$

giving $|P\rangle$ as an integral over the $|\xi'\rangle$'s, with the coefficient in the integrand again just the representative $\langle\xi'|P\rangle$ of $|P\rangle$. The conjugate imaginary equations to (25) and (26) would give the bra vector $\langle P|$ expanded in terms of the basic bras.

Our present mathematical methods enable us in the continuous case to expand any ket as an integral of eigenkets of ξ. If we do not use the δ function notation, the expansion of a general ket will consist of an integral plus a sum, as in equation (25) of § 10, but the δ function enables us to replace the sum by an integral in which the integrand consists of terms each containing a δ function as a factor. For example, the eigenket $|\xi''\rangle$ may be replaced by an integral of eigenkets, as is shown by the second of equations (23).

If $\langle Q|$ is any bra and $|P\rangle$ any ket we get, by further applications of (22) and (24),

$$\langle Q|P\rangle = \sum_{\xi'} \langle Q|\xi'\rangle\langle\xi'|P\rangle \tag{27}$$

for discrete ξ' and

$$\langle Q|P\rangle = \int \langle Q|\xi'\rangle \, d\xi' \, \langle\xi'|P\rangle \tag{28}$$

for continuous ξ'. These equations express the scalar product of $\langle Q|$ and $|P\rangle$ in terms of their representatives $\langle Q|\xi'\rangle$ and $\langle \xi'|P\rangle$. Equation (27) is just the usual formula for the scalar product of two vectors in terms of the coordinates of the vectors, and (28) is the natural modification of this formula for the case of continuous ξ', with an integral instead of a sum.

The generalization of the foregoing work to the case when ξ has both discrete and continuous eigenvalues is quite straightforward. Using ξ^r and ξ^s to denote discrete eigenvalues and ξ' and ξ'' to denote continuous eigenvalues, we have the set of equations

$$\langle \xi^r|\xi^s\rangle = \delta_{\xi^r\xi^s}, \quad \langle \xi^r|\xi'\rangle = 0, \quad \langle \xi'|\xi''\rangle = \delta(\xi' - \xi'') \tag{29}$$

as the generalization of (16) or (21). These equations express that the basic vectors are all orthogonal, that those belonging to discrete eigenvalues are normalized and those belonging to continuous eigenvalues have their lengths fixed by the same rule as led to (20). From (29) we can derive, as the generalization of (22) or (24),

$$\sum_{\xi^r} |\xi^r\rangle\langle \xi^r| + \int |\xi'\rangle\, d\xi'\, \langle \xi'| = 1, \tag{30}$$

the range of integration being the range of continuous eigenvalues. With the help of (30), we get immediately

$$|P\rangle = \sum_{\xi^r} |\xi^r\rangle\langle \xi^r|P\rangle + \int |\xi'\rangle\, d\xi'\, \langle \xi'|P\rangle \tag{31}$$

as the generalization of (25) or (26), and

$$\langle Q|P\rangle = \sum_{\xi^r} \langle Q|\xi^r\rangle\langle \xi^r|P\rangle + \int \langle Q|\xi'\rangle\, d\xi'\, \langle \xi'|P\rangle \tag{32}$$

as the generalization of (27) or (28).

Let us now pass to the general case when we have several commuting observables $\xi_1, \xi_2, \ldots, \xi_u$ forming a complete commuting set and set up an orthogonal representation in which the basic vectors are simultaneous eigenvectors of all of them, and are written $\langle \xi'_1 \cdots \xi'_u|, |\xi'_1 \cdots \xi'_u\rangle$. Let us suppose $\xi_1, \xi_2, \ldots, \xi_v$ ($v \leq u$) have discrete eigenvalues and $\xi_{v+1}, \xi_{v+2}, \ldots, \xi_u$ have continuous eigenvalues.

Consider the quantity $\langle \xi'_1 \cdots \xi'_v \xi''_{v+1} \cdots \xi''_u | \xi'_1 \cdots \xi'_v \xi'_{v+1} \cdots \xi'_u\rangle$. From the orthogonality theorem, it must vanish unless each $\xi''_s = \xi'_s$ for $s = v+1, v+2, \ldots, u$. By extending the work connected with expression (29) of § 10 to simultaneous eigenvectors of several commuting observables and extending also the axiom (30), we find that the $(u-v)$-fold integral of this quantity with respect to each ξ_s'' over a range extending through the value ξ_s' is a finite positive number. Calling this number c', the $'$ denoting that it is a function of $\xi_1', \xi_2', \ldots, \xi_v', \xi_{v+1}', \xi_{v+2}', \ldots, \xi_u'$, we can express our results by the equation

$$\langle \xi_1' \cdots \xi_v' \xi_{v+1}' \cdots \xi_u' | \xi_1' \cdots \xi_v' \xi_{v+1}'' \cdots \xi_u'' \rangle = c' \, \delta(\xi_{v+1}' - \xi_{v+1}'') \cdots \delta(\xi_u' - \xi_u''), \tag{33}$$

with one δ factor on the right-hand side for each value of s from $v+1$ to u. We now change the lengths of our basic vectors so as to make c' unity, by a procedure similar to that which led to (20). By a further use of the orthogonality theorem, we get finally

$$\langle \xi_1' \cdots \xi_u' | \xi_1'' \cdots \xi_u'' \rangle = \delta_{\xi_1' \xi_1''} \cdots \delta_{\xi_v' \xi_v''} \, \delta(\xi_{v+1}' - \xi_{v+1}'') \cdots \delta(\xi_u' - \xi_u''), \tag{34}$$

with a two-suffix δ symbol on the right-hand side for each ξ with discrete eigenvalues and a δ function for each ξ with continuous eigenvalues. This is the generalization of (16) or (21) to the case when there are several commuting observables in the complete set.

From (34) we can derive, as the generalization of (22) or (24)

$$\sum_{\xi_1' \cdots \xi_v'} \int \cdots \int |\xi_1' \cdots \xi_u'\rangle \, d\xi_{v+1}' \cdots d\xi_u' \, \langle \xi_1' \cdots \xi_u'| = 1, \tag{35}$$

the integral being a $(u-v)$-fold one over all the ξ''s with continuous eigenvalues and the summation being over all the ξ''s with discrete eigenvalues. Equations (34) and (35) give the fundamental properties of the basic vectors in the present case. From (35) we can immediately write down the generalization of (25) or (26) and of (27) or (28).

The case we have just considered can be further generalized by allowing some of the ξ's to have both discrete and continuous eigenvalues. The modifications required in the equations are quite straightforward, but will not be given here as they are rather cumbersome to write down in general form.

There are some problems in which it is convenient not to make the c' of equation (33) equal unity, but to make it equal to some definite function of the ξ''s instead. Calling this function of the ξ''s ρ'^{-1} we then have, instead of (34)

$$\langle \xi_1' \cdots \xi_u' | \xi_1'' \cdots \xi_u'' \rangle = \rho'^{-1} \delta_{\xi_1' \xi_1''} \cdots \delta_{\xi_v' \xi_v''} \, \delta(\xi_{v+1}' - \xi_{v+1}'') \cdots \delta(\xi_u' - \xi_u''), \tag{36}$$

and instead of (35) we get

$$\sum_{\xi'_1\cdots\xi'_v}\int\cdots\int|\xi'_1\cdots\xi'_u\rangle\rho'\,d\xi'_{v+1}\cdots d\xi'_u\langle\xi'_1\cdots\xi'_u|=1. \qquad (37)$$

ρ' is called the *weight function* of the representation, $\rho'\,d\xi'_{v+1}\cdots d\xi'_u$ being the 'weight' attached to a small volume element of the space of the variables $\xi'_{v+1},\xi'_{v+2},\ldots,\xi'_u$.

The representations we considered previously all had the weight function unity. The introduction of a weight function not unity is entirely a matter of convenience and does not add anything to the mathematical power of the representation. The basic bras $\langle\xi'_1\cdots\xi'^*_u|$ of a representation with the weight function ρ' are connected with the basic bras $\langle\xi'_1\cdots\xi'_u|$ of the corresponding representation with the weight function unity by

$$\langle\xi'_1\cdots\xi'^*_u|=\rho'^{-1/2}\langle\xi'_1\cdots\xi'_u|, \qquad (38)$$

as is easily verified. An example of a useful representation with non-unit weight function occurs when one has two ξ's which are the polar and azimuthal angles θ and ϕ giving a direction in three-dimensional space and one takes $\rho'=\sin\theta'$. One then has the element of solid angle $\sin\theta'\,d\theta'd\phi'$ occurring in (37).

17. The Representation of Linear Operators

In § 14 we saw how to represent ket and bra vectors by sets of numbers. We now have to do the same for linear operators, in order to have a complete scheme for representing all our abstract quantities by sets of numbers. The same basic vectors that we had in § 14 can be used again for this purpose.

Let us suppose the basic vectors are simultaneous eigenvectors of a complete set of commuting observables $\xi_1,\xi_2,\ldots,\xi_u$. If α is any linear operator, we take a general basic bra $\langle\xi'_1\cdots\xi'_u|$ and a general basic ket $|\xi''_1\cdots\xi''_u\rangle$ and form the numbers

$$\langle\xi'_1\cdots\xi'_u|\alpha|\xi''_1\cdots\xi''_u\rangle. \qquad (39)$$

These numbers are sufficient to determine α completely, since in the first place they determine the ket $\alpha|\xi''_1\cdots\xi''_u\rangle$ (as they provide the representative of this ket), and the value of this ket for all the basic kets $|\xi''_1\cdots\xi''_u\rangle$ determines α. The numbers (39) are called the *representative* of the linear operator α or of the dynamical variable α. They are more complicated than the representative of a ket or bra vector in that they involve the parameters that label two basic vectors instead of one.

Let us examine the form of these numbers in simple cases. Take first the case when there is only one ξ, forming a complete commuting set by itself, and suppose that it has discrete eigenvalues ξ'. The representative of α is then the discrete set of numbers $\langle \xi'|\alpha|\xi'' \rangle$. If one had to write out these numbers explicitly, the natural way of arranging them would be as a two-dimensional array, thus:

$$\begin{pmatrix} \langle \xi^1|\alpha|\xi^1\rangle & \langle \xi^1|\alpha|\xi^2\rangle & \langle \xi^1|\alpha|\xi^3\rangle & \dots \\ \langle \xi^2|\alpha|\xi^1\rangle & \langle \xi^2|\alpha|\xi^2\rangle & \langle \xi^2|\alpha|\xi^3\rangle & \dots \\ \langle \xi^3|\alpha|\xi^1\rangle & \langle \xi^3|\alpha|\xi^2\rangle & \langle \xi^3|\alpha|\xi^3\rangle & \dots \\ \dots & \dots & \dots & \dots \\ \dots & \dots & \dots & \dots \end{pmatrix} \qquad (40)$$

where $\xi^1, \xi^2, \xi^3, \dots$ are all the eigenvalues of ξ. Such an array is called a *matrix* and the numbers are called the *elements* of the matrix. We make the convention that the elements must always be arranged so that those in the same row refer to the same basic bra vector and those in the same column refer to the same basic ket vector.

An element $\langle \xi'|\alpha|\xi'\rangle$ referring to two basic vectors with the same label is called a *diagonal element* of the matrix, as all such elements lie on a diagonal. If we put α equal to unity, we have from (16) all the diagonal elements equal to unity and all the other elements equal to zero. The matrix is then called the *unit matrix*.

If α is real, we have

$$\langle \xi'|\alpha|\xi''\rangle = \overline{\langle \xi''|\alpha|\xi'\rangle}. \qquad (41)$$

The effect of these conditions on the matrix (40) is to make the diagonal elements all real and each of the other elements equal the conjugate complex of its mirror reflection in the diagonal. The matrix is then called a *Hermitian matrix*.

If we put α equal to ξ, we get for a general element of the matrix

$$\langle \xi'|\xi|\xi''\rangle = \xi'\langle \xi'|\xi''\rangle = \xi'\delta_{\xi'\xi''}. \qquad (42)$$

Thus all the elements not on the diagonal are zero. The matrix is then called a *diagonal matrix*. Its diagonal elements are just equal to the eigenvalues of ξ. More generally, if we put α equal to $f(\xi)$, a function of ξ, we get

$$\langle \xi'|f(\xi)|\xi''\rangle = f(\xi')\delta_{\xi'\xi''}, \qquad (43)$$

and the matrix is again a diagonal matrix.

Let us determine the representative of a product $\alpha\beta$ of two linear operators α and β in terms of the representatives of the factors. From equation (22) with ξ''' substituted for ξ' we obtain

$$\langle\xi'|\alpha\beta|\xi''\rangle = \langle\xi'|\alpha \sum_{\xi'''} |\xi'''\rangle\langle\xi'''|\beta|\xi''\rangle$$
$$= \sum_{\xi'''} \langle\xi'|\alpha|\xi'''\rangle\langle\xi'''|\beta|\xi''\rangle, \qquad (44)$$

which gives us the required result. Equation (44) shows that the matrix formed by the elements $\langle\xi'|\alpha\beta|\xi''\rangle$ equals the product of the matrices formed by the elements $\langle\xi'|\alpha|\xi''\rangle$ and $\langle\xi'|\beta|\xi''\rangle$ respectively, according to the usual mathematical rule for multiplying matrices. This rule gives for the element in the rth row and sth column of the product matrix the sum of the product of each element in the rth row of the first factor matrix with the corresponding element in the sth column of the second factor matrix. The multiplication of matrices is non-commutative, like the multiplication of linear operators.

We can summarize our results for the case when there is only one ξ and it has discrete eigenvalues as follows:

(i) *Any linear operator is represented by a matrix.*
(ii) *The unit operator is represented by the unit matrix.*
(iii) *A real linear operator is represented by a Hermitian matrix.*
(iv) *ξ and functions of ξ are represented by diagonal matrices.*
(v) *The matrix representing the product of two linear operators is the product of the matrices representing the two factors.*

Let us now consider the case when there is only one ξ and it has continuous eigenvalues. The representative of α is now $\langle\xi'|\alpha|\xi''\rangle$, a function of two variables ξ' and ξ'' which can vary continuously. It is convenient to call such a function a 'matrix', using this word in a generalized sense, in order that we may be able to use the same terminology for the discrete and continuous cases. One of these generalized matrices cannot, of course, be written out as a two-dimensional array like an ordinary matrix, since the number of its rows and columns is an infinity equal to the number of points on a line, and the number of its elements is an infinity equal to the number of points in an area.

We arrange our definitions concerning these generalized matrices so that the rules (i)–(v) which we had above for the discrete case hold also for the continuous case. The unit operator is represented by $\delta(\xi' - \xi'')$ and the generalized matrix formed by these elements we define to be the *unit matrix.* We still have equation (41) as the condition for α to be real and we define the generalized matrix formed by the elements $\langle\xi'|\alpha|\xi''\rangle$ to be *Hermitian* when it satisfies this condition. ξ is represented by

$$\langle\xi'|\xi|\xi''\rangle = \xi'\,\delta(\xi' - \xi'') \tag{45}$$

and $f(\xi)$ by

$$\langle\xi'|f(\xi)|\xi''\rangle = f(\xi')\,\delta(\xi' - \xi''), \tag{46}$$

and the generalized matrices formed by these elements we define to be *diagonal matrices*. From (11), we could equally well have ξ'' and $f(\xi'')$ as the coefficients of $\delta(\xi' - \xi'')$ on the right-hand sides of (45) and (46) respectively. Corresponding to equation (44) we now have, from (24)

$$\langle\xi'|\alpha\beta|\xi''\rangle = \int \langle\xi'|\alpha|\xi'''\rangle\,d\xi'''\,\langle\xi'''|\beta|\xi''\rangle, \tag{47}$$

with an integral instead of a sum, and we define the generalized matrix formed by the elements on the right-hand side here to be the product of the matrices formed by $\langle\xi'|\alpha|\xi''\rangle$ and $\langle\xi'|\beta|\xi''\rangle$. With these definitions we secure complete parallelism between the discrete and continuous cases and we have the rules (i)–(v) holding for both.

The question arises how a general diagonal matrix is to be defined in the continuous case, as so far we have only defined the right-hand sides of (45) and (46) to be examples of diagonal matrices. One might be inclined to define as diagonal any matrix whose (ξ', ξ'') elements all vanish except when ξ' differs infinitely little from ξ'', but this would not be satisfactory, because an important property of diagonal matrices in the discrete case is that they always commute with one another and we want this property to hold also in the continuous case. In order that the matrix formed by the elements $\langle\xi'|\omega|\xi''\rangle$ in the continuous case may commute with that formed by the elements on the right-hand side of (45) we must have, using the multiplication rule (47),

$$\int \langle\xi'|\omega|\xi'''\rangle\,d\xi'''\,\xi'''\,\delta(\xi''' - \xi'') = \int \xi'\,\delta(\xi' - \xi''')\,d\xi'''\,\langle\xi'''|\omega|\xi''\rangle.$$

With the help of formula (4), this reduces to

$$\langle\xi'|\omega|\xi''\rangle\xi'' = \xi'\langle\xi'|\omega|\xi''\rangle \tag{48}$$

or

$$(\xi' - \xi'')\langle\xi'|\omega|\xi''\rangle = 0.$$

This gives, according to the rule by which (13) follows from (12),

$$\langle\xi'|\omega|\xi''\rangle = c'\,\delta(\xi' - \xi'')$$

where c' is a number that may depend on ξ'. Thus $\langle \xi'|\omega|\xi''\rangle$ is of the form of the right-hand side of (46). For this reason *we define only matrices whose elements are of the form of the right-hand side of* (46) *to be diagonal matrices.* It is easily verified that these matrices all commute with one another. One can form other matrices whose (ξ', ξ'') elements all vanish when ξ' differs appreciably from ξ'' and have a different form of singularity when ξ' equals ξ'' [we shall later introduce the derivative $\delta'(x)$ of the δ function and $\delta'(\xi' - \xi'')$ will then be an example, see § 22 equation (19)], but these other matrices are not diagonal according to the definition.

Let us now pass on to the case when there is only one ξ and it has both discrete and continuous eigenvalues. Using ξ^r, ξ^s to denote discrete eigenvalues and ξ', ξ'' to denote continuous eigenvalues, we now have the representative of α consisting of four kinds of quantities, $\langle \xi^r|\alpha|\xi^s\rangle$, $\langle \xi^r|\alpha|\xi'\rangle$, $\langle \xi'|\alpha|\xi^r\rangle$, $\langle \xi'|\alpha|\xi''\rangle$. These quantities can all be put together and considered to form a more general kind of matrix having some discrete rows and columns and also a continuous range of rows and columns. We define unit matrix, Hermitian matrix, diagonal matrix, and the product of two matrices also for this more general kind of matrix so as to make the rules (i)–(v) still hold. The details are a straightforward generalization of what has gone before and need not be given explicitly.

Let us now go back to the general case of several ξ's, $\xi_1, \xi_2, \ldots, \xi_u$. The representative of α, expression (39), may still be looked upon as forming a matrix, with rows corresponding to different values of ξ'_1, $\xi'_2, \ldots, \xi'_u$ and columns corresponding to different values of $\xi''_1, \xi''_2, \ldots, \xi''_u$. Unless all the ξ's have discrete eigenvalues, this matrix will be of the generalized kind with continuous ranges of rows and columns. We again arrange our definitions so that the rules (i)–(v) hold, with rule (iv) generalized to:

(iv') *Each ξ_m ($m = 1, 2, \ldots, u$) and any function of them is represented by a diagonal matrix.*

A diagonal matrix is now defined as one whose general element $\langle \xi'_1 \cdots \xi'_u|\omega|\xi''_1 \cdots \xi''_u\rangle$ is of the form

$$\langle \xi'_1 \cdots \xi'_u|\omega|\xi''_1 \cdots \xi''_u\rangle = c' \delta_{\xi'_1 \xi''_1} \cdots \delta_{\xi'_v \xi''_v} \delta(\xi'_{v+1} - \xi''_{v+1}) \cdots \delta(\xi'_u - \xi''_u) \quad (49)$$

in the case when $\xi_1, \xi_2, \ldots, \xi_v$ have discrete eigenvalues and $\xi_{v+1}, \xi_{v+2}, \ldots, \xi_u$ have continuous eigenvalues, c' being any function of the ξ''s. This definition is the generalization of what we had with one ξ and makes diagonal matrices always commute with one another. The other definitions are straightforward and need not be given explicitly.

We now have a linear operator always represented by a matrix. The sum of two linear operators is represented by the sum of the matrices representing the operators and this, together with rule (v), means that *the matrices are subject to the same algebraic relations as the linear operators*. If any algebraic equation holds between certain linear operators, the same equation must hold between the matrices representing those operators.

The scheme of matrices can be extended to bring in the representatives of ket and bra vectors. The matrices representing linear operators are all square matrices with the same number of rows and columns, and with, in fact, a one-one correspondence between their rows and columns. We may look upon the representative of a ket $|P\rangle$ as *a matrix with a single column* by setting all the numbers $\langle \xi'_1 \cdots \xi'_u | P \rangle$ which form this representative one below the other. The number of rows in this matrix will be the same as the number of rows or columns in the square matrices representing linear operators. Such a single-column matrix can be multiplied on the left by a square matrix $\langle \xi'_1 \cdots \xi'_u | \alpha | \xi''_1 \cdots \xi''_u \rangle$ representing a linear operator, by a rule similar to that for the multiplication of two square matrices. The product is another single-column matrix with elements given by

$$\sum_{\xi''_1 \cdots \xi''_v} \int \cdots \int \langle \xi'_1 \cdots \xi'_u | \alpha | \xi''_1 \cdots \xi''_u \rangle \, d\xi''_{v+1} \cdots d\xi''_u \, \langle \xi''_1 \cdots \xi''_u | P \rangle.$$

From (35) this is just equal to $\langle \xi'_1 \cdots \xi'_u | \alpha | P \rangle$, the representative of $\alpha | P \rangle$. Similarly we may look upon the representative of a bra $\langle Q |$ as *a matrix with a single row* by setting all the numbers $\langle Q | \xi'_1 \cdots \xi'_u \rangle$ side by side. Such a single-row matrix may be multiplied on the right by a square matrix $\langle \xi'_1 \cdots \xi'_u | \alpha | \xi''_1 \cdots \xi''_u \rangle$, the product being another single-row matrix, which is just the representative of $\langle Q | \alpha$. The single-row matrix representing $\langle Q |$ may be multiplied on the right by the single-column matrix representing $|P\rangle$, the product being a matrix with just a single element, which is equal to $\langle Q | P \rangle$. Finally, the single-row matrix representing $\langle Q |$ may be multiplied on the left by the single-column matrix representing $|P\rangle$, the product being a square matrix, which is just the representative of $|P\rangle\langle Q|$. In this way all our abstract symbols, linear operators, bra vectors, and ket vectors, can be represented by matrices, which are subject to the same algebraic relations as the abstract symbols themselves.

18. Probability Amplitudes

Representations are of great importance in the physical interpretation of quantum mechanics as they provide a convenient method for obtaining the probabilities of observables having given values. In § 12 we obtained the probability of an observable having any specified value for a given state and in § 13 we generalized this result and obtained the probability of a set of commuting observables simultaneously having specified values for a given state. Let us now apply this result to a complete set of commuting observables, say the set of ξ's which we have been dealing with already. According to formula (51) of § 13, the probability of each ξ_r having the value ξ'_r for the state corresponding to the normalized ket vector $|x\rangle$ is

$$P_{\xi'_1 \ldots \xi'_u} = \langle x | \delta_{\xi_1 \xi'_1} \delta_{\xi_2 \xi'_2} \cdots \delta_{\xi_u \xi'_u} | x \rangle. \tag{50}$$

If the ξ's all have discrete eigenvalues, we can use (35) with $v = u$ and no integrals, and get

$$P_{\xi'_1 \ldots \xi'_u} = \sum_{\xi''_1 \ldots \xi''_u} \langle x | \delta_{\xi_1 \xi'_1} \delta_{\xi_2 \xi'_2} \cdots \delta_{\xi_u \xi'_u} | \xi''_1 \cdots \xi''_u \rangle \langle \xi''_1 \cdots \xi''_u | x \rangle$$

$$= \sum_{\xi''_1 \ldots \xi''_u} \langle x | \delta_{\xi''_1 \xi'_1} \delta_{\xi''_2 \xi'_2} \cdots \delta_{\xi''_u \xi'_u} | \xi''_1 \cdots \xi''_u \rangle \langle \xi''_1 \cdots \xi''_u | x \rangle$$

$$= \langle x | \xi'_1 \cdots \xi'_u \rangle \langle \xi'_1 \cdots \xi'_u | x \rangle$$

$$= |\langle \xi'_1 \cdots \xi'_u | x \rangle|^2. \tag{51}$$

We thus get the simple result that *the probability of the ξ's having the values ξ' is just the square of the modulus of the appropriate coordinate of the normalized ket vector corresponding to the state concerned.*

If the ξ's do not all have discrete eigenvalues, but if, say, $\xi_1, \xi_2, \ldots, \xi_v$ have discrete eigenvalues and $\xi_{v+1}, \xi_{v+2} \ldots, \xi_u$ have continuous eigenvalues, then to get something physically significant we must obtain the probability of each ξ_r ($r = 1, 2, \ldots, v$) having a specified value ξ'_r and each ξ_s ($s = v+1, v+2, \ldots, u$) lying in a specified small range ξ'_s to $\xi'_s + d\xi'_s$. For this purpose we must replace each factor $\delta_{\xi_s \xi'_s}$ in (50) by a factor χ_s, which is that function of the observable ξ_s which is equal to unity for ξ_s within the range ξ'_s to $\xi'_s + d\xi'_s$ and zero otherwise. Proceeding as before with the help of (35), we obtain for this probability

$$P_{\xi'_1 \ldots \xi'_u} d\xi'_{v+1} \cdots d\xi'_u = |\langle \xi'_1 \cdots \xi'_u | x \rangle|^2 d\xi'_{v+1} \cdots d\xi'_u. \tag{52}$$

Thus in every case *the probability distribution of values for the ξ's is given by the square of the modulus of the representative of the normalized ket vector corresponding to the state concerned.*

The numbers which form the representative of a normalized ket (or bra) may for this reason be called *probability amplitudes*. The square of the modulus of a probability amplitude is an ordinary probability, or a probability per unit range for those variables that have continuous ranges of values.

We may be interested in a state whose corresponding ket $|x\rangle$ cannot be normalized. This occurs, for example, if the state is an eigenstate of some observable belonging to an eigenvalue lying in a range of eigenvalues. The formula (51) or (52) can then still be used to give the relative probability of the ξ's having specified values or having values lying in specified small ranges, i.e. it will give correctly the ratios of the probabilities for different ξ''s. The numbers $\langle \xi'_1 \cdots \xi'_u | x \rangle$ may then be called *relative probability amplitudes*.

The representation for which the above results hold is characterized by the basic vectors being simultaneous eigenvectors of all the ξ's. It may also be characterized by the requirement that each of the ξ's shall be represented by a diagonal matrix, this condition being easily seen to be equivalent to the previous one. The latter characterization is usually the more convenient one. For brevity, we shall formulate it as each of the ξ's '*being diagonal in the representation*'.

Provided the ξ's form a *complete* set of commuting observables, the representation is completely determined by the characterization, apart from arbitrary phase factors in the basic vectors. Each basic bra $\langle \xi'_1 \cdots \xi'_u |$ may be multiplied by $e^{i\gamma'}$, where γ' is any real function of the variables $\xi'_1, \xi'_2, \ldots, \xi'_u$, without changing any of the conditions which the representation has to satisfy, i.e. the condition that the ξ's are diagonal or that the basic vectors are simultaneous eigenvectors of the ξ's, and the fundamental properties of the basic vectors (34) and (35). With the basic bras changed in this way, the representative $\langle \xi'_1 \cdots \xi'_u | P \rangle$ of a ket $|P\rangle$ gets multiplied by $e^{i\gamma'}$, the representative $\langle Q | \xi'_1 \cdots \xi'_u \rangle$ of a bra $\langle Q |$ gets multiplied by $e^{-i\gamma'}$ and the representative $\langle \xi'_1 \cdots \xi'_u | \alpha | \xi''_1 \cdots \xi''_u \rangle$ of a linear operator α gets multiplied by $e^{i(\gamma' - \gamma'')}$. The probabilities or relative probabilities (51), (52) are, of course, unaltered.

The probabilities that one calculates in practical problems in quantum mechanics are nearly always obtained from the squares of the moduli of probability amplitudes or relative probability amplitudes. Even when one is interested only in the probability of an incomplete set of commuting observables having specified values, it is usually necessary first to make the set a complete one by the introduction of some extra commuting observables and to obtain the probability of the complete set having specified values (as the square of the modulus of a probability amplitude), and then to sum or integrate over all possible values of the extra observables. A more direct application of formula (51) of § 13 is usually not practicable.

To introduce a representation in practice

(i) We look for observables which we would like to have diagonal, either because we are interested in their probabilities or for reasons of mathematical simplicity;

(ii) We must see that they all commute—a necessary condition since diagonal matrices always commute;

(iii) We then see that they form a complete commuting set, and if not we add some more commuting observables to them to make them into a complete commuting set;

(iv) We set up an orthogonal representation with this complete commuting set diagonal.

The representation is then completely determined except for the arbitrary phase factors. For most purposes the arbitrary phase factors are unimportant and trivial, so that we may count the representation as being completely determined by the observables that are diagonal in it. This fact is already implied in our notation, since the only indication in a representative of the representation to which it belongs are the letters denoting the observables that are diagonal.

It may be that we are interested in two representations for the same dynamical system. Suppose that in one of them the complete set of commuting observables $\xi_1, \xi_2, \ldots, \xi_u$ are diagonal and the basic bras are $\langle \xi'_1 \cdots \xi'_u |$ and in the other the complete set of commuting observables $\eta_1, \eta_2, \ldots, \eta_w$ are diagonal and the basic bras are $\langle \eta'_1 \cdots \eta'_w |$. A ket $|P\rangle$ will now have the two representatives $\langle \xi'_1 \cdots \xi'_u | P \rangle$ and $\langle \eta'_1 \cdots \eta'_w | P \rangle$. If $\xi_1, \xi_2, \ldots, \xi_v$ have discrete eigenvalues and $\xi_{v+1}, \xi_{v+2} \ldots, \xi_u$ have continuous eigenvalues and if $\eta_1, \eta_2, \ldots, \eta_x$ have discrete eigenvalues and $\eta_{x+1}, \eta_{x+2} \ldots, \eta_w$ have continuous eigenvalues, we get from (35)

$$\langle \eta'_1 \cdots \eta'_w | P \rangle$$

$$= \sum_{\xi'_1 \cdots \xi'_v} \int \cdots \int \langle \eta'_1 \cdots \eta'_w | \xi'_1 \cdots \xi'_u \rangle \, d\xi'_{v+1} \cdots d\xi'_u \langle \xi'_1 \cdots \xi'_u | P \rangle, \quad (53)$$

and interchanging ξ's and η's

$$\langle \xi'_1 \cdots \xi'_u | P \rangle$$
$$= \sum_{\eta'_1 \cdots \eta'_x} \int \cdots \int \langle \xi'_1 \cdots \xi'_u | \eta'_1 \cdots \eta'_w \rangle \, d\eta'_{x+1} \cdots d\eta'_w \langle \eta'_1 \cdots \eta'_w | P \rangle. \quad (54)$$

These are the transformation equations which give one representative of $|P\rangle$ in terms of the other. They show that either representative is expressible linearly in terms of the other, with the quantities

$$\langle \eta'_1 \cdots \eta'_w | \xi'_1 \cdots \xi'_u \rangle, \quad \langle \xi'_1 \cdots \xi'_u | \eta'_1 \cdots \eta'_w \rangle \quad (55)$$

as coefficients. These quantities are called the *transformation functions*. Similar equations may be written down to connect the two representatives of a bra vector or of a linear operator. The transformation functions (55) are in every case the means which enable one to pass from one representative to the other. Each of the transformation functions is the conjugate complex of the other, and they satisfy the conditions

$$\sum_{\xi'_1 \cdots \xi'_v} \int \cdots \int \langle \eta'_1 \cdots \eta'_w | \xi'_1 \cdots \xi'_u \rangle \, d\xi'_{v+1} \cdots d\xi'_u \langle \xi'_1 \cdots \xi'_u | \eta''_1 \cdots \eta''_w \rangle$$
$$= \delta_{\eta'_1 \eta''_1} \cdots \delta_{\eta'_x \eta''_x} \delta(\eta'_{x+1} - \eta''_{x+1}) \cdots \delta(\eta'_w - \eta''_w) \quad (56)$$

and the corresponding conditions with ξ's and η's interchanged, as may be verified from (35) and (34) and the corresponding equations for the η's.

Transformation functions are examples of probability amplitudes or relative probability amplitudes. Let us take the case when all the ξ's and all the η's have discrete eigenvalues. Then the basic ket $|\eta'_1 \cdots \eta'_w\rangle$ is normalized, so that its representative in the ξ-representation, $\langle \xi'_1 \cdots \xi'_u | \eta'_1 \cdots \eta'_w \rangle$, is a probability amplitude for each set of values for the ξ's. The state to which these probability amplitudes refer, namely the state corresponding to $|\eta'_1 \cdots \eta'_w\rangle$, is characterized by the condition that a simultaneous measurement of $\eta_1, \eta_2, \ldots, \eta_w$ is certain to lead to the results $\eta'_1, \eta'_2, \ldots, \eta'_w$. Thus $|\langle \xi'_1 \cdots \xi'_u | \eta'_1 \cdots \eta'_w \rangle|^2$ is the probability of the ξ's having the values $\xi'_1 \xi'_2 \cdots \xi'_u$ for the state for which the η's certainly have the values $\eta'_1 \eta'_2 \cdots \eta'_w$. Since

$$|\langle \xi'_1 \cdots \xi'_u | \eta'_1 \cdots \eta'_w \rangle|^2 = |\langle \eta'_1 \cdots \eta'_w | \xi'_1 \cdots \xi'_u \rangle|^2,$$

we have the theorem of reciprocity—*the probability of the ξ's having the values ξ' for the state for which the η's certainly have the values η' is equal to the probability of the η's having the values η' for the state for which the ξ's certainly have the values ξ'.*

If all the η's have discrete eigenvalues and some of the ξ's have continuous eigenvalues, $|\langle \xi'_1 \cdots \xi'_u | \eta'_1 \cdots \eta'_w \rangle|^2$ still gives the probability distribution of values for the ξ's for the state for which the η's certainly have the values η'. If some of the η's have continuous eigenvalues, $|\eta'_1 \cdots \eta'_w\rangle$ is not normalized and $|\langle \xi'_1 \cdots \xi'_u | \eta'_1 \ldots \eta'_w \rangle|^2$ then gives only the relative probability distribution of values for the ξ's for the state for which the η's certainly have the values η'.

19. Theorems about Functions of Observables

We shall illustrate the mathematical value of representations by using them to prove some theorems.

THEOREM 1. *A linear operator that commutes with an observable ξ commutes also with any function of ξ.*

The theorem is obviously true when the function is expressible as a power series. To prove it generally, let ω be the linear operator, so that we have the equation

$$\xi\omega - \omega\xi = 0. \tag{57}$$

Let us introduce a representation in which ξ is diagonal. If ξ by itself does not form a complete commuting set of observables, we must make it into a complete commuting set by adding certain observables, β say, to it, and then take the representation in which ξ and the β's are diagonal. (The case when ξ does form a complete commuting set by itself can be looked upon as a special case of the preceding one with the number of β variables zero.) In this representation equation (57) becomes

$$\langle \xi'\beta' | \xi\omega - \omega\xi | \xi''\beta'' \rangle = 0,$$

which reduces to

$$\xi'\langle \xi'\beta' | \omega | \xi''\beta'' \rangle - \langle \xi'\beta' | \omega | \xi''\beta'' \rangle \xi'' = 0.$$

In the case when the eigenvalues of ξ are discrete, this equation shows that all the matrix elements $\langle \xi'\beta' | \omega | \xi''\beta'' \rangle$ of ω vanish except those for which $\xi' = \xi''$. In the case when the eigenvalues of ξ are continuous it shows, like equation (48), that $\langle \xi'\beta' | \omega | \xi''\beta'' \rangle$ is of the form

$$\langle \xi'\beta' | \omega | \xi''\beta'' \rangle = c\,\delta(\xi' - \xi''),$$

where c is some function of ξ' and the β''s and β'''s. In either case we may say that the matrix representing ω '*is diagonal with respect to ξ*'. If $f(\xi)$ denotes any function of ξ in accordance with the general theory of § 11, which requires $f(\xi''')$ to be defined for ξ''' any eigenvalue of ξ, we can deduce in either case

$$f(\xi')\langle\xi'\beta'|\omega|\xi''\beta''\rangle - \langle\xi'\beta'|\omega|\xi''\beta''\rangle f(\xi'') = 0.$$

This gives

$$\langle\xi'\beta'|f(\xi)\omega - \omega f(\xi)|\xi''\beta''\rangle = 0,$$

so that

$$f(\xi)\omega - \omega f(\xi) = 0$$

and the theorem is proved.

As a special case of the theorem, we have the result that any observable that commutes with an observable ξ also commutes with any function of ξ. This result appears as a physical necessity when we identify, as in § 13, the condition of commutability of two observables with the condition of compatibility of the corresponding observations. Any observation that is compatible with the measurement of an observable ξ must also be compatible with the measurement of $f(\xi)$, since any measurement of ξ includes in itself a measurement of $f(\xi)$.

THEOREM 2. *A linear operator that commutes with each of a complete set of commuting observables is a function of those observables.*

Let ω be the linear operator and $\xi_1, \xi_2, \ldots, \xi_u$ the complete set of commuting observables, and set up a representation with these observables diagonal. Since ω commutes with each of the ξ's, the matrix representing it is diagonal with respect to each of the ξ's, by the argument we had above. This matrix is therefore a diagonal matrix and is of the form (49), involving a number c' which is a function of the ξ''s. It thus represents the function of the ξ's that c' is of the ξ''s, and hence ω equals this function of the ξ's.

THEOREM 3. *If an observable ξ and a linear operator g are such that any linear operator that commutes with ξ also commutes with g, then g is a function of ξ.*

This is the converse of Theorem 1. To prove it, we use the same representation with ξ diagonal as we had for Theorem 1. In the first place, we see that g must commute with ξ itself, and hence the representative of g must be diagonal with respect to ξ, i.e. it must be of the form

$$\langle\xi'\beta'|g|\xi''\beta''\rangle = a(\xi'\beta'\beta'')\delta_{\xi'\xi''} \quad \text{or} \quad a(\xi'\beta'\beta'')\,\delta(\xi' - \xi''),$$

according to whether ξ has discrete or continuous eigenvalues. Now let ω be any linear operator that commutes with ξ, so that its representative is of the form

$$\langle \xi'\beta'|\omega|\xi''\beta''\rangle = b(\xi'\beta'\beta'')\delta_{\xi'\xi''} \quad \text{or} \quad b(\xi'\beta'\beta'')\,\delta(\xi'-\xi'').$$

By hypothesis ω must also commute with g, so that

$$\langle \xi'\beta'|g\omega - \omega g|\xi''\beta''\rangle = 0. \tag{58}$$

If we suppose for definiteness that the β's have discrete eigenvalues, (58) leads, with the help of the law of matrix multiplication, to

$$\sum_{\beta'''} [a(\xi'\beta'\beta''')\,b(\xi'\beta'''\beta'') - b(\xi'\beta'\beta''')\,a(\xi'\beta'''\beta'')] = 0, \tag{59}$$

the left-hand side of (58) being equal to the left-hand side of (59) multiplied by $\delta_{\xi'\xi''}$ or $\delta(\xi'-\xi'')$. Equation (59) must hold for all functions $b(\xi'\beta'\beta'')$. We can deduce that

$$a(\xi'\beta'\beta'') = 0 \quad \text{for} \quad \beta' \neq \beta'',$$
$$a(\xi'\beta'\beta') = a(\xi'\beta''\beta'').$$

The first of these results shows that the matrix representing g is diagonal and the second shows that $a(\xi'\beta'\beta')$ is a function of ξ' only. We can now infer that g is that function of ξ which $a(\xi'\beta'\beta')$ is of ξ', so the theorem is proved. The proof is analogous if some of the β's have continuous eigenvalues.

Theorems 1 and 3 are still valid if we replace the observable ξ by any set of commuting observables $\xi_1, \xi_2, \ldots, \xi_r$, only formal changes being needed in the proofs.

20. Developments in Notation

The theory of representations that we have developed provides a general system for labelling kets and bras. In a representation in which the complete set of commuting observables $\xi_1, \xi_2, \ldots, \xi_u$ are diagonal any ket $|P\rangle$ will have a representative $\langle \xi'_1 \cdots \xi'_u | P\rangle$, or $\langle \xi'|p\rangle$ for brevity. This representative is a definite function of the variables ξ', say $\psi(\xi')$. The function ψ then determines the ket $|P\rangle$ completely, so it may be used to label this ket, to replace the arbitrary label P. In symbols, if

$$\langle \xi'|P\rangle = \psi(\xi') \tag{60}$$

we put

$$|P\rangle = |\psi(\xi)\rangle. \tag{60}$$

We must put $|P\rangle$ equal to $|\psi(\xi)\rangle$ and not $|\psi(\xi')\rangle$, since it does not depend on a particular set of eigenvalues for the ξ's, but only on the form of the function ψ.

With $f(\xi)$ any function of the observables $\xi_1, \xi_2, \ldots, \xi_u$, $f(\xi)|P\rangle$ will have as its representative

$$\langle \xi'|f(\xi)|P\rangle = f(\xi')\psi(\xi').$$

Thus according to (60) we put

$$f(\xi)|P\rangle = |f(\xi)\psi(\xi)\rangle.$$

With the help of the second of equations (60) we now get

$$f(\xi)|\psi(\xi)\rangle = |f(\xi)\psi(\xi)\rangle. \tag{61}$$

This is a general result holding for any functions f and ψ of the ξ's, and it shows that the vertical line | is not necessary with the new notation for a ket—either side of (61) may be written simply as $f(\xi)\psi(\xi)\rangle$. Thus the rule for the new notation becomes:—if

$$\langle \xi'|P\rangle = \psi(\xi') \tag{62}$$

we put

$$|P\rangle = \psi(\xi)\rangle. \tag{62}$$

We may further shorten $\psi(\xi)\rangle$ to $\psi\rangle$, leaving the variables ξ understood, if no ambiguity arises thereby.

The ket $\psi(\xi)\rangle$ may be considered as the product of the linear operator $\psi(\xi)$ with a ket which is denoted simply by $\rangle$ without a label. We call the ket $\rangle$ the *standard ket*. Any ket whatever can be expressed as a function of the ξ's multiplied into the standard ket. For example, taking $|P\rangle$ in (62) to be the basic ket $|\xi''\rangle$, we find

$$|\xi''\rangle = \delta_{\xi_1 \xi_1''} \cdots \delta_{\xi_v \xi_v''} \delta(\xi_{v+1} - \xi_{v+1}'') \cdots \delta(\xi_u - \xi_u'')\rangle \tag{63}$$

in the case when $\xi_1, \xi_2, \ldots, \xi_v$ have discrete eigenvalues and $\xi_{v+1}, \xi_{v+2}, \ldots, \xi_u$ have continuous eigenvalues. The standard ket is characterized by the condition that its representative $\langle \xi'|\rangle$ is unity over the whole domain of the variable ξ', as may be seen by putting $\psi = 1$ in (62).

A further contraction may be made in the notation, namely to leave the symbol $\rangle$ for the standard ket understood. A ket is then written simply as $\psi(\xi)$, a function of the observables ξ. A function of the ξ's used in this way to denote a ket is called a *wave function*.[1] The system of notation provided by wave functions is the one usually used by most authors for calculations in quantum mechanics. In using it one should remember that each wave function is understood to have the standard ket multiplied into it on the right, which prevents one from multiplying the wave function by any operator on the right. *Wave functions can be multiplied by operators only on the left.* This distinguishes them from ordinary functions of the ξ's, which are operators and can be multiplied by operators on either the left or the right. A wave function is just the representative of a ket expressed as a function of the observables ξ, instead of eigenvalues ξ' for those observables. The square of its modulus gives the probability (or the relative probability, if it is not normalized) of the ξ's having specified values, or lying in specified small ranges, for the corresponding state.

The new notation for bras may be developed in the same way as for kets. A bra $\langle Q|$ whose representative $\langle Q|\xi'\rangle$ is $\phi(\xi')$ we write $\langle \phi(\xi)|$. With this notation the conjugate imaginary to $|\psi(\xi)\rangle$ is $\langle \overline{\psi}(\xi)|$. Thus the rule that we have used hitherto, that a ket and its conjugate imaginary bra are both specified by the same label, must be extended to read—*if the labels of a ket involve complex numbers or complex functions, the labels of the conjugate imaginary bra involve the conjugate complex numbers or functions*. As in the case of kets we can show that $\langle \phi(\xi)|f(\xi)$ and $\langle \phi(\xi)\,f(\xi)|$ are the same, so that the vertical line can be omitted. We can consider $\langle \phi(\xi)|$ as the product of the linear operator $\phi(\xi)$ into the *standard bra* $\langle$, which is the conjugate imaginary of the standard ket $\rangle$. We may leave the standard bra understood, so that a general bra is written as $\phi(\xi)$, the conjugate complex of a wave function. The conjugate complex of a wave function can be multiplied by any linear operator on the right, but cannot be multiplied by a linear operator on the left. We can construct triple products of the form $\langle f(\xi)\rangle$. Such a triple product is a number, equal to $f(\xi)$ summed or integrated over the whole domain of eigenvalues for the ξ's,

$$\langle f(\xi)\rangle = \sum_{\xi'_1\ldots\xi'_v} \int \cdots \int f(\xi')\,d\xi'_{v+1}\cdots d\xi'_u \tag{64}$$

in the case when $\xi_1,\xi_2,\ldots,\xi_v$ have discrete eigenvalues and $\xi_{v+1},\xi_{v+2},\ldots,\xi_u$ have continuous eigenvalues.

---

[1] The reason for this name is that in the early days of quantum mechanics all the examples of these functions were of the form of waves. The name is not a descriptive one from the point of view of the modern general theory.

The standard ket and bra are defined with respect to a representation. If we carried through the above work with a different representation in which the complete set of commuting observables η are diagonal, or if we merely changed the phase factors in the representation with the ξ's diagonal, we should get a different standard ket and bra. In a piece of work in which more than one standard ket or bra appears one must, of course, distinguish them by giving them labels.

A further development of the notation which is of great importance for dealing with complicated dynamical systems will now be discussed. Suppose we have a dynamical system describable in terms of dynamical variables which can all be divided into two sets, set A and set B say, such that any member of set A commutes with any member of set B. A general dynamical variable must be expressible as a function of the A-variables and B-variables together. We may consider another dynamical system in which the dynamical variables are the A-variables only—let us call it the A-system. Similarly we may consider a third dynamical system in which the dynamical variables are the B-variables only—the B-system. The original system can then be looked upon as a combination of the A-system and the B-system in accordance with the mathematical scheme given below.

Let us take any ket $|a\rangle$ for the A-system and any ket $|b\rangle$ for the B-system. We assume that they have a product $|a\rangle|b\rangle$ for which the commutative and distributive axioms of multiplication hold, i.e.

$$|a\rangle|b\rangle = |b\rangle|a\rangle,$$
$$(c_1|a_1\rangle + c_2|a_2\rangle)|b\rangle = c_1|a_1\rangle|b\rangle + c_2|a_2\rangle|b\rangle,$$
$$|a\rangle(c_1|b_1\rangle + c_2|b_2\rangle) = c_1|a\rangle|b_1\rangle + c_2|a\rangle|b_2\rangle,$$

the c's being numbers. We can give a meaning to any A-variable operating on the product $|a\rangle|b\rangle$ by assuming that it operates only on the $|a\rangle$ factor and commutes with the $|b\rangle$ factor, and similarly we can give a meaning to any B-variable operating on this product by assuming that it operates only on the $|b\rangle$ factor and commutes with the $|a\rangle$ factor. (This makes every A-variable commute with every B-variable.) Thus any dynamical variable of the original system can operate on the product $|a\rangle|b\rangle$, so this product can be looked upon as a ket for the original system, and may then be written $|ab\rangle$, the two labels a and b being sufficient to specify it. In this way we get the fundamental equations

$$|a\rangle|b\rangle = |b\rangle|a\rangle = |ab\rangle. \tag{65}$$

20. DEVELOPMENTS IN NOTATION

The multiplication here is of quite a different kind from any that occurs earlier in the theory. The ket vectors $|a\rangle$ and $|b\rangle$ are in two different vector spaces and their product is in a third vector space, which may be called the product of the two previous vector spaces. The number of dimensions of the product space is equal to the product of the number of dimensions of each of the factor spaces. A general ket vector of the product space is not of the form (65), but is a sum or integral of kets of this form.

Let us take a representation for the A-system in which a complete set of commuting observables ξ_A of the A-system are diagonal. We shall then have the basic bras $\langle\xi'_A|$ for the A-system. Similarly, taking a representation for the B-system with the observables ξ_B diagonal, we shall have the basic bras $\langle\xi'_B|$ for the B-system. The products

$$\langle\xi'_A|\langle\xi'_B| = \langle\xi'_A\xi'_B| \tag{66}$$

will then provide the basic bras for a representation for the original system, in which representation the ξ_A's and the ξ_B's will be diagonal. The ξ_A's and ξ_B's will together form a complete set of commuting observables for the original system. From (65) and (66) we get

$$\langle\xi'_A|a\rangle\langle\xi'_B|b\rangle = \langle\xi'_A\xi'_B|ab\rangle, \tag{67}$$

showing that the representative of $|ab\rangle$ equals the product of the representatives of $|a\rangle$ and of $|b\rangle$ in their respective representations.

We can introduce the standard ket, $\rangle_A$ say, for the A-system, with respect to the representation with the ξ_A's diagonal, and also the standard ket $\rangle_B$ for the B-system, with respect to the representation with the ξ_B's diagonal. Their product $\rangle_A\rangle_B$ is then the standard ket for the original system, with respect to the representation with the ξ_A's and ξ_B's diagonal. Any ket for the original system may be expressed as

$$\psi(\xi_A\xi_B)\rangle_A\rangle_B. \tag{68}$$

It may be that in a certain calculation we wish to use a particular representation for the B-system, say the above representation with the ξ_B's diagonal, but do not wish to introduce any particular representation for the A-system. It would then be convenient to use the standard ket $\rangle_B$ for the B-system and no standard ket for the A-system. Under these circumstances we could write any ket for the original system as

$$|\xi_B\rangle\rangle_B, \tag{69}$$

in which $|\xi_B\rangle$ is a ket for the A-system and is also a function of the ξ_B's, i.e. it is a ket for the A-system for each set of values for the ξ_B's—in fact

(69) equals (68) if we take
$$|\xi_B\rangle = \psi(\xi_A\xi_B))_A.$$
We may leave the standard ket $\rangle_B$ in (69) understood, and then we have the general ket for the original system appearing as $|\xi_B\rangle$, a ket for the A-system and a wave function in the variables ξ_B of the B-system. An example of this notation will be used in § 66.

The above work can be immediately extended to a dynamical system describable in terms of dynamical variables which can be divided into three or more sets $A, B, C, \ldots$ such that any member of one set commutes with any member of another. Equation (65) gets generalized to
$$|a\rangle|b\rangle|c\rangle \cdots = |abc \cdots\rangle,$$
the factors on the left being kets for the component systems and the ket on the right being a ket for the original system. Equations (66), (67), and (68) get generalized to many factors in a similar way.

CHAPTER IV

The Quantum Conditions

21. Poisson Brackets

Our work so far has consisted in setting up a general mathematical scheme connecting states and observables in quantum mechanics. One of the dominant features of this scheme is that observables, and dynamical variables in general, appear in it as quantities which do not obey the commutative law of multiplication. It now becomes necessary for us to obtain equations to replace the commutative law of multiplication, equations that will tell us the value of $\xi\eta - \eta\xi$ when ξ and η are any two observables or dynamical variables. Only when such equations are known shall we have a complete scheme of mechanics with which to replace classical mechanics. These new equations are called *quantum conditions* or *commutation relations*.

The problem of finding quantum conditions is not of such a general character as those we have been concerned with up to the present. It is instead a special problem which presents itself with each particular dynamical system one is called upon to study. There is, however, a fairly general method of obtaining quantum conditions, applicable to a very large class of dynamical systems. This is the method of *classical analogy* and will form the main theme of the present chapter. Those dynamical systems to which this method is not applicable must be treated individually and special considerations used in each case.

The value of classical analogy in the development of quantum mechanics depends on the fact that classical mechanics provides a valid description of dynamical systems under certain conditions, when the particles and bodies composing the systems are sufficiently massive for the disturbance accompanying an observation to be negligible. Classical mechanics must therefore be a limiting case of quantum mechanics. We should thus expect to find that important concepts in classical mechanics correspond to important concepts in quantum mechanics, and, from an understanding of the general nature of the analogy between classical and quantum mechanics, we may hope to get laws and theorems in quantum mechanics appearing as simple generalizations of well-known results in classical mechanics; in particular we may hope to get the quantum conditions appearing as a simple generalization of the classical law that all dynamical variables commute.

Let us take a dynamical system composed of a number of particles in interaction. As independent dynamical variables for dealing with the system we may use the Cartesian coordinates of all the particles and the corresponding Cartesian components of velocity of the particles. It is, however, more convenient to work with the momentum components instead of the velocity components. Let us call the coordinates q_r, r going from 1 to three times the number of particles, and the corresponding momentum components p_r. The q's and p's are called *canonical coordinates and momenta*.

The method of Lagrange's equations of motion involves introducing coordinates q_r and momenta p_r in a more general way, applicable also for a system not composed of particles (e.g. a system containing rigid bodies). These more general q's and p's are also called canonical coordinates and momenta. Any dynamical variable is expressible in terms of a set of canonical coordinates and momenta.

An important concept in general dynamical theory is the *Poisson Bracket*. Any two dynamical variables u and v have a P.B. (Poisson Bracket) which we shall denote by $[u, v]$, defined by

$$[u, v] = \sum_r \left(\frac{\partial u}{\partial q_r} \frac{\partial v}{\partial p_r} - \frac{\partial u}{\partial p_r} \frac{\partial v}{\partial q_r} \right), \tag{1}$$

u and v being regarded as functions of a set of canonical coordinates and momenta q_r and p_r for the purpose of the differentiations. The right-hand side of (1) is independent of which set of canonical coordinates and momenta are used, this being a consequence of the general definition of canonical coordinates and momenta, so the P.B. $[u, v]$ is well defined.

The main properties of P.B.s, which follow at once from their definition (1), are

$$[u, v] = -[v, u], \tag{2}$$
$$[u, c] = 0, \tag{3}$$

where c is a number (which may be considered as a special case of a dynamical variable),

$$[u_1 + u_2, v] = [u_1, v] + [u_2, v],$$
$$[u, v_1 + v_2] = [u, v_1] + [u, v_2], \tag{4}$$

$$[u_1 u_2, v] = \sum_r \left[\left(\frac{\partial u_1}{\partial q_r} u_2 + u_1 \frac{\partial u_2}{\partial q_r} \right) \frac{\partial v}{\partial p_r} - \left(\frac{\partial u_1}{\partial p_r} u_2 + u_1 \frac{\partial u_2}{\partial p_r} \right) \frac{\partial v}{\partial q_r} \right]$$
$$= [u_1, v] u_2 + u_1 [u_2, v],$$
$$[u, v_1 v_2] = [u, v_1] v_2 + v_1 [u, v_2]. \tag{5}$$

Also the identity

$$[u, [v, w]] + [v, [w, u]] + [w, [u, v]] = 0 \tag{6}$$

is easily verified. Equations (4) express that the P.B. $[u, v]$ involves u and v linearly, while equations (5) correspond to the ordinary rules for differentiating a product.

Let us try to introduce a quantum P.B. which shall be the analogue of the classical one. We assume the quantum P.B. to satisfy all the conditions (2) to (6), it being now necessary that the order of the factors u_1 and u_2 in the first of equations (5) should be preserved throughout the equation, as in the way we have here written it, and similarly for the v_1 and v_2 in the second of equations (5). These conditions are already sufficient to determine the form of the quantum P.B. uniquely, as may be seen from the following argument. We can evaluate the P.B. $[u_1 u_2, v_1 v_2]$ in two different ways, since we can use either of the two formulas (5) first, thus,

$$[u_1 u_2, v_1 v_2] = [u_1, v_1 v_2] u_2 + u_1 [u_2, v_1 v_2]$$
$$= ([u_1, v_1] v_2 + v_1 [u_1, v_2]) u_2 + u_1 ([u_2, v_1] v_2 + v_1 [u_2, v_2])$$
$$= [u_1, v_1] v_2 u_2 + v_1 [u_1, v_2] u_2 + u_1 [u_2, v_1] v_2 + u_1 v_1 [u_2, v_2]$$

and

$$[u_1 u_2, v_1 v_2] = [u_1 u_2, v_1] v_2 + v_1 [u_1 u_2, v_2]$$
$$= [u_1, v_1] u_2 v_2 + u_1 [u_2, v_1] v_2 + v_1 [u_1, v_2] u_2 + v_1 u_1 [u_2, v_2].$$

Equating these two results, we obtain

$$[u_1, v_1](u_2 v_2 - v_2 u_2) = (u_1 v_1 - v_1 u_1)[u_2, v_2].$$

Since this condition holds with u_1 and v_1 quite independent of u_2 and v_2, we must have

$$u_1 v_1 - v_1 u_1 = i\hbar [u_1, v_1],$$
$$u_2 v_2 - v_2 u_2 = i\hbar [u_2, v_2],$$

where $\hbar$ must not depend on u_1 and v_1, nor on u_2 and v_2, and also must commute with $(u_1 v_1 - v_1 u_1)$. It follows that $\hbar$ must be simply a number. We want the P.B. of two real variables to be real, as in the classical theory, which requires from the work at the top of p. 28, that $\hbar$ shall be a real number when introduced, as here, with the coefficient i. We are thus led to the following *definition for the quantum P.B. $[u, v]$ of any two variables u and v*,

$$uv - vu = i\hbar[u, v], \tag{7}$$

in which $\hbar$ is a new universal constant. It has the dimensions of action. In order that the theory may agree with experiment, we must take $\hbar$ equal to $h/2\pi$, where h is the universal constant that was introduced by Planck, known as Planck's constant. It is easily verified that the quantum P.B. satisfies all the conditions (2), (3), (4), (5), and (6).

The problem of finding quantum conditions now reduces to the problem of determining P.B.s in quantum mechanics. The strong analogy between the quantum P.B. defined by (7) and the classical P.B. defined by (1) leads us to make the assumption that the quantum P.B.s, or at any rate the simpler ones of them, have the same values as the corresponding classical P.B.s. The simplest P.B.s are those involving the canonical coordinates and momenta themselves and have the following values in the classical theory:

$$[q_r, q_s] = 0, \quad [p_r, p_s] = 0, \quad [q_r, p_s] = \delta_{rs}. \tag{8}$$

We therefore assume that the corresponding quantum P.B.s also have the values given by (8). By eliminating the quantum P.B.s with the help of (7), we obtain the equations

$$q_r q_s - q_s q_r = 0, \quad p_r p_s - p_s p_r = 0, \quad q_r p_s - p_s q_r = i\hbar\delta_{rs}, \tag{9}$$

which are the *fundamental quantum conditions*. They show us where the lack of commutability among the canonical coordinates and momenta lies. They also provide us with a basis for calculating commutation relations between other dynamical variables. For instance, if ξ and η are any two functions of the q's and p's expressible as power series, we may express $\xi\eta - \eta\xi$ or $[\xi, \eta]$, by repeated applications of the laws (2), (3), (4), and (5), in terms of the elementary P.B.s given in (8) and so evaluate it. The result is often, in simple cases, the same as the classical result, or departs from the classical result only through requiring a special order for factors in a product, this order being, of course, unimportant in the classical theory. Even when ξ and η are more general functions of the q's and p's not expressible as power series, equations (9) are still sufficient to fix the value of $\xi\eta - \eta\xi$, as will become clear from the following work. Equations (9) thus give the solution of the problem of finding the quantum conditions, for all those dynamical systems which have a classical analogue and which are describable in terms of canonical coordinates and momenta. This does not include all possible systems in quantum mechanics.

Equations (7) and (9) provide the foundation for the analogy between quantum mechanics and classical mechanics. They show that *classical mechanics may be regarded as the limiting case of quantum mechanics when $\hbar$ tends to zero*. A P.B. in quantum mechanics is a purely algebraic notion and is thus a rather more fundamental concept than a classical P.B., which can be defined only with reference to a set of canonical coordinates and momenta. For this reason canonical coordinates and momenta are of less importance in quantum mechanics than in classical mechanics; in fact, we may have a system in quantum mechanics for which canonical coordinates and momenta do not exist and we can still give a meaning to P.B.s. Such a system would be one without a classical analogue and we should not be able to obtain its quantum conditions by the method here described.

From equations (9) we see that two variables with different suffixes r and s always commute. It follows that any function of q_r and p_r will commute with any function of q_s and p_s when s differs from r. Different values of r correspond to different degrees of freedom of the dynamical system, so we get the result that *dynamical variables referring to different degrees of freedom commute*. This law, as we have derived it from (9), is proved only for dynamical systems with classical analogues, but we assume it to hold generally. In this way we can make a start on the problem of finding quantum conditions for dynamical systems for which canonical coordinates and momenta do not exist, provided we can give a meaning to different degrees of freedom, as we may be able to do with the help of physical insight.

We can now see the physical meaning of the division, which was discussed in the preceding section, of the dynamical variables into sets, any member of one set commuting with any member of another. Each set corresponds to certain degrees of freedom, or possibly just one degree of freedom. The division may correspond to the physical process of resolving the dynamical system into its constituent parts, each constituent being capable of existing by itself as a physical system, and the various constituents having to be brought into interaction with one another to produce the original system. Alternatively the division may be merely a mathematical procedure of resolving the dynamical system into degrees of freedom which cannot be separated physically, e.g. the system consisting of a particle with internal structure may be divided into the degrees of freedom describing the motion of the centre of the particle and those describing the internal structure.

22. Schrödinger's Representation

Let us consider a dynamical system with n degrees of freedom having a classical analogue, and thus describable in terms of canonical coordinates and momenta q_r, p_r ($r = 1, 2, \ldots, n$). We assume that *the coordinates q_r are all observables and have continuous ranges of eigenvalues*, these assumptions being reasonable from the physical significance of the q's. Let us set up a representation with the q's diagonal. The question arises whether the q's form a complete commuting set for this dynamical system. It seems pretty obvious from inspection that they do. We shall here assume that they do, and the assumption will be justified later (see top of p. 93). With the q's forming a complete commuting set, the representation is fixed except for the arbitrary phase factors in it.

Let us consider first the case of $n = 1$, so that there is only one q and p, satisfying

$$qp - pq = i\hbar. \tag{10}$$

Any ket may be written in the standard ket notation $\psi(q)\rangle$. From it we can form another ket $d\psi/dq\rangle$, whose representative is the derivative of the original one. This new ket is a linear function of the original one and is thus the result of some linear operator applied to the original one. Calling this linear operator d/dq, we have

$$\frac{d}{dq}\psi\rangle = \frac{d\psi}{dq}\rangle. \tag{11}$$

Equation (11) holding for all functions ψ defines the linear operator d/dq. We have

$$\frac{d}{dq}\rangle = 0. \tag{12}$$

Let us treat the linear operator d/dq according to the general theory of linear operators of § 7. We should then be able to apply it to a bra $\langle\phi(q)|$, the product $\langle\phi d/dq$ being defined, according to (3) of § 7, by

$$\left(\langle\phi\frac{d}{dq}\right)\psi\rangle = \langle\phi\left(\frac{d}{dq}\psi\rangle\right) \tag{13}$$

for all functions $\psi(q)$. Taking representatives, we get

$$\int \langle\phi\frac{d}{dq}|q'\rangle \, dq' \, \psi(q') = \int \phi(q') \, dq' \, \frac{d\psi(q')}{dq'}. \tag{14}$$

22. SCHRÖDINGER'S REPRESENTATION

We can transform the right-hand side by partial integration and get

$$\int \langle \phi \frac{d}{dq} | q' \rangle \, dq' \, \psi(q') = -\int \frac{d\phi(q')}{dq'} \, dq' \, \psi(q'). \tag{15}$$

provided the contributions from the limits of integration vanish. This gives

$$\langle \phi \frac{d}{dq} | q' \rangle = -\frac{d\phi(q')}{dq'},$$

showing that

$$\langle \phi \frac{d}{dq} = -\langle \frac{d\phi}{dq}. \tag{16}$$

Thus d/dq operating to the left on the conjugate complex of a wave function has the meaning of minus differentiation with respect to q.

The validity of this result depends on our being able to make the passage from (14) to (15), which requires that we must restrict ourselves to bras and kets corresponding to wave functions that satisfy suitable boundary conditions. The conditions usually holding in practice are that they vanish at the boundaries. (Somewhat more general conditions will be given in the next section.) These conditions do not limit the physical applicability of the theory, but, on the contrary, are usually required also on physical grounds. For example, if q is a Cartesian coordinate of a particle, its eigenvalues run from $-\infty$ to ∞, and the physical requirement that the particle has zero probability of being at infinity leads to the condition that the wave function vanishes for $q = \pm\infty$.

The conjugate complex of the linear operator d/dq can be evaluated by noting that the conjugate imaginary of $d/dq \cdot \psi\rangle$ or $d\psi/dq\rangle$ is $\langle d\overline{\psi}/dq$, or $-\langle \overline{\psi} d/dq$ from (16). Thus the conjugate complex of d/dq is $-d/dq$, so *d/dq is a pure imaginary linear operator.*

To get the representative of d/dq we note that, from an application of formula (63) of § 20,

$$|q''\rangle = \delta(q - q'')\rangle, \tag{17}$$

so that

$$\frac{d}{dq}|q''\rangle = \frac{d}{dq} \delta(q - q'')\rangle, \tag{18}$$

and hence

$$\langle q' | \frac{d}{dq} | q'' \rangle = \frac{d}{dq'} \delta(q' - q''). \tag{19}$$

The representative of d/dq involves the derivative of the δ function.

Let us work out the commutation relation connecting d/dq with q. We have

$$\frac{d}{dq}q\psi\rangle = \frac{dq\psi}{dq}\rangle = q\frac{d}{dq}\psi\rangle + \psi\rangle. \tag{20}$$

Since this holds for any ket $\psi\rangle$, we have

$$\frac{d}{dq}q - q\frac{d}{dq} = 1. \tag{21}$$

Comparing this result with (10), we see that $-i\hbar d/dq$ *satisfies the same commutation relation with q that p does.*

To extend the foregoing work to the case of arbitrary n, we write the general ket as $\psi(q_1 \cdots q_n)\rangle = \psi\rangle$ and introduce the n linear operators $\partial/\partial q_r$ ($r = 1, 2, \ldots, n$), which can operate on it in accordance with the formula

$$\frac{\partial}{\partial q_r}\psi\rangle = \frac{\partial \psi}{\partial q_r}\rangle, \tag{22}$$

corresponding to (11). We have

$$\frac{\partial}{\partial q_r}\rangle = 0 \tag{23}$$

corresponding to (12). Provided we restrict ourselves to bras and kets corresponding to wave functions satisfying suitable boundary conditions, these linear operators can operate also on bras, in accordance with the formula

$$\langle\phi\frac{\partial}{\partial q_r} = -\langle\frac{\partial\phi}{\partial q_r}, \tag{24}$$

corresponding to (16). Thus $\partial/\partial q_r$ can operate to the left on the conjugate complex of a wave function, when it has the meaning of minus partial differentiation with respect to q_r. We find as before that each $\partial/\partial q_r$ is a pure imaginary linear operator. Corresponding to (21) we have the commutation relations

$$\frac{\partial}{\partial q_r}q_s - q_s\frac{\partial}{\partial q_r} = \delta_{rs}. \tag{25}$$

We have further

$$\frac{\partial}{\partial q_r}\frac{\partial}{\partial q_s}\psi\rangle = \frac{\partial^2 \psi}{\partial q_r \partial q_s}\rangle = \frac{\partial}{\partial q_s}\frac{\partial}{\partial q_r}\psi\rangle, \tag{26}$$

showing that

$$\frac{\partial}{\partial q_r}\frac{\partial}{\partial q_s} = \frac{\partial}{\partial q_s}\frac{\partial}{\partial q_r}. \tag{27}$$

Comparing (25) and (27) with (9), we see that *the linear operators* $-i\hbar \partial/\partial q_r$ *satisfy the same commutation relations with the q's and with each other that the p's do.*

It would be possible to take

$$p_r = -i\hbar \frac{\partial}{\partial q_r} \tag{28}$$

without getting any inconsistency. This possibility enables us to see that the q's must form a complete commuting set of observables, since it means that any function of the q's and p's could be taken to be a function of the q's and $-i\hbar\partial/\partial q$'s and then could not commute with all the q's unless it is a function of the q's only.

The equations (28) do not necessarily hold. But in any case the quantities $p_r + i\hbar\partial/\partial q_r$ each commute with all the q's, so each of them is a function of the q's, from Theorem 2 of § 19. Thus

$$p_r = -i\hbar \frac{\partial}{\partial q_r} + f_r(q). \tag{29}$$

Since p_r and $-i\hbar\partial/\partial q_r$ are both real, $f_r(q)$ must be real. For an function f of the q's we have

$$\frac{\partial}{\partial q_r} f |\psi\rangle = f \frac{\partial}{\partial q_r} |\psi\rangle + \frac{\partial f}{\partial q_r} |\psi\rangle,$$

showing that

$$\frac{\partial}{\partial q_r} f - f \frac{\partial}{\partial q_r} = \frac{\partial f}{\partial q_r}. \tag{30}$$

With the help of (29) we can now deduce the general formula

$$p_r f - f p_r = -i\hbar \frac{\partial f}{\partial q_r}. \tag{31}$$

This formula may be written in P.B. notation

$$[f, p_r] = \frac{\partial f}{\partial q_r}, \tag{32}$$

when it is the same as in the classical theory, as follows from (1). Multiplying (27) by $(-i\hbar)^2$ and substituting for $-i\hbar\partial/\partial q_r$ and $-i\hbar\partial/\partial q_s$ their values given by (29), we get

$$(p_r - f_r)(p_s - f_s) = (p_s - f_s)(p_r - f_r),$$

which reduces, with the help of the quantum condition $p_r p_s = p_s p_r$, to

$$p_r f_s + f_r p_s = p_s f_r + f_s p_r.$$

This reduces further, with the help of (31), to

$$\frac{\partial f_s}{\partial q_r} = \frac{\partial f_r}{\partial q_s}, \tag{33}$$

showing that the functions f_r are all of the form

$$f_r = \frac{\partial F}{\partial q_r} \tag{34}$$

with F independent of r. Equation (29) now becomes

$$p_r = -i\hbar \frac{\partial}{\partial q_r} + \frac{\partial F}{\partial q_r}. \tag{35}$$

We have been working with a representation which is fixed to the extent that the q's must be diagonal in it, but which contains arbitrary phase factors. If the phase factors are changed, the operators $\partial/\partial q_r$ get changed. It will now be shown that, by a suitable change in the phase factors, the function F in (35) can be made to vanish, so that equations (28) are made to hold.

Using stars to distinguish quantities referring to the new representation with the new phase factors, we shall have the new basic bras connected with the previous ones by

$$\langle q'_1 \cdots q'^*_n | = e^{i\gamma'} \langle q'_1 \cdots q'_n | \tag{36}$$

where $\gamma' = \gamma(q')$ is a real function of the q''s. The new representative of a ket is $e^{i\gamma'}$ times the old one, showing that $e^{i\gamma}\psi\rangle^* = \psi\rangle$, so we get

$$\rangle^* = e^{-i\gamma}\rangle \tag{37}$$

as the connexion between the new standard ket and the original one. The new linear operator $(\partial/\partial q_r)^*$ satisfies, corresponding to (22),

$$\left(\frac{\partial}{\partial q_r}\right)^* \psi\rangle^* = \frac{\partial \psi}{\partial q_r}\rangle^* = e^{-i\gamma}\frac{\partial \psi}{\partial q_r}\rangle$$

with the help of (37). Using (22), this gives

$$\left(\frac{\partial}{\partial q_r}\right)^* \psi\rangle^* = -e^{-i\gamma}\frac{\partial}{\partial q_r}\psi\rangle = e^{-i\gamma}\frac{\partial}{\partial q_r}e^{i\gamma}\psi\rangle^*,$$

showing that

$$\left(\frac{\partial}{\partial q_r}\right)^* = e^{-i\gamma}\frac{\partial}{\partial q_r}e^{i\gamma}, \tag{38}$$

or, with the help of (30),

$$\left(\frac{\partial}{\partial q_r}\right)^* = \frac{\partial}{\partial q_r} + i\frac{\partial \gamma}{\partial q_r}. \tag{39}$$

By choosing γ so that

$$F = \hbar\gamma + \text{a constant}, \tag{40}$$

(35) becomes

$$p_r = -i\hbar \left(\frac{\partial}{\partial q_r}\right)^*. \tag{41}$$

Equation (40) fixes γ except for an arbitrary constant, so the representation is fixed except for an arbitrary constant phase factor.

In this way we see that a representation can be set up in which the q's are diagonal and equations (28) hold. This representation is a very useful one for many problems. It will be called *Schrödinger's representation*, as it was the representation in terms of which Schrödinger gave his original formulation of quantum mechanics in 1926. Schrödinger's representation exists whenever one has canonical q's and p's, and is completely determined by these q's and p's except for an arbitrary constant phase factor. It owes its great convenience to its allowing one to express immediately any algebraic function of the q's and p's of the form of a power series in the p's as an operator of differentiation, e.g. if $f(q_1, \ldots, q_n, p_1, \ldots, p_n)$ is such a function, we have

$$f(q_1, \ldots, q_n, p_1, \ldots, p_n) = f\left(q_1, \ldots, q_n, -i\hbar\frac{\partial}{\partial q_1}, \ldots, -i\hbar\frac{\partial}{\partial q_n}\right), \tag{42}$$

provided we preserve the order of the factors in a product on substituting the $-i\hbar \partial/\partial q$'s for the p's.

From (23) and (28), we have

$$p_r\rangle = 0. \tag{43}$$

Thus the standard ket in Schrödinger's representation is characterized by the condition that it is a simultaneous eigenket of all the momenta belonging to the eigenvalues zero. Some properties of the basic vectors of Schrödinger's representation may also be noted. Equation (22) gives

$$\langle q'_1 \cdots q'_n | \frac{\partial}{\partial q_r} \psi\rangle = \langle q'_1 \cdots q'_n | \frac{\partial \psi}{\partial q_r}\rangle = \frac{\partial \psi(q'_1 \cdots q'_n)}{\partial q'_r} = \frac{\partial}{\partial q'_r} \langle q'_1 \cdots q'_n | \psi\rangle.$$

Hence

$$\langle q'_1 \cdots q'_n | \frac{\partial}{\partial q_r} = \frac{\partial}{\partial q'_r} \langle q'_1 \cdots q'_n |, \tag{44}$$

so that

$$\langle q'_1 \cdots q'_n | p_r = -i\hbar \frac{\partial}{\partial q'_r} \langle q'_1 \cdots q'_n |. \tag{45}$$

Similarly, equation (24) leads to

$$p_r | q'_1 \cdots q'_n \rangle = i\hbar \frac{\partial}{\partial q'_r} | q'_1 \cdots q'_n \rangle. \tag{46}$$

23. The Momentum Representation

Let us take a system with one degree of freedom, describable in terms of a q and p with the eigenvalues of q running from $-\infty$ to ∞, and let us take an eigenket $|p'\rangle$ of p. Its representative in the Schrödinger representation, $\langle q'|p'\rangle$, satisfies:

$$p' \langle q'|p'\rangle = \langle q'|p|p'\rangle = -i\hbar \frac{d}{dq'} \langle q'|p'\rangle,$$

with the help of (45) applied to the case of one degree of freedom. The solution of this differential equation for $\langle q'|p'\rangle$ is

$$\langle q'|p'\rangle = c' e^{ip'q'/\hbar}, \tag{47}$$

where $c' = c(p')$ is independent of q', but may involve p'.

The representative $\langle q'|p'\rangle$ does not satisfy the boundary conditions of vanishing at $q' = \pm\infty$. This gives rise to some difficulty, which shows itself up most directly in the failure of the orthogonality theorem. If we take a second eigenket $|p''\rangle$ of p with representative

$$\langle q'|p''\rangle = c'' e^{ip''q'/\hbar},$$

belonging to a different eigenvalue p'', we shall have

$$\langle p'|p''\rangle = \int_{-\infty}^{\infty} \langle p'|q'\rangle \, dq' \, \langle q'|p''\rangle = \overline{c'} c'' \int_{-\infty}^{\infty} e^{-i(p'-p'')q'/\hbar} \, dq'. \tag{48}$$

This integral does not converge according to the usual definition of convergence. To bring the theory into order, we adopt a new definition of convergence of an integral whose domain extends to infinity, analogous to the Cesàro definition of the sum of an infinite series. With this new definition, an integral whose value to the upper limit q' is of the form $\cos aq'$ or $\sin aq'$, with a a real number not zero, is counted as zero when q' tends to infinity, i.e. we take the mean value of the oscillations, and similarly for the lower limit of q' tending to minus infinity. This makes the right-hand side of (48) vanish for $p'' \neq p'$, so that the orthogonality theorem is restored. Also it makes the right-hand sides of (13) and (14) equal when $\langle \phi$

23. THE MOMENTUM REPRESENTATION

and $\psi\rangle$ are eigenvectors of p, so that eigenvectors of p become permissible vectors to use with the operator d/dq. Thus the boundary conditions that the representative of a permissible bra or ket has to satisfy become extended to allow the representative to oscillate like $\cos aq'$ or $\sin aq'$ as q' goes to infinity or minus infinity.

For p'' very close to p', the right-hand side of (48) involves a δ function. To evaluate it, we need the formula

$$\int_{-\infty}^{\infty} e^{iax}\, dx = 2\pi\, \delta(a) \qquad (49)$$

for real a, which may be proved as follows. The formula evidently holds for a different from zero, as both sides are then zero. Further we have, for any continuous function $f(a)$,

$$\int_{-\infty}^{\infty} f(a)\, da \int_{-g}^{g} e^{iax}\, dx = \int_{-\infty}^{\infty} f(a)\, da\, 2a^{-1} \sin ag = 2\pi\, f(0)$$

in the limit when g tends to infinity. A more complicated argument shows that we get the same result if instead of the limits g and $-g$ we put g_1 and $-g_2$, and then let g_1 and g_2 tend to infinity in different ways (not too widely different). This shows the equivalence of both sides of (49) as factors in an integrand, which proves the formula.

With the help of (49), (48) becomes

$$\langle p'|p''\rangle = \overline{c'}c''2\pi\, \delta\!\left(\frac{p'-p''}{\hbar}\right) = \overline{c'}c''h\, \delta(p'-p'')$$

$$= |c'|^2 h\, \delta(p'-p''). \qquad (50)$$

We have obtained an eigenket of p belonging to any real eigenvalue p', its representative being given by (47). Any ket $|X\rangle$ can be expanded in terms of these eigenkets of p, since its representative $\langle q'|X\rangle$ can be expanded in terms of the representatives (47) by Fourier analysis. It follows that *the momentum p is an observable*, in agreement with the experimental result that momenta can be observed.

A symmetry now appears between q and p. Each of them is an observable with eigenvalues extending from $-\infty$ to ∞, and the commutation relation connecting q and p, equation (10), remains invariant if we interchange q and p and write $-i$ for i. We have set up a representation in which q is diagonal and $p = -i\hbar d/dq$. It follows from the symmetry that we can also set up a representation in which p is diagonal and

$$q = i\hbar \frac{d}{dp}, \qquad (51)$$

the operator d/dp being defined by a procedure similar to that used for d/dq. This representation will be called the *momentum representation*. It is less useful than the previous Schrödinger representation because, while the Schrödinger representation enables one to express as an operator of differentiation any function of q and p that is a power series in p, the momentum representation enables one so to express any function of q and p that is a power series in q, and the important quantities in dynamics are almost always power series in p but are often not power series in q. All the same the momentum representation is of value for certain problems (see § 50).

Let us calculate the transformation function $\langle q'|p'\rangle$ connecting the two representations. The basic kets $|p'\rangle$ of the momentum representation are eigenkets of p and their Schrödinger representatives $\langle q'|p'\rangle$ are given by (47) with the coefficients c' suitably chosen. The phase factors of these basic kets must be chosen so as to make (51) hold. The easiest way to bring in this condition is to use the symmetry between q and p referred to above, according to which $\langle q'|p'\rangle$ must go over into $\langle p'|q'\rangle$ if we interchange q' and p' and write $-i$ for i. Now $\langle q'|p'\rangle$ is equal to the right-hand side of (47) and $\langle p'|q'\rangle$ to the conjugate complex expression, and hence c' must be independent of p'. Thus c' is just a number c. Further, we must have

$$\langle p'|p''\rangle = \delta(p' - p''),$$

which shows, on comparison with (50), that $|c| = h^{-1/2}$. We can choose the arbitrary constant phase factor in either representation so as to make $c = h^{-1/2}$, and we then get

$$\langle q'|p'\rangle = h^{-1/2} e^{ip'q'/\hbar} \tag{52}$$

for the transformation function.

The foregoing work may easily be generalized to a system with n degrees of freedom, describable in terms of n q's and p's, with the eigenvalues of each q running from $-\infty$ to ∞. Each p will then be an observable with eigenvalues running from $-\infty$ to ∞, and there will be symmetry between the set of q's and the set of p's, the commutation relations remaining invariant if we interchange each q_r with the corresponding p_r and write $-i$ for i. A momentum representation can be set up in which the p's are diagonal and each

$$q_r = i\hbar \frac{\partial}{\partial p_r}. \tag{53}$$

The transformation function connecting it with the Schrödinger representation will be given by the product of the transformation functions for each degree of freedom separately, as is shown by formula (67) of § 20, and will thus be

$$\langle q'_1 \cdots q'_n | p'_1 \cdots p'_n \rangle = \langle q'_1 | p'_1 \rangle \cdots \langle q'_n | p'_n \rangle$$
$$= h^{-n/2} e^{i(p'_1 q'_1 + \cdots + p'_n q'_n)/\hbar}. \qquad (54)$$

24. Heisenberg's Principle of Uncertainty

For a system with one degree of freedom, the Schrödinger and the momentum representatives of a ket $|X\rangle$ are connected by

$$\langle p'|X\rangle = h^{-1/2} \int_{-\infty}^{\infty} e^{-iq'p'/\hbar} \, dq' \, \langle q'|X\rangle,$$
$$\langle q'|X\rangle = h^{-1/2} \int_{-\infty}^{\infty} e^{iq'p'/\hbar} \, dp' \, \langle p'|X\rangle. \qquad (55)$$

These formulas have an elementary significance. They show that *either of the representatives is given, apart from numerical coefficients, by the amplitudes of the Fourier components of the other.*

It is interesting to apply (55) to a ket whose Schrödinger representative consists of what is called a *wave packet*. This is a function whose value is very small everywhere outside a certain domain, of width $\Delta q'$ say, and inside this domain is approximately periodic with a definite frequency.[1] If a Fourier analysis is made of such a wave packet, the amplitude of all the Fourier components will be small, except those in the neighbourhood of the definite frequency. The components whose amplitudes are not small will fill up a frequency[1] band whose width is of the order $1/\Delta q'$, since two components whose frequencies differ by this amount, if in phase in the middle of the domain $\Delta q'$, will be just out of phase and interfering at the ends of this domain. Now in the first of equations (55) the variable $(2\pi)^{-1} p'/\hbar = p'/h$ plays the part of frequency. Thus with $\langle q'|X\rangle$ of the form of a wave packet, the function $\langle p'|X\rangle$, being composed of the amplitudes of the Fourier components of the wave packet, will be small everywhere in the p'-space outside a certain domain of width $\Delta p' = h/\Delta q'$.

Let us now apply the physical interpretation of the square of the modulus of the representative of a ket as a probability. We find that our wave packet represents a state for which a measurement of q is almost certain

---
[1] Frequency here means reciprocal of wave-length.

to lead to a result lying in a domain of width $\Delta q'$ and a measurement of p is almost certain to lead to a result lying in a domain of width $\Delta p'$. We may say that for this state q has a definite value with an error of order $\Delta q'$ and p has a definite value with an error of order $\Delta p'$. The product of these two errors is

$$\Delta q' \Delta p' = h. \qquad (56)$$

Thus the more accurately one of the variables q, p has a definite value, the less accurately the other has a definite value. For a system with several degrees of freedom, equation (56) applies to each degree of freedom separately.

Equation (56) is known as *Heisenberg's Principle of Uncertainty*. It shows clearly the limitations in the possibility of simultaneously assigning numerical values, for any particular state, to two non-commuting observables, when those observables are a canonical coordinate and momentum, and provides a plain illustration of how observations in quantum mechanics may be incompatible. It also shows how classical mechanics, which assumes that numerical values can be assigned simultaneously to all observables, may be a valid approximation when h can be considered as small enough to be negligible. Equation (56) holds only in the most favourable case, which occurs when the representative of the state is of the form of a wave packet. Other forms of representative would lead to a $\Delta q'$ and $\Delta p'$ whose product is larger than h.

Heisenberg's principle of uncertainty shows that, in the limit when either q or p is completely determined, the other is completely undetermined. This result can also be obtained directly from the transformation function $\langle q'|p'\rangle$. According to the end of § 18, $|\langle q'|p'\rangle|^2 \, dq'$ is proportional to the probability of q having a value in the small range from q' to $q' + dq'$ for the state for which p certainly has the value p', and from (52) this probability is independent of q' for a given dq'. Thus if p certainly has a definite value p', all values of q are equally probable. Similarly, if q certainly has a definite value q', all values of p are equally probable.

It is evident physically that a state for which all values of q are equally probable, or one for which all values of p are equally probable, cannot be attained in practice, in the first case because of limitations of size and in the second because of limitations of energy. Thus an eigenstate of p or an eigenstate of q cannot be attained in practice. The argument at the end of § 12 already showed that such eigenstates are unattainable, because of the infinite precision that would be needed to set them up, and we now have another argument leading to the same conclusion.

25. Displacement Operators

We get a new insight into the meaning of some of the quantum conditions by making a study of displacement operators. These appear in the theory when we take into consideration that the scheme of relations between states and dynamical variables given in Chapter II is essentially a *physical* scheme, so that if certain states and dynamical variables are connected by some relation, on our displacing them all in a definite way (for example, displacing them all through a distance δx in the direction of the x-axis of Cartesian coordinates), the new states and dynamical variables would have to be connected by the same relation.

The displacement of a state or observable is a perfectly definite process physically. Thus to displace a state or observable through a distance δx in the direction of the x-axis, we should merely have to displace all the apparatus used in preparing the state, or all the apparatus required to measure the observable, through the distance δx in the direction of the x-axis, and the displaced apparatus would define the displaced state or observable. The displacement of a dynamical variable must be just as definite as the displacement of an observable, because of the close mathematical connexion between dynamical variables and observables. A displaced state or dynamical variable is uniquely determined by the undisplaced state or dynamical variable together with the direction and magnitude of the displacement.

The displacement of a ket vector is not such a definite thing though. If we take a certain ket vector, it will represent a certain state and we may displace this state and get a perfectly definite new state, but this new state will not determine our displaced ket, but only the direction of our displaced ket. We help to fix our displaced ket by requiring that it shall have the same length as the undisplaced ket, but even then it is not completely determined, but can still be multiplied by an arbitrary phase factor. One would think at first sight that each ket one displaces would have a different arbitrary phase factor, but with the help of the following argument, we see that it must be the same for them all. We make use of the law that superposition relationships between states remain invariant under the displacement. A superposition relationship between states is expressed mathematically by a linear equation between the kets corresponding to those states, for example

$$|R\rangle = c_1|A\rangle + c_2|B\rangle, \tag{57}$$

where c_1 and c_2 are numbers, and the invariance of the superposition relationship requires that the displaced states correspond to kets with the same linear equation between them—in our example they would correspond to $|Rd\rangle, |Ad\rangle, |Bd\rangle$ say, satisfying

$$|Rd\rangle = c_1|Ad\rangle + c_2|Bd\rangle. \tag{58}$$

We take these kets to be our displaced kets, rather than these kets multiplied by arbitrary independent phase factors, which latter kets would satisfy a linear equation with different coefficients c_1, c_2. The only arbitrariness now left in the displaced kets is that of a single arbitrary phase factor to be multiplied into all of them.

The condition that linear equations between the kets remain invariant under the displacement and that an equation such as (58) holds whenever the corresponding (57) holds, means that the displaced kets are linear functions of the undisplaced kets and thus each displaced ket $|Pd\rangle$ is the result of some linear operator applied to the corresponding undisplaced ket $|P\rangle$. In symbols,

$$|Pd\rangle = D|P\rangle, \tag{59}$$

where D is a linear operator independent of $|P\rangle$ and depending only on the displacement. The arbitrary phase factor by which all the displaced kets may be multiplied results in D being undetermined to the extent of an arbitrary numerical factor of modulus unity.

With the displacement of kets made definite in the above manner and the displacement of bras, of course, made equally definite, through their being the conjugate imaginaries of the kets, we can now assert that any symbolic equation between kets, bras, and dynamical variables must remain invariant under the displacement of every symbol occurring in it, on account of such an equation having some physical significance which will not get changed by the displacement.

Take as an example the equation

$$\langle Q|P\rangle = c,$$

c being a number. Then we must have

$$\langle Qd|Pd\rangle = c = \langle Q|P\rangle. \tag{60}$$

From the conjugate imaginary of (59) with Q instead of P,

$$\langle Qd| = \langle Q|\overline{D}. \tag{61}$$

Hence (60) gives

$$\langle Q|\overline{D}D|P\rangle = \langle Q|P\rangle.$$

25. DISPLACEMENT OPERATORS

Since this holds for arbitrary $\langle Q|$ and $|P\rangle$, we must have

$$\overline{D}D = 1, \qquad (62)$$

giving us a general condition which D has to satisfy.

Take as a second example the equation

$$v|P\rangle = |R\rangle,$$

where v is any dynamical variable. Then, using v_d to denote the displaced dynamical variable, we must have

$$v_d|Pd\rangle = |Rd\rangle.$$

With the help of (59) we get

$$v_d|Pd\rangle = D|R\rangle = Dv|P\rangle = DvD^{-1}|Pd\rangle.$$

Since $|Pd\rangle$ can be any ket, we must have

$$v_d = DvD^{-1} \qquad (63)$$

which shows that the linear operator D determines the displacement of dynamical variables as well as that of kets and bras. Note that the arbitrary numerical factor of modulus unity in D does not affect v_d, and also it does not affect the validity of (62).

Let us now pass to an infinitesimal displacement, i.e. taking the displacement through the distance δx in the direction of the x-axis, let us make $\delta x \to 0$. From physical continuity we should expect a displaced ket $|Pd\rangle$ to tend to the original $|P\rangle$ and we may further expect the limit

$$\lim_{\delta x \to 0} \frac{|Pd\rangle - |P\rangle}{\delta x} = \lim_{\delta x \to 0} \frac{D-1}{\delta x}|P\rangle$$

to exist. This requires that the limit

$$\lim_{\delta x \to 0} \frac{D-1}{\delta x} \qquad (64)$$

shall exist. This limit is a linear operator which we shall call the *displacement operator* for the x-direction and denote by d_x. The arbitrary numerical factor $e^{i\gamma}$ with γ real which we may multiply into D must be made to tend to unity as $\delta x \to 0$ and then introduces an arbitrariness in d_x, namely, d_x may be replaced by

$$\lim_{\delta x \to 0} \frac{De^{i\gamma} - 1}{\delta x} = \lim_{\delta x \to 0} \frac{D - 1 + i\gamma}{\delta x} = d_x + ia_x,$$

where a_x is the limit of $\gamma/\delta x$. Thus d_x contains an arbitrary additive pure imaginary number.

For δx small
$$D = 1 + \delta x d_x. \tag{65}$$
Substituting this into (62), we get
$$(1 + \delta x \overline{d}_x)(1 + \delta x d_x) = 1,$$
which reduces, with neglect of δx^2, to
$$\delta x (\overline{d}_x + d_x) = 0.$$
Thus d_x is a pure imaginary linear operator. Substituting (65) into (63) we get, with neglect of δx^2 again,
$$v_d = (1 + \delta x d_x) v (1 - \delta x d_x) = v + \delta x (d_x v - v d_x), \tag{66}$$
showing that
$$\lim_{\delta x \to 0} \frac{v_d - v}{\delta x} = d_x v - v d_x. \tag{67}$$

We may describe any dynamical system in terms of the following dynamical variables: the Cartesian coordinates x, y, z of the centre of mass of the system, the components p_x, p_y, p_z of the total momentum of the system, which are the canonical momenta conjugate to x, y, z respectively, and any dynamical variables needed for describing internal degrees of freedom of the system. If we suppose a piece of apparatus which has been set up to measure x, to be displaced a distance δx in the direction of the x-axis, it will measure $x - \delta x$, hence
$$x_d = x - \delta x.$$
Comparing this with (66) for $v = x$, we obtain
$$d_x x - x d_x = -1. \tag{68}$$
This is the quantum condition connecting d_x with x. From similar arguments we find that y, z, p_x, p_y, p_z and the internal dynamical variables, which are unaffected by the displacement, must commute with d_x. Comparing these results with (9), we see that $i\hbar d_x$ satisfies just the same quantum conditions as p_x. Their difference, $p_x - i\hbar d_x$, commutes with all the dynamical variables and must therefore be a number. This number, which is necessarily real since p_x and $i\hbar d_x$ are both real, may be made zero by a suitable choice of the arbitrary, pure imaginary number that can be added to d_x. We then have the result
$$p_x = i\hbar d_x, \tag{69}$$
or *the x-component of the total momentum of the system is $i\hbar$ times the displacement operator d_x.*

This is a fundamental result, which gives a new significance to displacement operators. There is a corresponding result, of course, also for the y and z displacement operators d_y and d_z. The quantum conditions which state that p_x, p_y and p_z commute with each other are now seen to be connected with the fact that displacements in different directions are commutable operations.

26. Unitary Transformations

Let U be any linear operator that has a reciprocal U^{-1} and consider the equation

$$\alpha^* = U\alpha U^{-1}, \tag{70}$$

α being an arbitrary linear operator. This equation may be regarded as expressing a transformation from any linear operator α to a corresponding linear operator α^*, and as such it has rather remarkable properties. In the first place it should be noted that each α^* has the same eigenvalues as the corresponding α; since, if α' is any eigenvalue of α and $|\alpha'\rangle$ is an eigenket belonging to it, we have

$$\alpha|\alpha'\rangle = \alpha'|\alpha'\rangle$$

and hence

$$\alpha^* U|\alpha'\rangle = U\alpha U^{-1} U|\alpha'\rangle = U\alpha|\alpha'\rangle = \alpha' U|\alpha'\rangle,$$

showing that $U|\alpha'\rangle$ is an eigenket of α^* belonging to the same eigenvalue α', and similarly any eigenvalue of α^* may be shown to be also an eigenvalue of α. Further, if we take several α's that are connected by algebraic equations and transform them all according to (70), the corresponding α^*'s will be connected by the same algebraic equations. This result follows from the fact that the fundamental algebraic processes of addition and multiplication are left invariant by the transformation (70), as is shown by the following equations:

$$(\alpha_1 + \alpha_2)^* = U(\alpha_1 + \alpha_2)U^{-1} = U\alpha_1 U^{-1} + U\alpha_2 U^{-1} = \alpha_1^* + \alpha_2^*,$$
$$(\alpha_1 \alpha_2)^* = U\alpha_1 \alpha_2 U^{-1} = U\alpha_1 U^{-1} U\alpha_2 U^{-1} = \alpha_1^* \alpha_2^*.$$

Let us now see what condition would be imposed on U by the requirement that any real α transforms into a real α^*. Equation (70) may be written

$$\alpha^* U = U\alpha. \tag{71}$$

Taking the conjugate complex of both sides in accordance with (5) of § 8 we find, if α and α^* are both real,

$$\overline{U}\alpha^* = \alpha\overline{U}. \tag{72}$$

Equation (71) gives us

$$\overline{U}\alpha^* U = \overline{U}U\alpha$$

and equation (72) gives us

$$\overline{U}\alpha^* U = \alpha\overline{U}U.$$

Hence

$$\overline{U}U\alpha = \alpha\overline{U}U.$$

Thus $\overline{U}U$ commutes with any real linear operator and therefore also with any linear operator whatever, since any linear operator can be expressed as one real one plus i times another. Hence $\overline{U}U$ is a number. It is obviously real, its conjugate complex according to (5) of § 8 being the same as itself, and further it must be a positive number, since for any ket $|P\rangle$, $\langle P|\overline{U}U|P\rangle$ is positive as well as $\langle P|P\rangle$. We can suppose it to be unity without any loss of generality in the transformation (70). We then have

$$\overline{U}U = 1. \tag{73}$$

Equation (73) is equivalent to any of the following

$$U = \overline{U}^{-1}, \quad \overline{U} = U^{-1}, \quad U^{-1}\overline{U}^{-1} = 1. \tag{74}$$

A matrix or linear operator U that satisfies (73) and (74) is said to be *unitary* and a transformation (70) with unitary U is called a *unitary transformation*. A unitary transformation transforms real linear operators into real linear operators and leaves invariant any algebraic equation between linear operators. It may be considered as applying also to kets and bras, in accordance with the equations

$$|P^*\rangle = U|P\rangle, \quad \langle P^*| = \langle P|\overline{U} = \langle P|U^{-1}, \tag{75}$$

and then it leaves invariant any algebraic equation between linear operators, kets, and bras. It transforms eigenvectors of α into eigenvectors of α^*. From this one can easily deduce that it transforms an observable into an observable and that it leaves invariant any functional relation between observables based on the general definition of a function given in § 11.

The inverse of a unitary transformation is also a unitary transformation, since from (74), if U is unitary, U^{-1} is also unitary. Further, if two

26. UNITARY TRANSFORMATIONS

unitary transformations are applied in succession, the result is a third unitary transformation, as may be verified in the following way. Let the two unitary transformations be (70) and

$$\alpha^\dagger = V\alpha^* V^{-1}.$$

The connexion between $\alpha^\dagger$ and α is then

$$\alpha^\dagger = VU\alpha U^{-1} V^{-1}$$
$$= (VU)\alpha(VU)^{-1} \tag{76}$$

from (42) of § 11. Now VU is unitary since

$$\overline{VU}VU = \overline{U}\,\overline{V}VU = \overline{U}U = 1,$$

and hence (76) is a unitary transformation.

The transformation given in the preceding section from undisplaced to displaced quantities is an example of a unitary transformation, as is shown by equations (62), (63), corresponding to equations (73), (70), and equations (59), (61), corresponding to equations (75).

In classical mechanics one can make a transformation from the canonical coordinates and momenta q_r, p_r ($r = 1, 2, \ldots, n$) to a new set of variables q_r^*, p_r^* ($r = 1, 2, \ldots, n$) satisfying the same P.B. relations as the q's and p's, i.e. equations (8) of § 21 with q^*'s and p^*'s replacing the q's and p's, and can express all dynamical variables in terms of the q^*'s and p^*'s. The q^*'s and p^*'s are then also called canonical coordinates and momenta and the transformation is called a *contact transformation*. One can easily verify that the P.B. of any two dynamical variables u and v is correctly given by formula (1) of § 21 with q^*'s and p^*'s instead of q's and p's, so that the P.B. relationship is invariant under a contact transformation. This results in the new canonical coordinates and momenta being on the same footing as the original ones for many purposes of general dynamical theory, even though the new coordinates q_r^* may not be a set of Lagrangian coordinates but may be functions of the Lagrangian coordinates and velocities.

It will now be shown that, for a quantum dynamical system that has a classical analogue, *unitary transformations in the quantum theory are the analogue of contact transformations in the classical theory*. Unitary transformations are more general than contact transformations, since the former can be applied to systems in quantum mechanics that have no classical analogue, but for those systems in quantum mechanics which are describable in terms of canonical coordinates and momenta, the analogy between the two kinds of transformation holds. To establish it, we note that a unitary transformation applied to the quantum variables q_r, p_r gives new variables q_r^*, p_r^* satisfying the same P.B. relations, since the P.B. relations are equivalent to the algebraic relations (9) of § 21 and algebraic relations are left invariant by a unitary transformation. Conversely, any real variables q_r^*, p_r^* satisfying the P.B. relations for canonical coordinates and momenta are connected with the q_r, p_r by a unitary transformation, as is shown by the following argument.

We use the Schrödinger representation, and write the basic ket $|q_1' \cdots q_n'\rangle$ as $|q'\rangle$ for brevity. Since we are assuming that the q_r^*, p_r^* satisfy the P.B. relations for canonical coordinates and momenta, we can set up a Schrödinger representation referring to them, with the q_r^* diagonal and each p_r^* equal to $-i\hbar \partial/\partial q_r^*$. The basic kets in this second Schrödinger representation will be $|q_1^{*\prime} \cdots q_n^{*\prime}\rangle$, which we write $|q^{*\prime}\rangle$ for brevity. Now introduce the linear operator U defined by

$$\langle q^{*\prime} | U | q' \rangle = \delta(q^{*\prime} - q'), \qquad (77)$$

where $\delta(q^{*\prime} - q')$ is short for

$$\delta(q^{*\prime} - q') = \delta(q_1^{*\prime} - q_1') \delta(q_2^{*\prime} - q_2') \cdots \delta(q_n^{*\prime} - q_n'). \qquad (78)$$

The conjugate complex of (77) is

$$\langle q' | \overline{U} | q^{*\prime} \rangle = \delta(q^{*\prime} - q'),$$

and hence[1]

$$\langle q' | \overline{U} U | q'' \rangle = \int \langle q' | \overline{U} | q^{*\prime} \rangle dq^{*\prime} \langle q^{*\prime} | U | q'' \rangle$$

$$= \int \delta(q^{*\prime} - q') dq^{*\prime} \delta(q^{*\prime} - q'')$$

$$= \delta(q' - q''),$$

so that

$$\overline{U} U = 1.$$

Thus U is a unitary operator. We have further

$$\langle q^{*\prime} | q_r^* U | q' \rangle = q_r^{*\prime} \delta(q^{*\prime} - q')$$

and

$$\langle q^{*\prime} | U q_r | q' \rangle = \delta(q^{*\prime} - q') q_r'.$$

---

[1] We use the notation of a single integral sign and $dq^{*\prime}$ to denote an integral over all the variables $q_1^{*\prime}, q_2^{*\prime}, \ldots, q_n^{*\prime}$. This abbreviation will be used also in future work.

The right-hand sides of these two equations are equal on account of the property of the δ function (11) of § 15, and hence

$$q_r^* U = U q_r$$

or

$$q_r^* = U q_r U^{-1}.$$

Again, from (45) and (46),

$$\langle q^{*\prime}|p_r^* U|q'\rangle = -i\hbar \frac{\partial}{\partial q_r^{*\prime}} \delta(q^{*\prime} - q'),$$

$$\langle q^{*\prime}|U p_r|q'\rangle = i\hbar \frac{\partial}{\partial q_r'} \delta(q^{*\prime} - q').$$

The right-hand sides of these two equations are obviously equal, and hence

$$p_r^* U = U p_r$$

or

$$p_r^* = U p_r U^{-1}.$$

Thus all the conditions for a unitary transformation are verified.

We get an infinitesimal unitary transformation by taking U in (70) to differ by an infinitesimal from unity. Put

$$U = 1 + i\epsilon F,$$

where ϵ is infinitesimal, so that its square can be neglected. Then

$$U^{-1} = 1 - i\epsilon F.$$

The unitary condition (73) or (74) requires that F shall be real. The transformation equation (70) now takes the form

$$\alpha^* = (1 + i\epsilon F)\alpha(1 - i\epsilon F),$$

which gives

$$\alpha^* - \alpha = i\epsilon(F\alpha - \alpha F). \tag{79}$$

It may be written in P.B. notation

$$\alpha^* - \alpha = \epsilon\hbar[\alpha, F]. \tag{80}$$

If α is a canonical coordinate or momentum, this is formally the same as a classical infinitesimal contact transformation.

CHAPTER V

The Equations of Motion

27. Schrödinger's Form for the Equations of Motion

Our work from § 5 onwards has all been concerned with one instant of time. It gave the general scheme of relations between states and dynamical variables for a dynamical system at one instant of time. To get a complete theory of dynamics we must consider also the connexion between different instants of time. When one makes an observation on the dynamical system, the state of the system gets changed in an unpredictable way, but in between observations causality applies, in quantum mechanics as in classical mechanics, and the system is governed by equations of motion which make the state at one time determine the state at a later time. These equations of motion we now proceed to study. They will apply so long as the dynamical system is left undisturbed by any observation or similar process.[1] Their general form can be deduced from the principle of superposition of Chapter I.

Let us consider a particular state of motion throughout the time during which the system is left undisturbed. We shall have the state at any time t corresponding to a certain ket which depends on t and which may be written $|t\rangle$. If we deal with several of these states of motion we distinguish them by giving them labels such as A, and we then write the ket which corresponds to the state at time t for one of them $|At\rangle$. The requirement that the state at one time determines the state at another time means that $|At_0\rangle$ determines $|At\rangle$ except for a numerical factor. The principle of superposition applies to these states of motion throughout the time during which the system is undisturbed, and means that if we take a superposition relation holding for certain states at time t_0 and giving rise to a linear equation between the corresponding kets, e.g. the equation

$$|Rt_0\rangle = c_1|At_0\rangle + c_2|Bt_0\rangle,$$

---

[1] The preparation of a state is a process of this kind. It often takes the form of making an observation and selecting the system when the result of the observation turns out to be a certain pre-assigned number.

the same superposition relation must hold between the states of motion throughout the time during which the system is undisturbed and must lead to the same equation between the kets corresponding to these states at any time t (in the undisturbed time interval), i.e. the equation

$$|Rt\rangle = c_1|At\rangle + c_2|Bt\rangle,$$

provided the arbitrary numerical factors by which these kets may be multiplied are suitably chosen. It follows that the $|Pt\rangle$'s are linear functions of the $|Pt_0\rangle$'s and each $|Pt\rangle$ is the result of some linear operator applied to $|Pt_0\rangle$. In symbols

$$|Pt\rangle = T|Pt_0\rangle, \tag{1}$$

where T is a linear operator independent of P and depending only on t (and t_0).

We now assume that each $|Pt\rangle$ has the same length as the corresponding $|Pt_0\rangle$. It is not necessarily possible to choose the arbitrary numerical factors by which the $|Pt\rangle$'s may be multiplied so as to make this so without destroying the linear dependence of the $|Pt\rangle$'s on the $|Pt_0\rangle$'s, so the new assumption is a physical one and not just a question of notation. It involves a kind of sharpening of the principle of superposition. The arbitrariness in $|Pt\rangle$ now becomes merely a phase factor, which must be independent of P in order that the linear dependence of the $|Pt\rangle$'s on the $|Pt_0\rangle$'s may be preserved. From the condition that the length of $c_1|Pt\rangle+c_2|Qt\rangle$ equals that of $c_1|Pt_0\rangle+c_2|Qt_0\rangle$ for any complex numbers c_1, c_2, we can deduce that

$$\langle Qt|Pt\rangle = \langle Qt_0|Pt_0\rangle. \tag{2}$$

The connexion between the $|Pt\rangle$'s and $|Pt_0\rangle$'s is formally similar to the connexion we had in § 25 between the displaced and undisplaced kets, with a process of time displacement instead of the space displacement of § 25. Equations (1) and (2) play the part of equations (59) and (60) of § 25. We can develop the consequences of these equations as in § 25 and can deduce that T contains an arbitrary numerical factor of modulus unity and satisfies

$$\overline{T}T = 1, \tag{3}$$

corresponding to (62) of § 25, so T *is unitary*. We pass to the infinitesimal case by making $t \to t_0$ and assume from physical continuity that the limit

$$\lim_{t \to t_0} \frac{|Pt\rangle - |Pt_0\rangle}{t - t_0}$$

27. SCHRÖDINGER'S FORM FOR THE EQUATIONS OF MOTION

exists. This limit is just the derivative of $|Pt_0\rangle$ with respect to t_0. From (1) it equals

$$\frac{d|Pt_0\rangle}{dt_0} = \left(\lim_{t \to t_0} \frac{T-1}{t-t_0}\right)|Pt_0\rangle. \tag{4}$$

The limit operator occurring here is, like (64) of § 25, a pure imaginary linear operator and is undetermined to the extent of an arbitrary additive pure imaginary number. Putting this limit operator multiplied by $i\hbar$ equal to H, or rather $H(t_0)$ since it may depend on t_0, equation (4) becomes, when written for a general t,

$$i\hbar \frac{d|Pt\rangle}{dt} = H(t)|Pt\rangle. \tag{5}$$

Equation (5) gives the general law for the variation with time of the ket corresponding to the state at any time. It is *Schrödinger's form for the equations of motion*. It involves just one real linear operator $H(t)$, which must be characteristic of the dynamical system under consideration. We assume that $H(t)$ *is the total energy of the system*. There are two justifications for this assumption, (i) the analogy with classical mechanics, which will be developed in the next section, and (ii) we have $H(t)$ appearing as $i\hbar$ times an operator of displacement in time similar to the operators of displacement in the x-, y-, and z-directions of § 25, so corresponding to (69) of § 25 we should have $H(t)$ equal to the total energy, since the theory of relativity puts energy in the same relation to time as momentum to distance.

We assume on physical grounds that the total energy of a system is always an observable. For an isolated system it is a constant, and may then be written H. Even when it is not a constant we shall often write it simply H, leaving its dependence on t understood. If the energy depends on t, it means the system is acted on by external forces. An action of this kind is to be distinguished from a disturbance caused by a process of observation, as the former is compatible with causality and equations of motion while the latter is not.

We can get a connexion between $H(t)$ and the T of equation (1) by substituting for $|Pt\rangle$ in (5) its value given by equation (1). This gives

$$i\hbar \frac{dT}{dt}|Pt_0\rangle = H(t)T|Pt_0\rangle.$$

Since $|Pt_0\rangle$ may be any ket, we have

$$i\hbar \frac{dT}{dt} = H(t)T. \tag{6}$$

Equation (5) is very important for practical problems, where it is usually used in conjunction with a representation. Introducing a representation with a complete set of commuting observables ξ diagonal and putting $\langle \xi' | Pt \rangle$ equal to $\psi(\xi' t)$, we have, passing to the standard ket notation,

$$|Pt\rangle = \psi(\xi t)\rangle.$$

Equation (5) now becomes

$$i\hbar \frac{\partial}{\partial t} \psi(\xi t)\rangle = H \psi(\xi t)\rangle. \tag{7}$$

Equation (7) is known as *Schrödinger's wave equation* and its solutions $\psi(\xi t)$ are *time-dependent wave functions*. Each solution corresponds to a state of motion of the system and the square of its modulus gives the probability of the ξ's having specified values at any time t. For a system describable in terms of canonical coordinates and momenta we may use Schrödinger's representation and can then take H to be an operator of differentiation in accordance with (42) of § 22.

28. Heisenberg's Form for the Equations of Motion

In the preceding section we set up a picture of the states of undisturbed motion by making each of them correspond to a moving ket, the state at any time corresponding to the ket at that time. We shall call this the *Schrödinger picture*. Let us apply to our kets the unitary transformation which makes each ket $|a\rangle$ go over into

$$|a^*\rangle = T^{-1}|a\rangle. \tag{8}$$

This transformation is of the form given by (75) of § 26 with T^{-1} for U, but *it depends on the time t* since T depends on t. It is thus to be pictured as the application of a continuous motion (consisting of rotations and uniform deformations) to the whole ket vector space. A ket which is originally fixed becomes a moving one, its motion being given by (8) with $|a\rangle$ independent of t. On the other hand, a ket which is originally moving to correspond to a state of undisturbed motion, i.e. in accordance with equation (1), becomes fixed, since on substituting $|Pt\rangle$ for $|a\rangle$ in (8) we get $|a^*\rangle$ independent of t. Thus *the transformation brings the kets corresponding to states of undisturbed motion to rest.*

The unitary transformation must be applied also to bras and linear operators, in order that equations between the various quantities may remain invariant. The transformation applied to bras is given by the conjugate imaginary of (8) and applied to linear operators it is given by (70) of

§ 26 with T^{-1} for U, i.e.

$$\alpha^* = T^{-1}\alpha T. \tag{9}$$

A linear operator which is originally fixed transforms into a moving linear operator in general. Now a dynamical variable corresponds to a linear operator which is originally fixed (because it does not refer to t at all), so after the transformation it corresponds to a moving linear operator. The transformation thus leads us to a new picture of the motion, in which the states correspond to fixed vectors and the dynamical variables to moving linear operators. We shall call this the *Heisenberg picture*.

The physical condition of the dynamical system at any time involves the relation of the dynamical variables to the state, and the change of the physical condition with time may be ascribed either to a change in the state, with the dynamical variables kept fixed, which gives us the Schrödinger picture, or to a change in the dynamical variables, with the state kept fixed, which gives us the Heisenberg picture.

In the Heisenberg picture there are equations of motion for the dynamical variables. Take a dynamical variable corresponding to the fixed linear operator v in the Schrödinger picture. In the Heisenberg picture it corresponds to a moving linear operator, which we write as v_t instead of v^*, to bring out its dependence on t, and which is given by

$$v_t = T^{-1}vT \tag{10}$$

or

$$Tv_t = vT.$$

Differentiating with respect to t, we get

$$\frac{dT}{dt}v_t + T\frac{dv_t}{dt} = v\frac{dT}{dt}.$$

With the help of (6), this gives

$$HTv_t + i\hbar T\frac{dv_t}{dt} = vHT$$

or

$$i\hbar\frac{dv_t}{dt} = T^{-1}vHT - T^{-1}HTv_t$$
$$= v_t H_t - H_t v_t, \tag{11}$$

where

$$H_t = T^{-1}HT. \tag{12}$$

Equation (11) may be written in P.B. notation

$$\frac{dv_t}{dt} = [v_t, H_t]. \tag{13}$$

Equation (11) or (13) shows how any dynamical variable varies with time in the Heisenberg picture and gives us *Heisenberg's form for the equations of motion*. These equations of motion are determined by the one linear operator H_t, which is just the transform of the linear operator H occurring in Schrödinger's form for the equations of motion and corresponds to the energy in the Heisenberg picture. We shall call the dynamical variables in the Heisenberg picture, where they vary with the time, *Heisenberg dynamical variables*, to distinguish them from the fixed dynamical variables of the Schrödinger picture, which we shall call *Schrödinger dynamical variables*. Each Heisenberg dynamical variable is connected with the corresponding Schrödinger dynamical variable by equation (10). Since this connexion is a unitary transformation, all algebraic and functional relationships are the same for both kinds of dynamical variable. We have $T = 1$ for $t = t_0$, so that $v_{t_0} = v$ and any Heisenberg dynamical variable at time t_0 equals the corresponding Schrödinger dynamical variable.

Equation (13) can be compared with classical mechanics, where we also have dynamical variables varying with the time. The equations of motion of classical mechanics can be written in the Hamiltonian form

$$\frac{dq_r}{dt} = \frac{\partial H}{\partial p_r}, \quad \frac{dp_r}{dt} = -\frac{\partial H}{\partial q_r}, \tag{14}$$

where the q's and p's are a set of canonical coordinates and momenta and H is the energy expressed as a function of them and possibly also of t. The energy expressed in this way is called the *Hamiltonian*. Equations (14) give, for v any function of the q's and p's that does not contain the time t explicitly,

$$\begin{aligned}\frac{dv}{dt} &= \sum_r \left(\frac{\partial v}{\partial q_r} \frac{dq_r}{dt} + \frac{\partial v}{\partial p_r} \frac{dp_r}{dt} \right) \\ &= \sum_r \left(\frac{\partial v}{\partial q_r} \frac{\partial H}{\partial p_r} - \frac{\partial v}{\partial p_r} \frac{\partial H}{\partial q_r} \right) \\ &= [v, H], \end{aligned} \tag{15}$$

with the classical definition of a P.B., equation (1) of § 21. This is of the same form as equation (13) in the quantum theory. We thus get an analogy between the classical equations of motion in the Hamiltonian form

28. HEISENBERG'S FORM FOR THE EQUATIONS OF MOTION

and the quantum equations of motion in Heisenberg's form. This analogy provides a justification for the assumption that the linear operator H introduced in the preceding section is the energy of the system in quantum mechanics.

In classical mechanics a dynamical system is defined mathematically when the Hamiltonian is given, i.e. when the energy is given in terms of a set of canonical coordinates and momenta, as this is sufficient to fix the equations of motion. In quantum mechanics a dynamical system is defined mathematically when the energy is given in terms of dynamical variables whose commutation relations are known, as this is then sufficient to fix the equations of motion, in both Schrödinger's and Heisenberg's form. We need to have either H expressed in terms of the Schrödinger dynamical variables or H_t expressed in terms of the corresponding Heisenberg dynamical variables, the functional relationship being, of course, the same in both cases. We call the energy expressed in this way the *Hamiltonian* of the dynamical system in quantum mechanics, to keep up the analogy with the classical theory.

A system in quantum mechanics always has a Hamiltonian, whether the system is one that has a classical analogue and is describable in terms of canonical coordinates and momenta or not. However, if the system does have a classical analogue, its connexion with classical mechanics is specially close and one can usually assume that the Hamiltonian is the same function of the canonical coordinates and momenta in the quantum theory as in the classical theory.[1] There would be a difficulty in this, of course, if the classical Hamiltonian involved a product of factors whose quantum analogues do not commute, as one would not know in which order to put these factors in the quantum Hamiltonian, but this does not happen for most of the elementary dynamical systems whose study is important for atomic physics. In consequence we are able also largely to use the same language for describing dynamical systems in the quantum theory as in the classical theory (e.g. to talk about particles with given masses moving through given fields of force), and when given a system in classical mechanics, can usually give a meaning to 'the same' system in quantum mechanics.

Equation (13) holds for v_t any function of the Heisenberg dynamical variables not involving the time explicitly, i.e. for v any constant linear

---

[1] This assumption is found in practice to be successful only when applied with the dynamical coordinates and momenta referring to a Cartesian system of axes and not to more general curvilinear coordinates.

operator in the Schrödinger picture. It shows that such a function v_t is constant if it commutes with H_t or if v commutes with H. We then have

$$v_t = v_{t_0} = v,$$

and we call v_t or v a *constant of the motion*. It is necessary that v shall commute with H at all times, which is usually possible only if H is constant. In this case we can substitute H for v in (13) and deduce that H_t is constant, showing that H itself is then a constant of the motion. Thus if the Hamiltonian is constant in the Schrödinger picture, it is also constant in the Heisenberg picture.

For an isolated system, a system not acted on by any external forces, there are always certain constants of the motion. One of these is the total energy or Hamiltonian. Others are provided by the displacement theory of § 25. It is evident physically that the total energy must remain unchanged if all the dynamical variables are displaced in a certain way, so equation (63) of § 25 must hold with $v_d = v = H$. Thus D commutes with H and is a constant of the motion. Passing to the case of an infinitesimal displacement, we see that the displacement operators d_x, d_y, and d_z are constants of the motion and hence, from (69) of § 25, the total momentum is a constant of the motion. Again, the total energy must remain unchanged if all the dynamical variables are subjected to a certain rotation. This leads, as will be shown in § 35, to the result that the total angular momentum is a constant of the motion. *The laws of conservation of energy, momentum, and angular momentum hold for an isolated system in the Heisenberg picture in quantum mechanics, as they hold in classical mechanics.*

Two forms for the equations of motion of quantum mechanics have now been given. Of these, the Schrödinger form is the more useful one for practical problems, as it provides the simpler equations. The unknowns in Schrödinger's wave equation are the numbers which form the representative of a ket vector, while Heisenberg's equation of motion for a dynamical variable, if expressed in terms of a representation, would involve as unknowns the numbers forming the representative of the dynamical variable. The latter are far more numerous and therefore more difficult to evaluate than the Schrödinger unknowns. Heisenberg's form for the equations of motion is of value in providing an immediate analogy with classical mechanics and enabling one to see how various features of classical theory, such as the conservation laws referred to above, are translated into quantum theory.

29. Stationary States

We shall here deal with a dynamical system whose energy is constant. Certain specially simple relations hold for this case. Equation (6) can be integrated[1] to give

$$T = e^{-iH(t-t_0)/\hbar},$$

with the help of the initial condition that $T = 1$ for $t = t_0$. This result substituted into (1) gives

$$|Pt\rangle = e^{-iH(t-t_0)/\hbar}|Pt_0\rangle, \tag{16}$$

which is the integral of Schrödinger's equation of motion (5), and substituted into (10) it gives

$$v_t = e^{iH(t-t_0)/\hbar}ve^{-iH(t-t_0)/\hbar}, \tag{17}$$

which is the integral of Heisenberg's equation of motion (11), H_t being now equal to H. Thus we have solutions of the equations of motion in a simple form. However, these solutions are not of much practical value, because of the difficulty involved in evaluating the operator $e^{-iH(t-t_0)/\hbar}$, unless H is particularly simple, and for practical purposes one usually has to fall back on Schrödinger's wave equation.

Let us consider a state of motion such that at time t_0 it is an eigenstate of the energy. The ket $|Pt_0\rangle$ corresponding to it at this time must be an eigenket of H. If H' is the eigenvalue to which it belongs, equation (16) gives

$$|Pt\rangle = e^{-iH'(t-t_0)/\hbar}|Pt_0\rangle,$$

showing that $|Pt\rangle$ differs from $|Pt_0\rangle$ only by a phase factor. Thus the state always remains an eigenstate of the energy, and further, it does not vary with the time at all, since the direction of the ket $|Pt\rangle$ does not vary with the time. Such a state is called a *stationary state*. The probability for any particular result of an observation on it is independent of the time when the observation is made. From our assumption that the energy is an observable, there are sufficient stationary states for an arbitrary state to be dependent on them.

The time-dependent wave function $\psi(\xi t)$ representing a stationary state of energy H' will vary with time according to the law

$$\psi(\xi t) = \psi_0(\xi)e^{-iH't/\hbar}, \tag{18}$$

---

[1] The integration can be carried out as though H were an ordinary algebraic variable instead of a linear operator, because there is no quantity that does not commute with H in the work.

and Schrödinger's wave equation (7) for it reduces to

$$H'\psi_0\rangle = H\psi_0\rangle. \tag{19}$$

This equation merely asserts that the state represented by ψ_0 is an eigenstate of H. We call a function ψ_0 satisfying (19) an *eigenfunction* of H, belonging to the eigenvalue H'.

In the Heisenberg picture the stationary states correspond to fixed eigenvectors of the energy. We can set up a representation in which all the basic vectors are eigenvectors of the energy and so correspond to stationary states in the Heisenberg picture. We call such a representation a *Heisenberg representation*. The first form of quantum mechanics, discovered by Heisenberg in 1925, was in terms of a representation of this kind. The energy is diagonal in the representation. Any other diagonal dynamical variable must commute with the energy and is therefore a constant of the motion. The problem of setting up a Heisenberg representation thus reduces to the problem of finding a complete set of commuting observables, each of which is a constant of the motion, and then making these observables diagonal. The energy must be a function of these observables, from Theorem 2 of § 19. It is sometimes convenient to take the energy itself as one of them.

Let α denote the complete set of commuting observables in a Heisenberg representation, so that the basic vectors are written $\langle\alpha'|, |\alpha''\rangle$. The energy is a function of these observables α, say $H = H(\alpha)$. From (17) we get

$$\langle\alpha'|v_t|\alpha''\rangle = \langle\alpha'|e^{iH(t-t_0)/\hbar}ve^{-iH(t-t_0)/\hbar}|\alpha''\rangle$$
$$= e^{i(H'-H'')(t-t_0)/\hbar}\langle\alpha'|v|\alpha''\rangle, \tag{20}$$

where $H' = H(\alpha')$ and $H'' = H(\alpha'')$. The factor $\langle\alpha'|v|\alpha''\rangle$ on the right-hand side here is independent of t, being an element of the matrix representing the fixed linear operator v. Formula (20) shows how the Heisenberg matrix elements of any Heisenberg dynamical variable vary with time, and it makes v_t satisfy the equation of motion (11), as is easily verified. The variation given by (20) is simply periodic with the frequency

$$\frac{|H'-H''|}{2\pi\hbar} = \frac{|H'-H''|}{h}, \tag{21}$$

depending only on the energy difference of the two stationary states to which the matrix element refers. This result is closely connected with the Combination Law of Spectroscopy and Bohr's Frequency Condition, according to which (21) is the frequency of the electromagnetic radiation emitted or absorbed when the system makes a transition under the influence of radiation between the stationary states α' and α'', the eigenvalues of H being Bohr's energy levels. These matters will be dealt with in § 45.

30. The Free Particle

The most fundamental and elementary application of quantum mechanics is to the system consisting merely of a free particle, or particle not acted on by any forces. For dealing with it we use as dynamical variables the three Cartesian coordinates x, y, z and their conjugate momenta p_x, p_y, p_z. The Hamiltonian is equal to the kinetic energy of the particle, namely

$$H = \frac{1}{2m}(p_x^2 + p_y^2 + p_z^2) \tag{22}$$

according to Newtonian mechanics, m being the mass. This formula is valid only if the velocity of the particle is small compared with c, the velocity of light. For a rapidly moving particle, such as we often have to deal with in atomic theory, (22) must be replaced by the relativistic formula

$$H = c(m^2c^2 + p_x^2 + p_y^2 + p_z^2)^{1/2}. \tag{23}$$

For small values of p_x, p_y, and p_z (23) goes over into (22), except for the constant term mc^2 which corresponds to the rest-energy of the particle in the theory of relativity and which has no influence on the equations of motion. Formulas (22) and (23) can be taken over directly into the quantum theory, the square root in (23) being now understood as the positive square root defined at the end of § 11. The constant term mc^2 by which (23) differs from (22) for small values of p_x, p_y, and p_z can still have no physical effects, since the Hamiltonian in the quantum theory, as introduced in § 27, is undefined to the extent of an arbitrary additive real constant.

We shall here work with the more accurate formula (23). We shall first solve the Heisenberg equations of motion. From the quantum conditions (9) of § 21, p_x commutes with p_y and p_z, and hence, from Theorem 1 of § 19 extended to a set of commuting observables, p_x commutes with any function of p_x, p_y, and p_z and therefore with H. It follows that p_x is a constant of the motion. Similarly p_y and p_z are constants of the motion. These results are the same as in the classical theory. Again, the equation of motion for a coordinate, x_t say, is, according to (11),

$$i\hbar \dot{x}_t = i\hbar \frac{dx_t}{dt} = x_t c(m^2c^2 + p_x^2 + p_y^2 + p_z^2)^{1/2} - c(m^2c^2 + p_x^2 + p_y^2 + p_z^2)^{1/2} x_t.$$

The right-hand side here can be evaluated by means of formula (31) of § 22 with the roles of coordinates and momenta interchanged, so that it reads

$$q_r f - f q_r = i\hbar \frac{\partial f}{\partial p_r}, \qquad (24)$$

f now being any function of the p's. This gives

$$\dot{x}_t = \frac{\partial}{\partial p_x} c(m^2 c^2 + p_x^2 + p_y^2 + p_z^2)^{1/2} = \frac{c^2 p_x}{H}. \qquad (25)$$

Similarly,

$$\dot{y}_t = \frac{c^2 p_y}{H}, \quad \dot{z}_t = \frac{c^2 p_z}{H}. \qquad (25)$$

The magnitude of the velocity is

$$v = (\dot{x}_t^2 + \dot{y}_t^2 + \dot{z}_t^2)^{1/2} = \frac{c^2 (p_x^2 + p_y^2 + p_z^2)^{1/2}}{H}. \qquad (26)$$

Equations (25) and (26) are just the same as in the classical theory.

Let us consider a state that is an eigenstate of the momenta, belonging to the eigenvalues p'_x, p'_y, p'_z. This state must be an eigenstate of the Hamiltonian, belonging to the eigenvalue

$$H' = c(m^2 c^2 + p'^2_x + p'^2_y + p'^2_z)^{1/2}, \qquad (27)$$

and must therefore be a stationary state. The possible values for H' are all numbers from mc^2 to ∞, as in the classical theory. The wave function $\psi(xyz)$ representing this state at any time in Schrödinger's representation must satisfy

$$p'_x \psi(xyz)\rangle = p_x \psi(xyz)\rangle = -i\hbar \frac{\partial \psi(xyz)}{\partial x}\rangle,$$

with similar equations for p_y and p_z. These equations show that $\psi(xyz)$ is of the form

$$\psi(xyz) = a e^{i(p'_x x + p'_y y + p'_z z)/\hbar} \qquad (28)$$

where a is independent of x, y, and z. From (18) we see now that the time-dependent wave function $\psi(xyzt)$ is of the form

$$\psi(xyzt) = a_0 e^{i(p'_x x + p'_y y + p'_z z - H' t)/\hbar}, \qquad (29)$$

where a_0 is independent of x, y, z, and t.

30. THE FREE PARTICLE

The function (29) of x, y, z, and t describes plane waves in space-time. We see from this example the suitability of the terms 'wave function' and 'wave equation'. The frequency of the waves is

$$\nu = \frac{H'}{h}, \tag{30}$$

their wavelength is

$$\lambda = \frac{h}{(p_x'^2 + p_y'^2 + p_z'^2)^{1/2}} = \frac{h}{P'}, \tag{31}$$

P' being the length of the vector (p_x', p_y', p_z'), and their motion is in the direction specified by the vector (p_x', p_y', p_z') with the velocity

$$\lambda \nu = \frac{H'}{P'} = \frac{c^2}{v'}, \tag{32}$$

v' being the velocity of the particle corresponding to the momentum (p_x', p_y', p_z') as given by formula (26). Equations (30), (31), and (32) are easily seen to hold in all Lorentz frames of reference, the expression on the right-hand side of (29) being, in fact, relativistically invariant with p_x', p_y', p_z' and H' as the components of a 4-vector. These properties of relativistic invariance led de Broglie, before the discovery of quantum mechanics, to postulate the existence of waves of the form (29) associated with the motion of any particle. They are therefore known as *de Broglie waves*.

In the limiting case when the mass m is made to tend to zero, the classical velocity of the particle v becomes equal to c and hence, from (32), the wave velocity also becomes c. The waves are then like the light-waves associated with a photon, with the difference that they contain no reference to the polarization and involve a complex exponential instead of sines and cosines. Formulas (30) and (31) are still valid, connecting the frequency of the light-waves with the energy of the photon and the wavelength of the light-waves with the momentum of the photon.

For the state represented by (29), the probability of the particle being found in any specified small volume when an observation of its position is made is independent of where the volume is. This provides an example of Heisenberg's principle of uncertainty, the state being one for which the momentum is accurately given and for which, in consequence, the position is completely unknown. Such a state is, of course, a limiting case which never occurs in practice. The states usually met with in practice are those represented by wave packets, which may be formed by superposing a number of waves of the type (29) belonging to slightly different values of (p'_x, p'_y, p'_z), as discussed in § 24. The ordinary formula in hydrodynamics for the velocity of such a wave packet, i.e. the *group velocity* of the waves, is

$$\frac{dv}{d(1/\lambda)} \qquad (33)$$

which gives, from (30) and (31)

$$\frac{dH'}{dP'} = c\frac{d}{dP'}(m^2c^2 + P'^2)^{1/2} = \frac{c^2 P'}{H'} = v'. \qquad (34)$$

This is just the velocity of the particle. The wave packet moves in the same direction and with the same velocity as the particle moves in classical mechanics.

31. The Motion of Wave Packets

The result just deduced for a free particle is an example of a general principle. For any dynamical system with a classical analogue, a state for which the classical description is valid as an approximation is represented in quantum mechanics by a wave packet, all the coordinates and momenta having approximate numerical values, whose accuracy is limited by Heisenberg's principle of uncertainty. Now Schrödinger's wave equation fixes how such a wave packet varies with time, so in order that the classical description may remain valid, the wave packet should remain a wave packet and should move according to the laws of classical dynamics. We shall verify that this is so.

We take a dynamical system having a classical analogue and let its Hamiltonian be $H(q_r, p_r)$ ($r = 1, 2, \ldots, n$). The corresponding classical dynamical system will have as Hamiltonian $H_c(q_r, p_r)$ say, obtained by putting ordinary algebraic variables for the q_r and p_r in $H(q_r, p_r)$ and making $\hbar \to 0$ if it occurs in $H(q_r, p_r)$. The classical Hamiltonian H_c is, of course, a real function of its variables. It is usually a quadratic function of the momenta p_r, but not always so, the relativistic theory of a free particle being an example where it is not. The following argument is valid for H_c any algebraic function of the p's.

We suppose that the time-dependent wave function in Schrödinger's representation is of the form

$$\psi(qt) = Ae^{iS/\hbar}, \qquad (35)$$

where A and S are real functions of the q's and t which do not vary very rapidly with their arguments. The wave function is then of the form of waves, with A and S determining the amplitude and phase respectively. Schrödinger's wave equation (7) gives

$$i\hbar \frac{\partial}{\partial t} A e^{iS/\hbar} \rangle = H(q_r, p_r) A e^{iS/\hbar} \rangle$$

or

$$\left(i\hbar \frac{\partial A}{\partial t} - A \frac{\partial S}{\partial t} \right) \rangle = e^{-iS/\hbar} H(q_r, p_r) A e^{iS/\hbar} \rangle. \tag{36}$$

Now $e^{-iS/\hbar}$ is evidently a unitary linear operator and may be used for U in equation (70) of § 26 to give us a unitary transformation. The q's remain unchanged by this transformation, each p_r goes over into

$$e^{-iS/\hbar} p_r e^{iS/\hbar} = p_r + \frac{\partial S}{\partial q_r},$$

with the help of (31) of § 22, and H goes over into

$$e^{-iS/\hbar} H(q_r, p_r) e^{iS/\hbar} = H\left(q_r, p_r + \frac{\partial S}{\partial q_r} \right),$$

since algebraic relations are preserved by the transformation. Thus (36) becomes

$$\left(i\hbar \frac{\partial A}{\partial t} - A \frac{\partial S}{\partial t} \right) \rangle = H\left(q_r, p_r + \frac{\partial S}{\partial q_r} \right) A \rangle. \tag{37}$$

Let us now suppose that $\hbar$ can be counted as small and let us neglect terms involving $\hbar$ in (37). This involves neglecting the p_r's that occur in H in (37), since each p_r is equivalent to the operator $-i\hbar \partial / \partial q_r$ operating on the functions of the q's to the right of it. The surviving terms give

$$-\frac{\partial S}{\partial t} = H_c\left(q_r, \frac{\partial S}{\partial q_r} \right). \tag{38}$$

This is a differential equation which the phase function S has to satisfy. The equation is determined by the classical Hamiltonian function H_c and is known as the *Hamilton-Jacobi equation* in classical dynamics. It allows S to be real and so shows that the assumption of the wave form (35) does not lead to an inconsistency.

To obtain an equation for A, we must retain the terms in (37) which are linear in $\hbar$ and see what they give. A direct evaluation of these terms is rather awkward in the case of a general function H, and we can get the result we require more easily by first multiplying both sides of (37) by the bra vector $\langle Af$, where f is an arbitrary real function of the q's. This gives

$$\langle Af\left(i\hbar\frac{\partial A}{\partial t} - A\frac{\partial S}{\partial t}\right)\rangle = \langle AfH\left(q_r, p_r + \frac{\partial S}{\partial q_r}\right)A\rangle.$$

The conjugate complex equation is

$$\langle Af\left(-i\hbar\frac{\partial A}{\partial t} - A\frac{\partial S}{\partial t}\right)\rangle = \langle AH\left(q_r, p_r + \frac{\partial S}{\partial q_r}\right)fA\rangle.$$

Subtracting and dividing out by $i\hbar$, we obtain

$$2\langle Af\frac{\partial A}{\partial t}\rangle = \langle A\left[f, H\left(q_r, p_r + \frac{\partial S}{\partial q_r}\right)\right]A\rangle. \tag{39}$$

We now have to evaluate the P.B.

$$\left[f, H\left(q_r, p_r + \frac{\partial S}{\partial q_r}\right)\right].$$

Our assumption that $\hbar$ can be counted as small enables us to expand $H(q_r, p_r + \partial S/\partial q_r)$ as a power series in the p's. The terms of zero degree will contribute nothing to the P.B. The terms of the first degree in the p's give a contribution to the P.B. which can be evaluated most easily with the help of the classical formula (1) of § 21 (this formula being valid also in the quantum theory if u is independent of the p's and v is linear in the p's). The amount of this contribution is

$$\sum_s \frac{\partial f}{\partial q_s}\left[\frac{\partial H(q_r, p_r)}{\partial p_s}\right]_{p_r=\partial S/\partial q_r},$$

the notation meaning that we must substitute $\partial S/\partial q_r$ for each p_r in the function [ ] of the q's and p's, so as to obtain a function of the q's only. The terms of higher degree in the p's give contributions to the P.B. which vanish when $\hbar \to 0$. Thus (39) becomes, with neglect of terms involving $\hbar$, which is equivalent to the neglect of $\hbar^2$ in (37),

$$\langle f\frac{\partial A^2}{\partial t}\rangle = \langle A^2 \sum_s \frac{\partial f}{\partial q_s}\left[\frac{\partial H_c(q_r, p_r)}{\partial p_s}\right]_{p_r=\partial S/\partial q_r}\rangle. \tag{40}$$

Now if $a(q)$ and $b(q)$ are any two functions of the q's, formula (64) of § 20 gives

$$\langle a(q)b(q)\rangle = \int a(q')\,dq'\,b(q'),$$

and so

$$\langle a(q)\frac{\partial b(q)}{\partial q_r}\rangle = -\langle \frac{\partial a(q)}{\partial q_r}b(q)\rangle, \tag{41}$$

provided $a(q)$ and $b(q)$ satisfy suitable boundary conditions, as discussed in §§ 22 and 23. Hence (40) may be written

$$\langle f \frac{\partial A^2}{\partial t} \rangle = -\langle f \sum_s \frac{\partial}{\partial q_s} \left\{ A^2 \left[\frac{\partial H_c(q_r, p_r)}{\partial p_s} \right]_{p_r = \partial S/\partial q_r} \right\} \rangle.$$

Since this holds for an arbitrary real function f, we must have

$$\frac{\partial A^2}{\partial t} = -\sum_s \frac{\partial}{\partial q_s} \left\{ A^2 \left[\frac{\partial H_c(q_r, p_r)}{\partial p_s} \right]_{p_r = \partial S/\partial q_r} \right\}. \qquad (42)$$

This is the equation for the amplitude A of the wave function. To get an understanding of its significance, let us suppose we have a fluid moving in the space of the variables q, the density of the fluid at any point and time being A^2 and its velocity being

$$\frac{dq_s}{dt} = \left[\frac{\partial H_c(q_r, p_r)}{\partial p_s} \right]_{p_r = \partial S/\partial q_r}. \qquad (43)$$

Equation (42) is then just the equation of conservation for such a fluid. The motion of the fluid is determined by the function S satisfying (38), there being one possible motion for each solution of (38)

For a given S, let us take a solution of (42) for which at some definite time the density A^2 vanishes everywhere outside a certain small region. We may suppose this region to move with the fluid, its velocity at each point being given by (43), and then the equation of conservation (42) will require the density always to vanish outside the region. There is a limit to how small the region may be, imposed by the approximation we made in neglecting $\hbar$ in (39). This approximation is valid only provided

$$\hbar \frac{\partial}{\partial q_r} A \ll \frac{\partial S}{\partial q_r} A,$$

or

$$\frac{1}{A} \frac{\partial A}{\partial q_r} \ll \frac{1}{\hbar} \frac{\partial S}{\partial q_r},$$

which requires that A shall vary by an appreciable fraction of itself only through a range of the q's in which S varies by many times $\hbar$, i.e. a range consisting of many wavelengths of the wave function (35). Our solution is then a wave packet of the type discussed in § 24 and remains so for all time.

We thus get a wave function representing a state of motion for which the coordinates and momenta have approximate numerical values throughout all time. Such a state of motion in quantum theory corresponds to the states with which classical theory deals. The motion of our wave packet is determined by equations (38) and (43). From these we get, defining p_s as $\partial S/\partial q_s$,

$$\begin{aligned}\frac{dp_s}{dt} &= \frac{d}{dt}\frac{\partial S}{\partial q_s} = \frac{\partial^2 S}{\partial t \partial q_s} + \sum_u \frac{\partial^2 S}{\partial q_u \partial q_s}\frac{dq_u}{dt} \\ &= -\frac{\partial}{\partial q_s}H_c\left(q_r, \frac{\partial S}{\partial q_r}\right) + \sum_u \frac{\partial^2 S}{\partial q_u \partial q_s}\frac{\partial H_c(q_r, p_r)}{\partial p_u} \\ &= -\frac{\partial H_c(q_r, p_r)}{\partial q_s},\end{aligned} \qquad (44)$$

where in the last line the p's are counted as independent of the q's before the partial differentiation. Equations (43) and (44) are just the classical equations of motion in Hamiltonian form and show that the wave packet moves according to the laws of classical mechanics. We see in this way how the classical equations of motion are derivable from the quantum theory as a limiting case.

By a more accurate solution of the wave equation one can show that the accuracy with which the coordinates and momenta simultaneously have numerical values cannot remain permanently as favourable as the limit allowed by Heisenberg's principle of uncertainty, equation (56) of § 24, but if it is initially so it will become less favourable, the wave packet undergoing a spreading.[1]

32. The Action Principle[2]

Equation (10) shows that the Heisenberg dynamical variables at time t, v_t, are connected with their values at time t_0, v_{t_0}, or v, by a unitary transformation. The Heisenberg variables at time $t + \delta t$ are connected with their values at time t by an infinitesimal unitary transformation, as is shown by the equation of motion (11) or (13), which gives the connexion between $v_{t+\delta t}$ and v_t of the form of (79) or (80) of § 26 with H_t for F and $\delta t/\hbar$ for ϵ. The variation with time of the Heisenberg dynamical variables may thus be looked upon as the continuous unfolding of a unitary transformation. In classical mechanics the dynamical variables at time $t + \delta t$ are connected with their values at time t by an infinitesimal contact transformation and the whole motion may be looked upon as

---

[1] See Kennard, *Z. f. Physik*, **44** (1927), 344; Darwin, *Proc. Roy. Soc.* A, **117** (1927), 258.

[2] This section may be omitted by the student who is not specially concerned with higher dynamics.

32. THE ACTION PRINCIPLE

the continuous unfolding of a contact transformation. We have here the mathematical foundation of the analogy between the classical and quantum equations of motion, and can develop it to bring out the quantum analogue of all the main features of the classical theory of dynamics.

Suppose we have a representation in which the complete set of commuting observables ξ are diagonal, so that a basic bra is $\langle \xi' |$. We can introduce a second representation in which the basic bras are

$$\langle \xi'^* | = \langle \xi' | T. \tag{45}$$

The new basic bras depend on the time t and give us a moving representation, like a moving system of axes in an ordinary vector space. Comparing (45) with the conjugate imaginary of (8), we see that the new basic vectors are just the transforms in the Heisenberg picture of the original basic vectors in the Schrödinger picture, and hence they must be conneced with the Heisenberg dynamical variables v_t in the same way in which the original basic vectors are connected with the Schrödinger dynamical variables v. In particular, each $\langle \xi'^* |$ must be an eigenvector of the ξ_t's belonging to the eigenvalues ξ'. It may therefore be written $\langle \xi'_t |$, with the understanding that the numbers ξ'_t are the same eigenvalues of the ξ_t's that the ξ'''s are of the ξ's. From (45) we get

$$\langle \xi'_t | \xi''_t \rangle = \langle \xi' | T | \xi'' \rangle, \tag{46}$$

showing that the transformation function is just the representative of T in the original representation.

Differentiating (45) with respect to t and using (6), we get

$$i\hbar \frac{d}{dt} \langle \xi'_t | = i\hbar \langle \xi' | \frac{dT}{dt} = \langle \xi' | HT = \langle \xi'_t | H_t$$

with the help of (12). Multiplying on the right by any ket $|a\rangle$ independent of t, we get

$$i\hbar \frac{d}{dt} \langle \xi'_t | a \rangle = \langle \xi'_t | H_t | a \rangle = \int \langle \xi'_t | H_t | \xi''_t \rangle \, d\xi''_t \, \langle \xi''_t | a \rangle, \tag{47}$$

if we take for definiteness the case of continuous eigenvalues for the ξ's. Now equation (5), written in terms of representatives, reads

$$i\hbar \frac{d}{dt} \langle \xi' | Pt \rangle = \int \langle \xi' | H | \xi'' \rangle \, d\xi'' \, \langle \xi'' | Pt \rangle. \tag{48}$$

Since $\langle \xi'_t | H_t | \xi''_t \rangle$ is the same function of the variables ξ'_t and ξ''_t that $\langle \xi' | H | \xi'' \rangle$ is of ξ' and ξ'', equations (47) and (48) are of precisely the same form, with the variables ξ'_t, ξ''_t in (47) playing the role of the variables ξ' and ξ'' in (48) and the function $\langle \xi'_t | a \rangle$ playing the role of the function $\langle \xi'|Pt\rangle$. We can thus look upon (47) as a form of Schrödinger's wave equation, with the function $\langle \xi'_t|a\rangle$ of the variables ξ'_t as the wave function. In this way *Schrödinger's wave equation appears in a new light, as the condition on the representative, in the moving representation with the Heisenberg variables ξ_t diagonal, of the fixed ket corresponding to a state in the Heisenberg picture*. The function $\langle \xi'_t|a\rangle$ owes its variation with time to its left factor $\langle \xi'_t|$, in contradistinction to the function $\langle \xi'|Pt\rangle$, which owes its variation with time to its right factor $|Pt\rangle$.

If we put $|a\rangle = |\xi''\rangle$ in (47), we get

$$i\hbar \frac{d}{dt}\langle \xi'_t|\xi''\rangle = \int \langle \xi'_t|H_t|\xi'''_t\rangle \, d\xi'''_t \, \langle \xi'''_t|\xi''\rangle, \qquad (49)$$

showing that the transformation function $\langle \xi'_t|\xi''\rangle$ satisfies Schrödinger's wave equation. Now $\xi_{t_0} = \xi$, so we must have

$$\langle \xi'_{t_0}|\xi''\rangle = \delta(\xi'_{t_0} - \xi''), \qquad (50)$$

the δ function here being understood as the product of a number of factors, one for each ξ-variable, such as occurs for the variables ξ_{v+1}, $\xi_{v+2}, \ldots, \xi_u$ on the right-hand side of equation (34) of § 16. Thus the transformation function $\langle \xi'_t|\xi''\rangle$ is that solution of Schrödinger's wave equation for which the ξ's certainly have the values ξ'' at time t_0. The square of its modulus, $|\langle \xi'_t|\xi''\rangle|^2$, is the relative probability of the ξ's having the values ξ'_t at time $t > t_0$ if they certainly have the values ξ'' at time t_0. We may write $\langle \xi'_t|\xi''\rangle$ as $\langle \xi'_t|\xi''_{t_0}\rangle$ and consider it as depending on t_0 as well as on t. To get its dependence on t_0 we take the conjugate complex of equation (49), interchange t and t_0 and also interchange single primes and double primes. This gives

$$-i\hbar \frac{d}{dt_0}\langle \xi'_t|\xi''_{t_0}\rangle = \int \langle \xi'_t|\xi'''_{t_0}\rangle \, d\xi'''_{t_0} \, \langle \xi'''_{t_0}|H_{t_0}|\xi''_{t_0}\rangle. \qquad (51)$$

The foregoing discussion of the transformation function $\langle \xi'_t|\xi''\rangle$ is valid with the ξ's any complete set of commuting observables. The equations were written down for the case of the ξ's having continuous eigenvalues, but they would still be valid if any of the ξ's have discrete eigenvalues, provided the necessary formal changes are made in them. Let us now take a dynamical system having a classical analogue and let us take the ξ's to be the coordinates q. Put

$$\langle q'_t|q''\rangle = e^{iS/\hbar} \qquad (52)$$

32. THE ACTION PRINCIPLE

and so define the function S of the variables q'_t, q''. This function also depends explicitly on t. (52) is a solution of Schrödinger's wave equation and, if $\hbar$ can be counted as small, it can be handled in the same way as (35) was. The S of (52) differs from the S of (35) on account of there being no A in (52), which makes the S of (52) complex, but the real part of this S equals the S of (35) and its pure imaginary part is of the order $\hbar$. Thus, in the limit $\hbar \to 0$, the S of (52) will equal that of (35) and will therefore satisfy, corresponding to (38),

$$-\frac{\partial S}{\partial t} = H_c(q'_{rt}, p'_{rt}), \tag{53}$$

where

$$p'_{rt} = \frac{\partial S}{\partial q'_{rt}}, \tag{54}$$

and H_c is the Hamiltonian of the classical analogue of our quantum dynamical system. But (52) is also a solution of (51) with q's for ξ's, which is the conjugate complex of Schrödinger's wave equation in the variables q'' or q''_{t_0}. This causes S to satisfy also[1]

$$\frac{\partial S}{\partial t_0} = H_c(q''_r, p''_r), \tag{55}$$

where

$$p''_r = -\frac{\partial S}{\partial q''_r}. \tag{56}$$

The solution of the Hamilton-Jacobi equations (53), (55) is the action function of classical mechanics for the time interval t_0 to t, i.e. it is the time integral of the Lagrangian L,

$$S = \int_{t_0}^{t} L(t')\,dt'. \tag{57}$$

Thus *the S defined by (52) is the quantum analogue of the classical action function and equals it in the limit $\hbar \to 0$*. To get the quantum analogue of the classical Lagrangian, we pass to the case of an infinitesimal time interval by putting $t = t_0 + \delta t$ and we then have $\langle q'_{t_0+\delta t} | q''_{t_0} \rangle$ as the analogue of $e^{iL(t_0)\delta t/\hbar}$. For the sake of the analogy, one should consider $L(t_0)$ as a function of the coordinates q' at time $t_0 + \delta t$ and the coordinates q'' at time t_0, rather than as a function of the coordinates and velocities at time t_0, as one usually does.

---

[1] For a more accurate comparison of transformation functions with classical theory, see Van Vleck, *Proc. Nat. Acad.* **14**, 178.

The principle of least action in classical mechanics says that the action function (57) remains stationary for small variations of the trajectory of the system which do not alter the end points, i.e. for small variations of the q's at all intermediate times between t_0 and t with q_{t_0} and q_t fixed. Let us see what it corresponds to in the quantum theory.

Put

$$\exp\left[i\int_{t_a}^{t_b} L(t)\,dt/\hbar\right] = \exp\left[\frac{i\,S(t_b, t_a)}{\hbar}\right] = B(t_b, t_a), \qquad (58)$$

so that $B(t_b, t_a)$ corresponds to $\langle q'_{t_b}|q'_{t_a}\rangle$ in the quantum theory. (We here allow q'_{t_a} and q'_{t_b} to denote different eigenvalues of q_{t_a} and q_{t_b}, to save having to introduce a large number of primes into the analysis.) Now suppose the time interval $t_0 \to t$ to be divided up into a large number of small time intervals $t_0 \to t_1, t_1 \to t_2, \ldots, t_{m-1} \to t_m, t_m \to t$, by the introduction of a sequence of intermediate times $t_1, t_2, \ldots, t_m$. Then

$$B(t, t_0) = B(t, t_m)\,B(t_m, t_{m-1}) \cdots B(t_2, t_1)\,B(t_1, t_0). \qquad (59)$$

The corresponding quantum equation, which follows from the property of basic vectors (35) of § 16, is

$$\langle q'_t|q'_0\rangle = \iint \cdots \int \langle q'_t|q'_m\rangle\,dq'_m\,\langle q'_m|q'_{m-1}\rangle\,dq'_{m-1}\cdots \langle q'_2|q'_1\rangle\,dq'_1\,\langle q'_1|q'_0\rangle, \qquad (60)$$

q'_k being written for q'_{t_k} for brevity. At first sight there does not seem to be any close correspondence between (59) and (60). We must, however, analyse the meaning of (59) rather more carefully. We must regard each factor B as a function of the q's at the two ends of the time interval to which it refers. This makes the right-hand side of (59) a function, not only of q_t and q_{t_0}, but also of all the intermediate q's. Equation (59) is valid only when we substitute for the intermediate q's in its right-hand side their values for the real trajectory, small variations in which values leave S stationary and therefore also, from (58), leave $B(t, t_0)$ stationary. It is the process of substituting these values for the intermediate q's which corresponds to the integrations over all values for the intermediate q''s in (60). The quantum analogue of the action principle is thus absorbed in the composition law (60) and the classical requirement that the values of the intermediate q's shall make S stationary corresponds to the condition in quantum mechanics that all values of the intermediate q''s are important in proportion to their contribution to the integral in (60).

Let us see how (59) can be a limiting case of (60) for $\hbar$ small. We must suppose the integrand in (60) to be of the form $e^{iF/\hbar}$, where F is a function of $q'_0, q'_1, q'_2, \ldots, q'_m, q'_t$ which remains continuous as $\hbar$ tends to zero, so that the integrand is a rapidly oscillating function when $\hbar$ is small. The integral of such a rapidly oscillating function will be extremely small, except for the contribution arising from a region in the domain of integration where comparatively large variations in the q'_k produce only very small variations in F. Such a region must be the neighbourhood of a point where F is stationary for small variations of the q'_k. Thus the integral in (60) is determined essentially by the value of the integrand at a point where the integrand is stationary for small variations of the intermediate q''s, and so (60) goes over into (59).

Equations (54) and (56) express that the variables q'_t, p'_t are connected with the variables q'', p'' by a contact transformation and are one of the standard forms of writing the equations of a contact transformation. There is an analogous form for writing the equations of a unitary transformation in quantum mechanics. We get from (52), with the help of (45) of § 22,

$$\langle q'_t | p_{rt} | q'' \rangle = -i\hbar \frac{\partial}{\partial q'_{rt}} \langle q'_t | q'' \rangle = \frac{\partial S(q'_t, q'')}{\partial q'_{rt}} \langle q'_t | q'' \rangle. \tag{61}$$

Similarly, with the help of (46) of § 22,

$$\langle q'_t | p_r | q'' \rangle = i\hbar \frac{\partial}{\partial q''_r} \langle q'_t | q'' \rangle = -\frac{\partial S(q'_t, q'')}{\partial q''_r} \langle q'_t | q'' \rangle. \tag{62}$$

From the general definition of functions of commuting observables, we have

$$\langle q'_t | f(q_t) g(q) | q'' \rangle = f(q'_t) g(q'') \langle q'_t | q'' \rangle, \tag{63}$$

where $f(q_t)$ and $g(q)$ are functions of the q_t's and q's respectively. Let $G(q_t, q)$ be any function of the q_t's and q's consisting of a sum or integral of terms each of the form $f(q_t) g(q)$, so that all the q_t's in G occur to the left of all the q's Such a function we call *well ordered*. Applying (63) to each of the terms in G and adding or integrating, we get

$$\langle q'_t | G(q_t, q) | q'' \rangle = G(q'_t, q'') \langle q'_t | q'' \rangle.$$

Now let us suppose each p_{rt} and p_r can be expressed as a well-ordered function of the q_t's and q's and write these functions $p_{rt}(q_t, q), p_r(q_t, q)$. Putting these functions for G, we get

$$\langle q'_t | p_{rt} | q'' \rangle = p_{rt}(q'_t, q'') \langle q'_t | q'' \rangle,$$
$$\langle q'_t | p_r | q'' \rangle = p_r(q'_t, q'') \langle q'_t | q'' \rangle.$$

Comparing these equations with (61) and (62) respectively, we see that
$$p_{rt}(q_t', q'') = \frac{\partial S(q_t', q'')}{\partial q_{rt}'}, \quad p_r(q_t', q'') = -\frac{\partial S(q_t', q'')}{\partial q_r''}.$$

This means that
$$p_{rt} = \frac{\partial S(q_t, q)}{\partial q_{rt}}, \quad p_r = -\frac{\partial S(q_t, q)}{\partial q_r}, \tag{64}$$

provided the right-hand sides of (64) are written as well-ordered functions.

These equations are of the same form as (54) and (56), but refer to the non-commuting quantum variables q_t, q instead of the ordinary algebraic variables q_t', q''. They show how the conditions for a unitary transformation between quantum variables are analogous to the conditions for a contact transformation between classical variables. The analogy is not complete, however, because the classical S must be real and there is no simple condition corresponding to this for the S of (64).

33. The Gibbs Ensemble

In our work up to the present we have been assuming all along that our dynamical system at each instant of time is in a definite state, that is to say, its motion is specified as completely and accurately as is possible without conflicting with the general principles of the theory. In the classical theory this would mean, of course, that all the coordinates and momenta have specified values. Now we may be interested in a motion which is specified to a lesser extent than this maximum possible. The present section will be devoted to the methods to be used in such a case.

The procedure in classical mechanics is to introduce what is called a *Gibbs ensemble*, the idea of which is as follows. We consider all the dynamical coordinates and momenta as Cartesian coordinates in a certain space, the *phase space*, whose number of dimensions is twice the number of degrees of freedom of the system. Any state of the system can then be represented by a point in this space. This point will move according to the classical equations of motion (14). Suppose, now, that we are not given that the system is in a definite state at any time, but only that it is in one or other of a number of possible states according to a definite probability law. We should then be able to represent it by a fluid in the phase space, the mass of fluid in any volume of the phase space being the total probability of the system being in any state whose representative point lies in that volume. Each particle of the fluid will be moving according to the equations of motion (14). If we introduce the density ρ of the fluid at any point, equal to the probability per unit volume of phase space of the system being in the neighbourhood of the corresponding state, we shall have the equation of conservation

$$\frac{\partial \rho}{\partial t} = -\sum_r \left[\frac{\partial}{\partial q_r} \left(\rho \frac{dq_r}{dt} \right) + \frac{\partial}{\partial p_r} \left(\rho \frac{dp_r}{dt} \right) \right]$$

$$= -\sum_r \left[\frac{\partial}{\partial q_r} \left(\rho \frac{\partial H}{\partial p_r} \right) - \frac{\partial}{\partial p_r} \left(\rho \frac{\partial H}{\partial q_r} \right) \right]$$

$$= -[\rho, H]. \tag{65}$$

This may be considered as the equation of motion for the fluid, since it determines the density ρ for all time if ρ is given initially as a function of the q's and p's. It is, apart from the minus sign, of the same form as the ordinary equation of motion (15) for a dynamical variable.

The requirement that the total probability of the system being in any state shall be unity gives us a normalizing condition for ρ

$$\iint \rho \, dq \, dp = 1, \tag{66}$$

the integration being over the whole of phase space and the single differential dq or dp being written to denote the product of all the dq's or dp's. If β denotes any function of the dynamical variables, the average value of β will be

$$\iint \beta \rho \, dq \, dp. \tag{67}$$

It makes only a trivial alteration in the theory, but often facilitates discussion, if we work with a density ρ differing from the above one by a positive constant factor, k say, so that we have instead of (66)

$$\iint \rho \, dq \, dp = k.$$

With this density we can picture the fluid as representing a number k of similar dynamical systems, all following through their motions independently in the same place, without any mutual disturbance or interaction. The density at any point would then be the probable or average number of systems in the neighbourhood of any state per unit volume of phase space, and expression (67) would give the average total value of β for all the systems. Such a set of dynamical systems, which is the ensemble introduced by Gibbs, is usually not realizable in practice, except as a rough approximation, but it forms all the same a useful theoretical abstraction.

We shall now see that there exists a corresponding density ρ in quantum mechanics, having properties analogous to the above. It was first introduced by von Neumann. Its existence is rather surprising in view of the fact that phase space has no meaning in quantum mechanics, there being no possibility of assigning numerical values simultaneously to the q's and p's.

We consider a dynamical system which is at a certain time in one or other of a number of possible states according to some given probability law. These states may be either a discrete set or a continuous range, or both together. We shall here take for definiteness the case of a discrete set and suppose them labelled by a parameter m. Let the normalized ket vectors corresponding to them be $|m\rangle$ and let the probability of the system being in the mth state be P_m. We then define the quantum density ρ by

$$\rho = \sum_m |m\rangle P_m \langle m|. \tag{68}$$

Let ρ' be any eigenvalue of ρ and $|\rho'\rangle$ an eigenket belonging to this eigenvalue. Then

$$\sum_m |m\rangle P_m \langle m|\rho'\rangle = \rho|\rho'\rangle = \rho'|\rho'\rangle$$

so that

$$\sum_m \langle \rho'|m\rangle P_m \langle m|\rho'\rangle = \rho'\langle \rho'|\rho'\rangle$$

or

$$\sum_m P_m |\langle m|\rho'\rangle|^2 = \rho'\langle \rho'|\rho'\rangle.$$

Now P_m, being a probability, can never be negative. It follows that ρ' cannot be negative. Thus ρ has no negative eigenvalues, in analogy with the fact that the classical density ρ is never negative.

Let us now obtain the equation of motion for our quantum ρ. In Schrödinger's picture the kets and bras in (68) will vary with the time in accordance with Schrödinger's equation (5) and the conjugate imaginary of this equation, while the P_m's will remain constant, since the system, so long as it is left undisturbed, cannot change over from a state corresponding to one ket satisfying Schrödinger's equation to a state corresponding to another. We thus have

$$i\hbar \frac{d\rho}{dt} = \sum_m i\hbar \left(\frac{d|m\rangle}{dt} P_m \langle m| + |m\rangle P_m \frac{d\langle m|}{dt} \right)$$
$$= \sum_m (H|m\rangle P_m \langle m| - |m\rangle P_m \langle m|H)$$

$$= H\rho - \rho H. \tag{69}$$

This is the quantum analogue of the classical equation of motion (65). Our quantum ρ, like the classical one, is determined for all time if it is given initially.

From the assumption of § 12, the average value of any observable β when the system is in the state m is $\langle m|\beta|m\rangle$. Hence if the system is distributed over the various states m according to the probability law P_m, the average value of β will be $\sum_m P_m\langle m|\beta|m\rangle$. If we introduce a representation with a discrete set of basic ket vectors $|\xi'\rangle$ say, this equals

$$\sum_{m\xi'} P_m\langle m|\xi'\rangle\langle\xi'|\beta|m\rangle = \sum_{\xi'm}\langle\xi'|\beta|m\rangle P_m\langle m|\xi'\rangle$$
$$= \sum_{\xi'}\langle\xi'|\beta\rho|\xi'\rangle = \sum_{\xi'}\langle\xi'|\rho\beta|\xi'\rangle, \tag{70}$$

the last step being easily verified with the law of matrix multiplication, equation (44) of § 17. The expressions (70) are the analogue of the expression (67) of the classical theory. Whereas in the classical theory we have to multiply β by ρ and take the integral of the product over all phase space, in the quantum theory we have to multiply β by ρ, with the factors in either order, and take the diagonal sum of the product in a representation. If the representation involves a continuous range of basic vectors $|\xi'\rangle$, we get instead of (70)

$$\int \langle\xi'|\beta\rho|\xi'\rangle\,d\xi' = \int \langle\xi'|\rho\beta|\xi'\rangle\,d\xi', \tag{71}$$

so that we must carry through a process of 'integrating along the diagonal' instead of summing the diagonal elements. We shall define (71) to be the diagonal sum of $\beta\rho$ in the continuous case. It can easily be verified, from the properties of transformation functions (56) of § 18, that the diagonal sum is the same for all representations.

From the condition that the $|m\rangle$'s are normalized we get, with discrete ξ''s

$$\sum_{\xi'}\langle\xi'|\rho|\xi'\rangle = \sum_{\xi'm}\langle\xi'|m\rangle P_m\langle m|\xi'\rangle = \sum_m P_m = 1, \tag{72}$$

since the total probability of the system being in any state is unity. This is the analogue of equation (66). The probability of the system being in the state ξ', or the probability of the observables ξ which are diagonal in the representation having the values ξ', is, according to the rule for interpreting representatives of kets (51) of § 18,

$$\sum_m |\langle \xi'|m\rangle|^2 P_m = \langle \xi'|\rho|\xi'\rangle, \tag{73}$$

which gives us a meaning for each term in the sum on the left-hand side of (72). For continuous ξ''s, the right-hand side of (73) gives the probability of the ξ's having values in the neighbourhood of ξ' per unit range of variation of the values ξ'.

As in the classical theory, we may take a density equal to k times the above ρ and consider it as representing a Gibbs ensemble of k similar dynamical systems, between which there is no mutual disturbance or interaction. We shall then have k on the right-hand side of (72), and (70) or (71) will give the total average β for all the members of the ensemble, while (73) will give the total probability of a member of the ensemble having values for its ξ's equal to ξ' or in the neighbourhood of ξ' per unit range of variation of the values ξ'.

An important application of the Gibbs ensemble is to a dynamical system in thermodynamic equilibrium with its surroundings at a given temperature T. Gibbs showed that such a system is represented in classical mechanics by the density

$$\rho = ce^{-H/kT}, \tag{74}$$

H being the Hamiltonian, which is now independent of the time, k being Boltzmann's constant, and c being a number chosen to make the normalizing condition (66) hold. This formula may be taken over unchanged into the quantum theory. At high temperatures, (74) becomes $\rho = c$, which gives, on being substituted into the right-hand side of (73), $c\langle \xi'|\xi'\rangle = c$ in the case of discrete ξ''s. This shows that *at high temperatures all discrete states are equally probable.*

CHAPTER VI

Elementary Applications

34. The Harmonic Oscillator

A simple and interesting example of a dynamical system in quantum mechanics is the harmonic oscillator. This example is of importance for general theory, because it forms a corner-stone in the theory of radiation. The dynamical variables needed for describing the system are just one coordinate q and its conjugate momentum p. The Hamiltonian in classical mechanics is

$$H = \frac{1}{2m}(p^2 + m^2\omega^2 q^2), \tag{1}$$

where m is the mass of the oscillating particle and ω is 2π times the frequency. We assume the same Hamiltonian in quantum mechanics. This Hamiltonian, together with the quantum condition (10) of § 22, define the system completely.

The Heisenberg equations of motion are

$$\dot{q}_t = [q_t, H] = \frac{p_t}{m},$$
$$\dot{p}_t = [p_t, H] = -m\omega^2 q_t. \tag{2}$$

It is convenient to introduce the dimensionless complex dynamical variable

$$\eta = (2m\hbar\omega)^{-1/2}(p + im\omega q). \tag{3}$$

The equations of motion (2) give

$$\dot{\eta}_t = (2m\hbar\omega)^{-1/2}(-m\omega^2 q_t + i\omega p_t) = i\omega\eta_t.$$

This equation can be integrated to give

$$\eta_t = \eta_0 e^{i\omega t}, \tag{4}$$

where η_0 is a linear operator independent of t, and is equal to the value of η_t at time $t = 0$. The above equations are all as in the classical theory.

We can express q and p in terms of η and its conjugate complex $\bar{\eta}$ and may thus work entirely in terms of η and $\bar{\eta}$. We have

$$\hbar\omega\eta\bar{\eta} = (2m)^{-1}(p+im\omega q)(p-im\omega q)$$
$$= (2m)^{-1}[p^2 + m^2\omega^2 q^2 + im\omega(qp-pq)]$$
$$= H - \frac{1}{2}\hbar\omega \tag{5}$$

and similarly

$$\hbar\omega\bar{\eta}\eta = H + \frac{1}{2}\hbar\omega. \tag{6}$$

Thus

$$\bar{\eta}\eta - \eta\bar{\eta} = 1. \tag{7}$$

Equation (5) or (6) gives H in terms of η and $\bar{\eta}$ and (7) gives the commutation relation connecting η and $\bar{\eta}$. From (5)

$$\hbar\omega\bar{\eta}\eta\bar{\eta} = \bar{\eta}H - \frac{1}{2}\hbar\omega\bar{\eta}$$

and from (6)

$$\hbar\omega\bar{\eta}\eta\bar{\eta} = H\bar{\eta} + \frac{1}{2}\hbar\omega\bar{\eta}.$$

Thus

$$\bar{\eta}H - H\bar{\eta} = \hbar\omega\bar{\eta}. \tag{8}$$

Also, (7) leads to

$$\bar{\eta}\eta^n - \eta^n\bar{\eta} = n\eta^{n-1} \tag{9}$$

for any positive integer n, as may be verified by induction, since, by multiplying (9) by η on the left, we can deduce (9) with $n+1$ for n.

Let H' be an eigenvalue of H and $|H'\rangle$ an eigenket belonging to it. From (5)

$$\hbar\omega\langle H'|\eta\bar{\eta}|H'\rangle = \langle H'|H - \frac{1}{2}\hbar\omega|H'\rangle = \left(H' - \frac{1}{2}\hbar\omega\right)\langle H'|H'\rangle.$$

Now $\langle H'|\eta\bar{\eta}|H'\rangle$ is the square of the length of the ket $\bar{\eta}|H'\rangle$, and hence

$$\langle H'|\eta\bar{\eta}|H'\rangle \geq 0,$$

the case of equality occurring only if $\bar{\eta}|H'\rangle = 0$. Also $\langle H'|H'\rangle > 0$. Thus

$$H' \geq \frac{1}{2}\hbar\omega, \tag{10}$$

the case of equality occurring only if $\bar{\eta}|H'\rangle = 0$. From the form (1) of H as a sum of squares, we should expect its eigenvalues to be all positive or zero (since the average value of H for any state must be positive or zero). We now have the more stringent condition (10).

From (8)
$$H\bar{\eta}|H'\rangle = (\bar{\eta}H - \hbar\omega\bar{\eta})|H'\rangle = (H' - \hbar\omega)\bar{\eta}|H'\rangle. \tag{11}$$

Now if $H' \neq (1/2)\hbar\omega$, $\bar{\eta}|H'\rangle$ is not zero and is then according to (11) an eigenket of H belonging to the eigenvalue $H' - \hbar\omega$. Thus, with H' any eigenvalue of H not equal to $(1/2)\hbar\omega$, $H' - \hbar\omega$ is another eigenvalue of H. We can repeat the argument and infer that, if $H' - \hbar\omega \neq (1/2)\hbar\omega$, $H' - 2\hbar\omega$ is another eigenvalue of H. Continuing in this way, we obtain the series of eigenvalues H', $H' - \hbar\omega$, $H' - 2\hbar\omega$, $H' - 3\hbar\omega$, ..., which cannot extend to infinity, because then it would contain eigenvalues contradicting (10), and can terminate only with the value $(1/2)\hbar\omega$. Again, from the conjugate complex of equation (8)
$$H\eta|H'\rangle = (\eta H + \hbar\omega\eta)|H'\rangle = (H' + \hbar\omega)\eta|H'\rangle,$$

showing that $H' + \hbar\omega$ is another eigenvalue of H, with $\eta|H'\rangle$ as an eigenket belonging to it, unless $\eta|H'\rangle = 0$. The latter alternative can be ruled out, since it would lead to
$$0 = \hbar\omega\bar{\eta}\eta|H'\rangle = \left(H + \frac{1}{2}\hbar\omega\right)|H'\rangle = \left(H' + \frac{1}{2}\hbar\omega\right)|H'\rangle,$$

which contradicts (10). Thus $H' + \hbar\omega$ is always another eigenvalue of H, and so are $H' + 2\hbar\omega$, $H' + 3\hbar\omega$ and so on. Hence the eigenvalues of H are the series of numbers
$$\frac{1}{2}\hbar\omega, \quad \frac{3}{2}\hbar\omega, \quad \frac{5}{2}\hbar\omega, \quad \frac{7}{2}\hbar\omega, \quad \ldots, \tag{12}$$

extending to infinity. These are the possible energy values for the harmonic oscillator.

Let $|0\rangle$ be an eigenket of H belonging to the lowest eigenvalue $(1/2)\hbar\omega$, so that
$$\bar{\eta}|0\rangle = 0, \tag{13}$$

and form the sequence of kets
$$|0\rangle, \quad \eta|0\rangle, \quad \eta^2|0\rangle, \quad \eta^3|0\rangle, \quad \ldots. \tag{14}$$

These kets are all eigenkets of H, belonging to the sequence of eigenvalues (12) respectively. From (9) and (13)
$$\bar{\eta}\eta^n|0\rangle = n\eta^{n-1}|0\rangle \tag{15}$$

for any non-negative integer n. Thus the set of kets (14) is such that η or $\bar{\eta}$ applied to any one of the set gives a ket dependent on the set. Now all the dynamical variables in our problem are expressible in terms of η and $\bar{\eta}$, so the kets (14) must form a complete set (otherwise there would be some more dynamical variables). There is just one of these kets for each eigenvalue (12) of H, so H by itself forms a complete commuting set of observables. The kets (14) correspond to the various stationary states of the oscillator. The stationary state with energy $[n + (1/2)]\hbar\omega$, corresponding to $\eta^n|0\rangle$, is called the nth quantum state.

The square of the length of the ket $\eta^n|0\rangle$ is

$$\langle 0|\bar{\eta}^n\eta^n|0\rangle = n\langle 0|\bar{\eta}^{n-1}\eta^{n-1}|0\rangle$$

with the help of (15). By induction, we find that

$$\langle 0|\bar{\eta}^n\eta^n|0\rangle = n! \tag{16}$$

provided $|0\rangle$ is normalized. Thus the kets (14) multiplied by the coefficients $n!^{-1/2}$ with $n = 0, 1, 2, \ldots$, respectively form the basic kets of a representation, namely the representation with H diagonal. Any ket $|x\rangle$ can be expanded in the form

$$|x\rangle = \sum_{0}^{\infty} x_n \eta^n |0\rangle, \tag{17}$$

where the x_n's are numbers. In this way the ket $|x\rangle$ is put into correspondence with a power series $\sum x_n \eta^n$ in the variable η, the various terms in the power series corresponding to the various stationary states. If $|x\rangle$ is normalized, it defines a state for which the probability of the oscillator being in the nth quantum state, i.e. the probability of H having the value $[n + (1/2)]\hbar\omega$, is

$$P_n = n!|x_n|^2, \tag{18}$$

as follows from the same argument which led to (51) of § 18.

We may consider the ket $|0\rangle$ as a standard ket and the power series in η as a wave function, since any ket can be expressed as such a wave function multiplied into this standard ket. We get a kind of wave function differing from the usual kind, introduced by equations (62) of § 20, in that it is a function of the complex dynamical variable η instead of observables. It was first introduced by V. Fock, so we shall call the representation Fock's representation. It is for many purposes the most convenient representation for describing states of the harmonic oscillator. The standard ket $|0\rangle$ satisfies the condition (13), which replaces the conditions (43) of § 22 for the standard ket in Schrödinger's representation.

Let us introduce Schrödinger's representation with q diagonal and obtain the representatives of the stationary states. From (13) and (3)

$$(p - im\omega q)|0\rangle = 0,$$

so

$$\langle q'|p - im\omega q|0\rangle = 0.$$

With the help of (45) of § 22, this gives

$$\hbar \frac{\partial}{\partial q'} \langle q'|0\rangle + m\omega q' \langle q'|0\rangle = 0. \tag{19}$$

The solution of this differential equation is

$$\langle q'|0\rangle = \left(\frac{m\omega}{\pi\hbar}\right)^{1/4} e^{-m\omega q'^2/2\hbar}, \tag{20}$$

the numerical coefficient being chosen so as to make $|0\rangle$ normalized. We have here the representative of the *normal state*, as the state of lowest energy is called. The representatives of the other stationary states can be obtained from it. We have from (3)

$$\langle q'|\eta^n|0\rangle = (2m\hbar\omega)^{-n/2} \langle q'|(p + im\omega q)^n|0\rangle$$

$$= (2m\hbar\omega)^{-n/2} i^n \left(-\hbar \frac{\partial}{\partial q'} + m\omega q'\right)^n \langle q'|0\rangle$$

$$= i^n (2m\hbar\omega)^{-n/2} \left(\frac{m\omega}{\pi\hbar}\right)^{1/4} \left(-\hbar \frac{\partial}{\partial q'} + m\omega q'\right)^n e^{-m\omega q'^2/2\hbar}. \tag{21}$$

This may easily be worked out for small values of n. The result is of the form of $e^{-m\omega q'^2/2\hbar}$ times a power series of degree n in q'. A further factor $n!^{-1/2}$ must be inserted in (21) to get the normalized representative of the nth quantum state. The phase factor i^n may be discarded.

35. Angular Momentum

Let us consider a particle described by the three Cartesian coordinates x, y, z and their conjugate momenta p_x, p_y, p_z. Its angular momentum about the origin is defined as in the classical theory, by

$$m_x = yp_z - zp_y \quad m_y = zp_x - xp_z \quad m_z = xp_y - yp_x \tag{22}$$

or by the vector equation

$$\mathbf{m} = \mathbf{x} \times \mathbf{p}.$$

We must evaluate the P.B.s of the angular momentum components with the dynamical variables x, p_x, etc., and with each other. This we can do most conveniently with the help of the laws (4) and (5) of § 21, thus

$$[m_z, x] = [xp_y - yp_x, x] = -y[p_x, x] = y,$$
$$[m_z, y] = [xp_y - yp_x, y] = x[p_y, y] = -x, \tag{23}$$
$$[m_z, z] = [xp_y - yp_x, z] = 0, \tag{24}$$

and similarly,

$$[m_z, p_x] = p_y, \quad [m_z, p_y] = -p_x, \tag{25}$$
$$[m_z, p_z] = 0, \tag{26}$$

with corresponding relations for m_x and m_y. Again

$$[m_y, m_z] = [zp_x - xp_z, m_z] = z[p_x, m_z] - [x, m_z]p_z$$
$$= -zp_y + yp_z = m_x,$$
$$[m_z, m_x] = m_y, \quad [m_x, m_y] = m_z. \tag{27}$$

These results are all the same as in the classical theory. The sign in the results (23), (25), and (27) may easily be remembered from the rule that the + sign occurs when the three dynamical variables, consisting of the two in the P.B. on the left-hand side and the one forming the result on the right, are in the cyclic order (xyz) and the - sign occurs otherwise. Equations (27) may be put in the vector form

$$\mathbf{m} \times \mathbf{m} = i\hbar \mathbf{m}. \tag{28}$$

Now suppose we have several particles with angular momenta $\mathbf{m}_1$, $\mathbf{m}_2, \ldots$. Each of these angular momentum vectors will satisfy (28), thus

$$\mathbf{m}_r \times \mathbf{m}_r = i\hbar \mathbf{m}_r,$$

and any one of them will commute with any other, so that

$$\mathbf{m}_r \times \mathbf{m}_s + \mathbf{m}_s \times \mathbf{m}_r = 0 \quad (r \neq s).$$

Hence if $\mathbf{M} = \sum_r \mathbf{m}_r$ is the total angular momentum,

$$\mathbf{M} \times \mathbf{M} = \sum_{rs} \mathbf{m}_r \times \mathbf{m}_s = \sum_r \mathbf{m}_r \times \mathbf{m}_r + \sum_{r<s} (\mathbf{m}_r \times \mathbf{m}_s + \mathbf{m}_s \times \mathbf{m}_r)$$
$$= i\hbar \sum_r \mathbf{m}_r = i\hbar \mathbf{M}. \tag{29}$$

This result is of the same form as (28), so that the components of the total angular momentum $\mathbf{M}$ of any number of particles satisfy the same commutation relations as those of the angular momentum of a single particle.

35. ANGULAR MOMENTUM

Let A_x, A_y, A_z denote the three coordinates of any one of the particles, or else the three components of momentum of one of the particles. The A's will commute with the angular momenta of the other particles, and hence from (23), (24), (25), and (26)

$$[M_z, A_x] = A_y, \quad [M_z, A_y] = -A_x, \quad [M_z, A_z] = 0. \tag{30}$$

If B_x, B_y, B_z are a second set of three quantities denoting the coordinates or momentum components of one of the particles, they will satisfy similar relations to (30). We shall then have

$$[M_z, A_x B_x + A_y B_y + A_z B_z]$$
$$= [M_z, A_x]B_x + A_x[M_z, B_x] + [M_z, A_y]B_y + A_y[M_z, B_y]$$
$$= A_y B_x + A_x B_y - A_x B_y - A_y B_x = 0.$$

Thus the scalar product $A_x B_x + A_y B_y + A_z B_z$ commutes with M_z, and similarly with M_x and M_y. Introduce the vector product

$$\mathbf{A} \times \mathbf{B} = \mathbf{C}$$

or

$$A_y B_z - A_z B_y = C_x, \quad A_z B_x - A_x B_z = C_y, \quad A_x B_y - A_y B_x = C_z.$$

We have

$$[M_z, C_x] = -A_x B_z + A_z B_x = C_y$$

and similarly

$$[M_z, C_y] = -C_x, \quad [M_z, C_z] = 0.$$

These equations are again of the form (30), with $\mathbf{C}$ for $\mathbf{A}$. We can conclude from this work that equations of the form (30) hold for the three components of any vector that we can construct from our dynamical variables, and that any scalar commutes with $\mathbf{M}$.

We can introduce linear operators R referring to rotations about the origin in the same way in which we introduced the linear operators D in § 25 referring to displacements. Taking a rotation through an angle $\delta\phi$ about the z-axis and making $\delta\phi$ infinitesimal, we can obtain the limit operator corresponding to (64) of § 25,

$$\lim_{\delta\phi \to 0} \frac{R - 1}{\delta\phi},$$

which we shall call the *rotation operator* about the z-axis and denote by r_z. Like the displacement operators, r_z is a pure imaginary linear operator and is undetermined to the extent of an arbitrary additive pure imaginary number. Corresponding to (66) of § 25, the change in any dynamical variable v caused by a rotation through a small angle $\delta\phi$ about the z-axis is

$$\delta\phi(r_z v - v r_z), \tag{31}$$

to the first order in $\delta\phi$. Now the changes produced in the three components A_x, A_y, A_z of a vector by a (right-handed) rotation $\delta\phi$ about the z-axis applied to all measuring apparatus are $\delta\phi A_y$, $-\delta\phi A_x$, and 0 respectively, and any scalar quantity is unchanged by the rotation. Equating these changes to (31), we find that

$$r_z A_x - A_x r_z = A_y, \quad r_z A_y - A_y r_z = -A_x,$$
$$r_z A_z - A_z r_z = 0,$$

and r_z commutes with any scalar. Comparing these results with (30), we see that $i\hbar r_z$ satisfies the same commutation relations as M_z. Their difference, $M_z - i\hbar r_z$, commutes with all the dynamical variables and must therefore be a number. This number, which is necessarily real since M_z and $i\hbar r_z$ are real, may be made zero by a suitable choice of the arbitrary pure imaginary number that can be added to r_z. We then have the result

$$M_z = i\hbar r_z. \tag{32}$$

Similar equations hold for M_x and M_y. They are the analogues of (69) of § 25. Thus *the total angular momentum is connected with the rotation operators as the total momentum is connected with the displacement operators*. This conclusion is valid for any point as origin.

The above argument applies to the angular momentum arising from the motion of particles, defined by (22) for each particle. There is another kind of angular momentum occurring in atomic theory, *spin angular momentum*. The former kind of angular momentum will be called *orbital angular momentum*, to distinguish it. The spin angular momentum of a particle should be pictured as due to some internal motion of the particle, so that it is associated with different degrees of freedom from those describing the motion of the particle as a whole, and hence the dynamical variables that describe the spin must commute with x, y, z, p_x, p_y, and p_z. The spin does not correspond very closely to anything in classical mechanics, so the method of classical analogy is not suitable for studying it. However, we can build up a theory of the spin simply from the assumption that the components of the spin angular momentum are connected with the rotation operators in the same way as we had above for orbital angular momentum, i.e. equation (32) holds with M_z as the z-component

35. ANGULAR MOMENTUM

of the spin angular momentum of a particle and r_z as the rotation operator about the z-axis referring to states of spin of that particle. With this assumption, the commutation relations connecting the components of the spin angular momentum **M** with any vector **A** referring to the spin must be of the standard form (30), and hence, taking **A** to be the spin angular momentum itself, we have equation (29) holding also for the spin. We now have (29) holding quite generally, for any sum of spin and orbital angular momenta, and also (30) will hold generally, for **M** the total spin and orbital angular momentum and **A** any vector dynamical variable, and the connexion between angular momentum and rotation operators will be always valid.

As an immediate consequence of this connexion, we can deduce the *law of conservation of angular momentum*. For an isolated system, the Hamiltonian must be unchanged by any rotation about the origin, in other words it must be a scalar, so it must commute with the angular momentum about the origin. Thus the angular momentum is a constant of the motion. For this argument the origin may be any point.

As a second immediate consequence, we can deduce that *a state with zero total angular momentum is spherically symmetrical*. The state will correspond to a ket $|S\rangle$, say, satisfying

$$M_x|S\rangle = M_y|S\rangle = M_z|S\rangle = 0, \tag{33}$$

and hence

$$r_x|S\rangle = r_y|S\rangle = r_z|S\rangle = 0.$$

This shows that the ket $|S\rangle$ is unaltered by infinitesimal rotations, and it must therefore be unaltered by finite rotations, since the latter can be built up from infinitesimal ones. Thus the state is spherically symmetrical. The result may be understood in this way: if a state has zero total angular momentum, the dynamical system is equally likely to have any orientation, and hence spherical symmetry occurs. It is analogous to stating that if a state has zero total linear momentum, the system is equally likely to be anywhere in space.

The converse result is also true, *a spherically symmetrical state has zero total angular momentum*. This is obvious physically, since angular momentum is of the nature of a vector and, if it is not zero, its existence must destroy the spherical symmetry.

It should be noted that in (33) we have a ket $|S\rangle$ that is a simultaneous eigenket for non-commuting observables. This is usually not possible, but it is possible in the present special case, because the three equations (33) together with the commutation relations (29) do not lead to any inconsistency.

36. Properties of Angular Momentum

There are some general properties of angular momentum, deducible simply from the commutation relations between the three components. These properties must hold equally for spin and orbital angular momentum. Let m_x, m_y, m_z be the three components of an angular momentum, and introduce the quantity β defined by

$$\beta = m_x^2 + m_y^2 + m_z^2.$$

Since β is a scalar it must commute with m_x, m_y, and m_z. Let us suppose we have a dynamical system for which m_x, m_y, m_z are the only dynamical variables. Then β commutes with everything and must be a number. We can study this dynamical system on much the same lines as we used for the harmonic oscillator in § 34.

Put

$$m_x - im_y = \eta.$$

From the commutation relations (27) we get

$$\bar{\eta}\eta = (m_x + im_y)(m_x - im_y) = m_x^2 + m_y^2 - i(m_x m_y - m_y m_x)$$
$$= \beta - m_z^2 + \hbar m_z \tag{34}$$

and similarly

$$\eta\bar{\eta} = \beta - m_z^2 - \hbar m_z. \tag{35}$$

Thus

$$\bar{\eta}\eta - \eta\bar{\eta} = 2\hbar m_z. \tag{36}$$

Also

$$m_z \eta - \eta m_z = i\hbar m_y - \hbar m_x = -\hbar\eta. \tag{37}$$

We assume that the components of an angular momentum are observables and thus m_z has eigenvalues. Let m_z' be one of them, and $|m_z'\rangle$ an eigenket belonging to it. From (34)

$$\langle m_z'|\bar{\eta}\eta|m_z'\rangle = \langle m_z'|\beta - m_z^2 + \hbar m_z|m_z'\rangle = (\beta - m_z'^2 + \hbar m_z')\langle m_z'|m_z'\rangle.$$

The left-hand side here is the square of the length of the ket $\eta|m_z'\rangle$ and is thus greater than or equal to zero, the case of equality occurring if and only if $\eta|m_z'\rangle = 0$. Hence

$$\beta - m_z'^2 + \hbar m_z' \geq 0,$$

or
$$\beta + \frac{1}{4}\hbar^2 \geq \left(m'_z - \frac{1}{2}\hbar\right)^2. \tag{38}$$

Thus
$$\beta + \frac{1}{4}\hbar^2 \geq 0.$$

Defining the number k by
$$k + \frac{1}{2}\hbar = \left(\beta + \frac{1}{4}\hbar^2\right)^{1/2} = \left(m_x^2 + m_y^2 + m_z^2 + \frac{1}{4}\hbar^2\right)^{1/2}, \tag{39}$$

so that $k \geq -(1/2)\hbar$, the inequality (38) becomes
$$k + \frac{1}{2}\hbar \geq \left|m'_z - \frac{1}{2}\hbar\right|$$

or
$$k + \hbar \geq m'_z \geq -k. \tag{40}$$

An equality occurs if and only if $\eta|m'_z\rangle = 0$. Similarly from (35)
$$\langle m'_z|\eta\bar{\eta}|m'_z\rangle = (\beta - m'^2_z - \hbar m'_z)\langle m'_z|m'_z\rangle,$$

showing that
$$\beta - m'^2_z - \hbar m'_z \geq 0$$

or
$$k \geq m'_z \geq -k - \hbar,$$

with an equality occurring if and only if $\bar{\eta}|m'_z\rangle = 0$. This result combined with (40) shows that $k \geq 0$ and
$$k \geq m'_z \geq -k, \tag{41}$$

with $m'_z = k$ if $\bar{\eta}|m'_z\rangle = 0$ and $m'_z = -k$ if $\eta|m'_z\rangle = 0$.

From (37)
$$m_z\eta|m'_z\rangle = (\eta m_z - \hbar\eta)|m'_z\rangle = (m'_z - \hbar)\eta|m'_z\rangle.$$

Now if $m'_z \neq -k$, $\eta|m'_z\rangle$ is not zero and is then an eigenket of m_z belonging to the eigenvalue $m'_z - \hbar$. Similarly, if $m'_z - \hbar \neq -k$, $m'_z - 2\hbar$ is another eigenvalue of m_z, and so on. We get in this way a series of eigenvalues $m'_z, m'_z - \hbar, m'_z - 2\hbar, \ldots$, which must terminate from (41), and can terminate only with the value $-k$. Again, from the conjugate complex of equation (37)
$$m_z\bar{\eta}|m'_z\rangle = (\bar{\eta}m_z + \hbar\bar{\eta})|m'_z\rangle = (m'_z + \hbar)\bar{\eta}|m'_z\rangle,$$

showing that $m'_z + \hbar$ is another eigenvalue of m_z unless $\bar{\eta}|m'_z\rangle = 0$, in which case $m'_z = k$. Continuing in this way we get a series of eigenvalues $m'_z, m'_z + \hbar, m'_z + 2\hbar, \ldots$, which must terminate from (41), and can terminate only with the value k. We can conclude that $2k$ is an integral multiple of $\hbar$ and that the eigenvalues of m_z are

$$k, \quad k - \hbar, \quad k - 2\hbar, \quad \ldots, \quad -k + \hbar, \quad -k. \tag{42}$$

The eigenvalues of m_x and m_y are the same, from symmetry. These eigenvalues are all integral or half odd integral multiples of $\hbar$, according to whether $2k$ is an even or odd multiple of $\hbar$.

Let $|\max\rangle$ be an eigenket of m_z belonging to the maximum eigenvalue k, so that

$$\bar{\eta}|\max\rangle = 0, \tag{43}$$

and form the sequence of kets

$$|\max\rangle, \quad \eta|\max\rangle, \quad \eta^2|\max\rangle, \quad \ldots, \quad \eta^{2k/\hbar}|\max\rangle. \tag{44}$$

These kets are all eigenkets of m_z, belonging to the sequence of eigenvalues (42) respectively. The set of kets (44) is such that the operator η applied to any one of them gives a ket dependent on the set (η applied to the last gives zero), and from (36) and (43) one sees that $\bar{\eta}$ applied to any one of the set also gives a ket dependent on the set. All the dynamical variables for the system we are now dealing with are expressible in terms of η and $\bar{\eta}$, so the set of kets (44) is a complete set. There is just one of these kets for each eigenvalue (42) of m_z, so m_z by itself forms a complete commuting set of observables.

It is convenient to define the magnitude of the angular momentum vector **m** to be k, given by (39), rather than $\beta^{1/2}$, because the possible values for k are

$$0, \quad \frac{1}{2}\hbar, \quad \hbar, \quad \frac{3}{2}\hbar, \quad 2\hbar, \quad \ldots \tag{45}$$

extending to infinity, while the possible values for $\beta^{1/2}$ are a more complicated set of numbers.

For a dynamical system involving other dynamical variables besides m_x, m_y, and m_z, there may be variables that do not commute with β. Then β is no longer a number, but a general linear operator. This happens for any orbital angular momentum (22), as x, y, z, p_x, p_y, and p_z do not commute with β. We shall assume that β is always an observable, and k can then be defined by (39) with the positive square root function and is also an observable. We shall call k so defined the magnitude of the angular momentum vector **m** in the general case. The above analysis by which we obtained the eigenvalues of m_z is still valid if we replace $|m'_z\rangle$ by a simultaneous eigenket $|k'm'_z\rangle$ of the commuting observables k and m_z, and leads to the result that the possible eigenvalues for k are the numbers (45), and for each eigenvalue k' of k the eigenvalues of m_z are the numbers (42) with k' substituted for k. We have here an example of a phenomenon which we have not met with previously, namely that with two commuting observables, the eigenvalues of one depend on what eigenvalue we assign to the other. This phenomenon may be understood as the two observables being not altogether independent, but partially functions of one another. The number of independent simultaneous eigenkets of k and m_z belonging to the eigenvalues k' and m'_z must be independent of m'_z, since for each independent $|k'm'_z\rangle$ we can obtain an independent $|k'm''_z\rangle$, for any m''_z in the sequence (42), by multiplying $|k'm'_z\rangle$ by a suitable power of η or $\bar{\eta}$.

As an example let us consider a dynamical system with two angular momenta $\mathbf{m}_1$ and $\mathbf{m}_2$, which commute with one another. If there are no other dynamical variables, then all the dynamical variables commute with the magnitudes k_1 and k_2 of $\mathbf{m}_1$ and $\mathbf{m}_2$, so k_1 and k_2 are numbers. However, the magnitude K of the resultant angular momentum $\mathbf{M} = \mathbf{m}_1 + \mathbf{m}_2$ is not a number (it does not commute with the components of $\mathbf{m}_1$ and $\mathbf{m}_2$) and it is interesting to work out the eigenvalues of K. This can be done most simply by a method of counting independent kets. There is one independent simultaneous eigenket of m_{1z} and m_{2z} belonging to any eigenvalue m'_{1z} having one of the values $k_1, k_1 - \hbar, k_1 - 2\hbar, \ldots, -k_1$ and any eigenvalue m'_{2z} having one of the values $k_2, k_2 - \hbar, k_2 - 2\hbar, \ldots, -k_2$, and this ket is an eigenket of M_z belonging to the eigenvalue $M'_z = m'_{1z} + m'_{2z}$. The possible values of M'_z are thus $k_1 + k_2, k_1 + k_2 - \hbar, k_1 + k_2 - 2\hbar, \ldots, -k_1 - k_2$, and the number of times each of them occurs is given by the following scheme (if we assume for definiteness that $k_1 \geq k_2$),

$$\begin{array}{cccccc} k_1+k_2, & k_1+k_2-\hbar, & k_1+k_2-2\hbar, & \ldots, & k_1-k_2, & k_1-k_2-\hbar, \ldots \\ 1 & 2 & 3 & \ldots & 2k_2+1 & 2k_2+1 \quad \ldots \\ \ldots & -k_1+k_2, & -k_1+k_2-\hbar, & \ldots, & -k_1-k_2 & \\ \ldots & 2k_2+1 & 2k_2 & \ldots & 1 & \end{array} \quad (46)$$

Now each eigenvalue K' of K will be associated with the eigenvalues $K', K' - \hbar, K' - 2\hbar, \ldots, -K'$ for M_z, with the same number of independent simultaneous eigenkets of K and M_z for each of them. The total number of independent eigenkets of M_z belonging to any eigenvalue M'_z must be the same, whether we take them to be simultaneous eigenkets of m_{1z} and m_{2z} or simultaneous eigenkets of K and M_z, i.e. it is always given by the scheme (46). It follows that the eigenvalues for K are

$$k_1 + k_2, \quad k_1 + k_2 - \hbar, \quad k_1 + k_2 - 2\hbar, \quad \ldots, \quad k_1 - k_2, \qquad (47)$$

and that for each of these eigenvalues for K and an eigenvalue for M_z going with it there is just one independent simultaneous eigenket of K and M_z.

The effect of rotations on eigenkets of angular momentum variables should be noted. Take any eigenket $|M_z'\rangle$ of the z-component of total angular momentum for any dynamical system, and apply to it a small rotation through an angle $\delta\phi$ about the z-axis. It will change into

$$(1 + \delta\phi r_z)|M_z'\rangle = \left(1 - \frac{i\delta\phi M_z'}{\hbar}\right)|M_z'\rangle$$

with the help of (32). This equals

$$\left(1 - \frac{i\delta\phi M_z'}{\hbar}\right)|M_z'\rangle = e^{-i\delta\phi M_z'/\hbar}|M_z'\rangle$$

to the first order in $\delta\phi$. Thus $|M_z'\rangle$ gets multiplied by the numerical factor $e^{-i\delta\phi M_z'/\hbar}$. By applying a succession of these small rotations, we find that the application of a finite rotation through an angle ϕ about the z-axis causes $|M_z'\rangle$ to get multiplied by $e^{-i\phi M_z'/\hbar}$. Putting $\phi = 2\pi$, we find that an application of one revolution about the z-axis leaves $|M_z'\rangle$ unchanged if the eigenvalue M_z' is an integral multiple of $\hbar$ and causes $|M_z'\rangle$ to change sign if M_z' is half an odd integral multiple of $\hbar$. Now consider an eigenket $|K'\rangle$ of the magnitude K of the total angular momentum. If the eigenvalue K' is an integral multiple of $\hbar$, the possible eigenvalues of M_z are all integral multiples of $\hbar$ and the application of one revolution about the z-axis must leave $|K'\rangle$ unchanged. Conversely, if K' is half an odd integral multiple of $\hbar$, the possible eigenvalues of M_z are all half odd integral multiples of $\hbar$ and the revolution must change the sign of $|K'\rangle$. From symmetry, the application of a revolution about any other axis must have the same effect on $|K'\rangle$ as one about the z-axis. We thus get the general result, *the application of one revolution about any axis leaves a ket unchanged or changes its sign according to whether it belongs to eigenvalues of the magnitude of the total angular momentum which are integral or half odd integral multiples of* $\hbar$. A state, of course, is always unaffected by the revolution, since a state is unaffected by a change of sign of the ket corresponding to it.

For a dynamical system involving only orbital angular momenta, a ket must be unchanged by a revolution about an axis, since we can set up Schrödinger's representation, with the coordinates of all the particles diagonal, and the Schrödinger representative of a ket will get brought back to its original value by the revolution. It follows that *the eigenvalues of the magnitude of an orbital angular momentum are always integral multiples of* $\hbar$. The eigenvalues of a component of an orbital angular momentum are also always integral multiples of $\hbar$. For a spin angular momentum, Schrödinger's representation does not exist and both kinds of eigenvalue are possible.

37. The Spin of the Electron

Electrons, and also some of the other fundamental particles (protons, neutrons) have a spin whose magnitude is $(1/2)\hbar$. This is found from experimental evidence, and also there are theoretical reasons showing that this spin value is more elementary than any other, even spin zero (see Chapter XI). The study of this particular spin is therefore of special importance.

For dealing with an angular momentum **m** whose magnitude is $(1/2)\hbar$, it is convenient to put

$$\mathbf{m} = \frac{1}{2}\hbar\boldsymbol{\sigma}. \qquad (48)$$

The components of the vector $\boldsymbol{\sigma}$ then satisfy, from (27),

$$\sigma_y\sigma_z - \sigma_z\sigma_y = 2i\sigma_x,$$
$$\sigma_z\sigma_x - \sigma_x\sigma_z = 2i\sigma_y, \qquad (49)$$
$$\sigma_x\sigma_y - \sigma_y\sigma_x = 2i\sigma_z.$$

The eigenvalues of m_z are $(1/2)\hbar$ and $-(1/2)\hbar$, so the eigenvalues of σ_z are 1 and -1, and σ_z^2 has just the one eigenvalue 1. It follows that σ_z^2 must equal 1, and similarly for σ_x^2 and σ_y^2, i.e.

$$\sigma_x^2 = \sigma_y^2 = \sigma_z^2 = 1. \qquad (50)$$

We can get equations (49) and (50) into a simpler form by means of some straightforward non-commutative algebra. From (50)

$$\sigma_y^2\sigma_z - \sigma_z\sigma_y^2 = 0$$

or

$$\sigma_y(\sigma_y\sigma_z - \sigma_z\sigma_y) + (\sigma_y\sigma_z - \sigma_z\sigma_y)\sigma_y = 0$$

or
$$\sigma_y\sigma_x + \sigma_x\sigma_y = 0$$
with the help of the first of equations (49). This means $\sigma_x\sigma_y = -\sigma_y\sigma_x$. Two dynamical variables or linear operators like these which satisfy the commutative law of multiplication except for a minus sign will be said to *anticommute*. Thus σ_x anticommutes with σ_y. From symmetry each of the three dynamical variables $\sigma_x, \sigma_y, \sigma_z$ must anticommute with any other. Equations (49) may now be written

$$\sigma_y\sigma_z = i\sigma_x = -\sigma_z\sigma_y,$$
$$\sigma_z\sigma_x = i\sigma_y = -\sigma_x\sigma_z, \tag{51}$$
$$\sigma_x\sigma_y = i\sigma_z = -\sigma_y\sigma_x,$$

and also from (50)
$$\sigma_x\sigma_y\sigma_z = i. \tag{52}$$

Equations (50), (51), (52) are the fundamental equations satisfied by the spin variables σ describing a spin whose magnitude is $(1/2)\hbar$.

Let us set up a matrix representation for the σ's and let us take σ_z to be diagonal. If there are no other independent dynamical variables besides the m's or σ's in our dynamical system, then σ_z by itself forms a complete set of commuting observables, since the form of equations (50) and (51) is such that we cannot construct out of σ_x, σ_y, and σ_z any new dynamical variable that commutes with σ_z. The diagonal elements of the matrix representing σ_z being the eigenvalues 1 and -1 of σ_z, the matrix itself will be
$$\begin{pmatrix} 1 & 0 \\ 0 & -1 \end{pmatrix}.$$

Let σ_x be represented by
$$\begin{pmatrix} a_1 & a_2 \\ a_3 & a_4 \end{pmatrix}.$$
This matrix must be Hermitian, so that a_1 and a_4 must be real and a_2 and a_3 conjugate complex numbers. The equation $\sigma_z\sigma_x = -\sigma_x\sigma_z$ gives us
$$\begin{pmatrix} a_1 & a_2 \\ -a_3 & -a_4 \end{pmatrix} = -\begin{pmatrix} a_1 & -a_2 \\ a_3 & -a_4 \end{pmatrix},$$
so that $a_1 = a_4 = 0$. Hence σ_x is represented by a matrix of the form
$$\begin{pmatrix} 0 & a_2 \\ a_3 & 0 \end{pmatrix}.$$

The equation $\sigma_x^2 = 1$ now shows that $a_2 a_3 = 1$. Thus a_2 and a_3, being conjugate complex numbers, must be of the form $e^{i\alpha}$ and $e^{-i\alpha}$ respectively, where α is a real number, so that σ_x is represented by a matrix of the form

$$\begin{pmatrix} 0 & e^{i\alpha} \\ e^{-i\alpha} & 0 \end{pmatrix}.$$

Similarly it may be shown that σ_y is also represented by a matrix of this form. By suitably choosing the phase factors in the representation, which is not completely determined by the condition that σ_z shall be diagonal, we can arrange that σ_x shall be represented by the matrix

$$\begin{pmatrix} 0 & 1 \\ 1 & 0 \end{pmatrix}.$$

The representative of σ_y is then determined by the equation $\sigma_y = i\sigma_x\sigma_z$. We thus obtain finally the three matrices

$$\begin{pmatrix} 0 & 1 \\ 1 & 0 \end{pmatrix}, \quad \begin{pmatrix} 0 & -i \\ i & 0 \end{pmatrix}, \quad \begin{pmatrix} 1 & 0 \\ 0 & -1 \end{pmatrix}, \tag{53}$$

to represent σ_x, σ_y, and σ_z respectively, which matrices satisfy all the algebraic relations (49), (50), (51), (52). The component of the vector σ in an arbitrary direction specified by the direction cosines l, m, n, namely $l\sigma_x + m\sigma_y + n\sigma_z$, is represented by

$$\begin{pmatrix} n & l - im \\ l + im & -n \end{pmatrix}. \tag{54}$$

The representative of a ket vector will consist of just two numbers, corresponding to the two values $+1$ and -1 for σ_z'. These two numbers form a function of the variable σ_z' whose domain consists of only the two points $+1$ and -1. The state for which σ_z has the value unity will be represented by the function, $f_\alpha(\sigma_z')$ say, consisting of the pair of numbers $1, 0$ and that for which σ_z has the value -1 will be represented by the function, $f_\beta(\sigma_z')$ say, consisting of the pair $0, 1$. Any function of the variable σ_z', i.e. any pair of numbers, can be expressed as a linear combination of these two. Thus *any state can be obtained by superposition of the two states for which σ_z equals $+1$ and -1 respectively*. For example, the state for which the component of σ in the direction l, m, n, represented by (54), has the value $+1$ is represented by the pair of numbers a, b which satisfy

$$\begin{pmatrix} n & l - im \\ l + im & -n \end{pmatrix} \begin{pmatrix} a \\ b \end{pmatrix} = \begin{pmatrix} a \\ b \end{pmatrix}$$

or

$$na + (l - im)b = a,$$
$$(l + im)a - nb = b.$$

Thus

$$\frac{a}{b} = \frac{l - im}{1 - n} = \frac{1 + n}{l + im}.$$

This state can be regarded as a superposition of the two states for which σ_z equals $+1$ and -1, the relative weights in the superposition process being as

$$|a|^2 : |b|^2 = |l - im|^2 : (1 - n)^2 = 1 + n : 1 - n. \tag{55}$$

For the complete description of an electron [or other elementary particle with spin $(1/2)\hbar$] we require the spin dynamical variables σ, whose connexion with the spin angular momentum is given by (48), together with the Cartesian coordinates x, y, z and momenta p_x, p_y, p_z. The spin dynamical variables commute with these coordinates and momenta. Thus a complete set of commuting observables for a system consisting of a single electron will be x, y, z, σ_z, In a representation in which these are diagonal, the representative of any state will be a function of four variables x', y', z', σ'_z. Since σ'_z has a domain consisting of only two points, namely 1 and -1, this function of four variables is the same as two functions of three variables, namely the two functions

$$\langle x'y'z'|\rangle_+ = \langle x', y', z', +1|\rangle, \quad \langle x'y'z'|\rangle_- = \langle x', y', z', -1|\rangle. \tag{56}$$

Thus *the presence of the spin may be considered either as introducing a new variable into the representative of a state or as giving this representative two components.*

38. Motion in a Central Field of Force

An atom consists of a massive positively charged nucleus together with a number of electrons moving round, under the influence of the attractive force of the nucleus and their own mutual repulsions. An exact treatment of this dynamical system is a very difficult mathematical problem. One can, however, gain some insight into the main features of the system by making the rough approximation of regarding each electron as moving independently in a certain *central* field of force, namely that of the nucleus, assumed fixed, together with some kind of average of the forces due to the other electrons. Thus our present problem of the motion of a particle in a central field of force forms a corner-stone in the theory of the atom.

Let the Cartesian coordinates of the particle, referred to a system of axes with the centre of force as origin, be x, y, z and the corresponding components of momentum p_x, p_y, p_z. The Hamiltonian, with neglect of relativistic mechanics, will be of the form

$$H = \frac{1}{2m}(p_x^2 + p_y^2 + p_z^2) + V, \tag{57}$$

where V, the potential energy, is a function only of $(x^2 + y^2 + z^2)$. To develop the theory it is convenient to introduce polar dynamical variables. We introduce first the radius r, defined as the positive square root

$$r = (x^2 + y^2 + z^2)^{1/2}.$$

Its eigenvalues go from 0 to ∞. If we evaluate its P.B.s with p_x, p_y, and p_z, we obtain, with the help of formula (32) of § 22,

$$[r, p_x] = \frac{\partial r}{\partial x} = \frac{x}{r}, \quad [r, p_y] = \frac{y}{r}, \quad [r, p_z] = \frac{z}{r},$$

the same as in the classical theory. We introduce also the dynamical variable p_r defined by

$$p_r = \frac{xp_x + yp_y + zp_z}{r}. \tag{58}$$

Its P.B. with r is given by

$$r[r, p_r] = [r, rp_r] = [r, xp_x + yp_y + zp_z]$$
$$= x[r, p_x] + y[r, p_y] + z[r, p_z]$$
$$= x\frac{x}{r} + y\frac{y}{r} + z\frac{z}{r} = r.$$

Hence

$$[r, p_r] = 1$$

or

$$rp_r - p_r r = i\hbar.$$

The commutation relation between r and p_r is just the one for a canonical coordinate and momentum, namely equation (10) of § 22. This makes p_r like the momentum conjugate to the r coordinate, but it is not exactly equal to this momentum because it is not real, its conjugate complex being

$$\bar{p}_r = \frac{p_x x + p_y y + p_z z}{r}$$

$$= \frac{xp_x + yp_y + zp_z - 3i\hbar}{r}$$

$$= \frac{rp_r - 3i\hbar}{r} = p_r - \frac{2i\hbar}{r}. \tag{59}$$

Thus $p_r - i\hbar r^{-1}$ is real and is the true momentum conjugate to r.

The angular momentum **m** of the particle about the origin is given by (22) and its magnitude k is given by (39). Since r and p_r are scalars, they commute with **m**, and therefore also with k.

We can express the Hamiltonian in terms of r, p_r, and k. We have, if $\sum_{xyz}$ denotes a sum over cyclic permutations of the suffixes x, y, z,

$$k(k+\hbar)$$
$$= \sum_{xyz} m_z^2 = \sum_{xyz} (xp_y - yp_x)^2$$
$$= \sum_{xyz} (xp_y xp_y + yp_x yp_x - xp_y yp_x - yp_x xp_y)$$
$$= \sum_{xyz} (x^2 p_y^2 + y^2 p_x^2 - xp_x p_y y - yp_y p_x x + x^2 p_x^2 - xp_x p_x x - 2i\hbar xp_x)$$
$$= (x^2 + y^2 + z^2)(p_x^2 + p_y^2 + p_z^2)$$
$$\quad - (xp_x + yp_y + zp_z)(p_x x + p_y y + p_z z + 2i\hbar)$$
$$= r^2(p_x^2 + p_y^2 + p_z^2) - rp_r(\bar{p}_r r + 2i\hbar) = r^2(p_x^2 + p_y^2 + p_z^2) - rp_r^2 r.$$

from (59). Hence

$$H = \frac{1}{2m}\left[\frac{1}{r}p_r^2 r + \frac{k(k+\hbar)}{r^2}\right] + V. \tag{60}$$

This form for H is such that k commutes not only with H, as is necessary since k is a constant of the motion, but also with every dynamical variable occurring in H, namely r, p_r, and V, which is a function of r. In consequence, a simple treatment becomes possible, namely, we may consider an eigenstate of k belonging to an eigenvalue k' and then we can substitute k' for k in (60) and get a problem in one degree of freedom r.

Let us introduce Schrödinger's representation with x, y, z diagonal. Then p_x, p_y, p_z are equal to the operators $-i\hbar\, \partial/\partial x, -i\hbar\, \partial/\partial y, -i\hbar\, \partial/\partial z$ respectively. A state is represented by a wave function $\psi(xyzt)$ satisfying Schrödinger's wave equation (7) of § 27, which now reads, with H given by (57)

$$i\hbar \frac{\partial \psi}{\partial t} = \left[-\frac{\hbar^2}{2m}\left(\frac{\partial^2}{\partial x^2} + \frac{\partial^2}{\partial y^2} + \frac{\partial^2}{\partial z^2}\right) + V\right]\psi. \tag{61}$$

We may pass from the Cartesian coordinates x, y, z to the polar coordinates r, θ, ϕ by means of the equations

$$x = r \sin \theta \cos \phi,$$
$$y = r \sin \theta \sin \phi, \qquad (62)$$
$$z = r \cos \theta,$$

and may express the wave function in terms of the polar coordinates, so that it reads $\psi(r\theta\phi t)$. The equations (62) give the operator equation

$$\frac{\partial}{\partial r} = \frac{\partial x}{\partial r}\frac{\partial}{\partial x} + \frac{\partial y}{\partial r}\frac{\partial}{\partial y} + \frac{\partial z}{\partial r}\frac{\partial}{\partial z} = \frac{x}{r}\frac{\partial}{\partial x} + \frac{y}{r}\frac{\partial}{\partial y} + \frac{z}{r}\frac{\partial}{\partial z},$$

which shows, on being compared with (58), that $p_r = -i\hbar \partial/\partial r$. Thus Schrödinger's wave equation reads, with the form (60) for H,

$$i\hbar \frac{\partial \psi}{\partial t} = \left[\frac{\hbar^2}{2m}\left(-\frac{1}{r}\frac{\partial^2}{\partial r^2}r + \frac{k(k+\hbar)}{\hbar^2 r^2}\right) + V \right]\psi. \qquad (63)$$

Here k is a certain linear operator which, since it commutes with r and $\partial/\partial r$, can involve only θ, ϕ, $\partial/\partial\theta$, and $\partial/\partial\phi$. From the formula

$$k(k+\hbar) = m_x^2 + m_y^2 + m_z^2, \qquad (64)$$

which comes from (39), and from (62) one can work out the form of $k(k+\hbar)$ and one finds

$$\frac{k(k+\hbar)}{\hbar^2} = -\frac{1}{\sin\theta}\frac{\partial}{\partial\theta}\sin\theta\frac{\partial}{\partial\theta} - \frac{1}{\sin^2\theta}\frac{\partial^2}{\partial\phi^2}. \qquad (65)$$

This operator is well known in mathematical physics. Its eigenfunctions are called *spherical harmonics* and its eigenvalues are $n(n+1)$ where n is an integer. Thus the theory of spherical harmonics provides an alternative proof that the eigenvalues of k are integral multiples of $\hbar$.

For an eigenstate of k belonging to the eigenvalue $n\hbar$ (n a nonnegative integer) the wave function will be of the form

$$\psi = \frac{\chi(rt)\, S_n(\theta\phi)}{r}, \qquad (66)$$

where $S_n(\theta\phi)$ satisfies

$$k(k+\hbar) S_n(\theta\phi) = n(n+1)\hbar^2 S_n(\theta\phi), \qquad (67)$$

i.e. from (65) S_n is a spherical harmonic of order n. The factor r^{-1} is inserted in (66) for convenience. Substituting (66) into (63), we get as the equation for χ

$$i\hbar\frac{\partial\chi}{\partial t} = \left\{\frac{\hbar^2}{2m}\left[-\frac{\partial^2}{\partial r^2} + \frac{n(n+1)}{r^2}\right] + V\right\}\chi. \tag{68}$$

If the state is a stationary state belonging to the energy value H', χ will be of the form

$$\chi(rt) = \chi_0(r)e^{-iH't/\hbar}$$

and (68) will reduce to

$$H'\chi_0 = \left\{\frac{\hbar^2}{2m}\left[-\frac{d^2}{dr^2} + \frac{n(n+1)}{r^2}\right] + V\right\}\chi_0. \tag{69}$$

This equation may be used to determine the energy-levels H' of the system. For each solution χ_0 of (69), arising from a given n, there will be $2n+1$ independent states, because there are $2n+1$ independent solutions of (67) corresponding to the $2n+1$ different values that a component of the angular momentum, m_z say, can take on.

The probability of the particle being in an element of volume $dx\,dy\,dz$ is proportional to $|\psi|^2\,dx\,dy\,dz$. With ψ of the form (66) this becomes $r^{-2}|\chi|^2|S_n|^2\,dx\,dy\,dz$. The probability of the particle being in a spherical shell between r and $r+dr$ is then proportional to $|\chi|^2\,dr$. It now becomes clear that, in solving equation (68) or (69), we must impose a boundary condition on the function χ at $r = 0$, namely the function must be such that the integral to the origin $\int_0 |\chi|^2\,dr$ is convergent. If this integral were not convergent, the wave function would represent a state for which the chances are infinitely in favour of the particle being at the origin and such a state would not be physically admissible.

The boundary condition at $r = 0$ obtained by the above consideration of probabilities is, however, not sufficiently stringent. We get a more stringent condition by verifying that the wave function obtained by solving the wave equation in polar coordinates (63) really satisfies the wave equation in Cartesian coordinates (61). Let us take the case of $V = 0$, giving us the problem of the free particle. Applied to a stationary state with energy $H' = 0$, equation (61) gives

$$\nabla^2\psi = 0, \tag{70}$$

where ∇^2 is written for the Laplacian operator $\partial^2/\partial x^2 + \partial^2/\partial y^2 + \partial^2/\partial z^2$, and equation (63) gives

$$\left[\frac{1}{r}\frac{\partial^2}{\partial r^2}r - \frac{k(k+\hbar)}{\hbar^2 r^2}\right]\psi = 0. \tag{71}$$

A solution of (71) for $k = 0$ is $\psi = r^{-1}$. This does not satisfy (70), since, although $\nabla^2 r^{-1}$ vanishes for any finite value of r, its integral through a volume containing the origin is -4π (as may be verified by transforming this volume integral to a surface integral by means of Gauss's theorem), and hence

$$\nabla^2 r^{-1} = -4\pi\, \delta(x)\, \delta(y)\, \delta(z). \tag{72}$$

Thus not every solution of (71) gives a solution of (70), and more generally, not every solution of (63) is a solution of (61). We must impose on the solution of (63) the condition that it shall not tend to infinity as rapidly as r^{-1} when $r \to 0$ in order that, when substituted into (61), it shall not give a δ function on the right like the right-hand side of (72). Only when equation (63) is supplemented with this condition does it become equivalent to equation (61). We thus have the boundary condition $r\psi \to 0$ or $\chi \to 0$ as $r \to 0$.

There are also boundary conditions for the wave function at $r = \infty$. If we are interested only in 'closed' states, i.e. states for which the particle does not go off to infinity, we must restrict the integral to infinity $\int^\infty |\chi(r)|^2\, dr$ to be convergent. These closed states, however, are not the only ones that are physically permissible, as we can also have states in which the particle arrives from infinity, is scattered by the central field of force, and goes off to infinity again. For these states the wave function may remain finite as $r \to \infty$. Such states will be dealt with in Chapter VIII under the heading of collision problems. In any case the wave function must not tend to infinity as $r \to \infty$, or it will represent a state that has no physical meaning.

39. Energy-levels of the Hydrogen Atom

The above analysis may be applied to the problem of the hydrogen atom with neglect of relativistic mechanics and the spin of the electron. The potential energy V is now[1] $-e^2/r$, so that equation (69) becomes

$$\left[\frac{d^2}{dr^2} - \frac{n(n+1)}{r^2} + \frac{2me^2}{\hbar^2}\frac{1}{r}\right]\chi_0 = -\frac{2mH'}{\hbar^2}\chi_0. \tag{73}$$

A thorough investigation of this equation has been given by Schrödinger.[2] We shall here obtain its eigenvalues H' by an elementary argument.

---

[1] The e here, denoting minus the charge on an electron, is, of course, to be distinguished from the e denoting the base of exponentials.

[2] Schrödinger, *Ann, d. Physik*, **79** (1926), 361.

It is convenient to put
$$\chi_0 = f(r)e^{-r/a}, \tag{74}$$
introducing the new function $f(r)$, where a is one or other of the square roots
$$a = \pm\sqrt{-\frac{\hbar^2}{2mH'}}. \tag{75}$$
Equation (73) now becomes
$$\left[\frac{d^2}{dr^2} - \frac{2}{a}\frac{d}{dr} - \frac{n(n+1)}{r^2} + \frac{2me^2}{\hbar^2}\frac{1}{r}\right] f(r) = 0. \tag{76}$$
We look for a solution of this equation in the form of a power series
$$f(r) = \sum_s c_s r^s, \tag{77}$$
in which consecutive values for s differ by unity although these values themselves need not be integers. On substituting (77) in (76) we obtain
$$\sum_s c_s \left[s(s-1)r^{s-2} - \frac{2s}{a}r^{s-1} - n(n+1)r^{s-2} + \frac{2me^2}{\hbar^2}r^{s-1}\right] = 0,$$
which gives, on equating to zero the coefficient of r^{s-2}, the following relation between successive coefficients c_s,
$$c_s[s(s-1) - n(n+1)] = c_{s-1}\left[\frac{2(s-1)}{a} - \frac{2me^2}{\hbar^2}\right]. \tag{78}$$
We saw in the preceding section that only those eigenfunctions χ are allowed that tend to zero with r and hence, from (74), $f(r)$ must tend to zero with r. The series (77) must therefore terminate on the side of small s and the minimum value of s must be greater than zero. Now the only possible minimum values of s are those that make the coefficient of c_s in (78) vanish, i.e. $n+1$ and $-n$, and the second of these is negative or zero. Thus the minimum value of s must be $n+1$. Since n is always an integer, the values of s will all be integers. The series (77) will in general extend to infinity on the side of large s. For large values of s the ratio of successive terms is
$$\frac{c_s}{c_{s-1}}r = \frac{2r}{sa}$$
according to (78). Thus the series (77) will always converge, as the ratios of the higher terms to one another are the same as for the series
$$\sum_s \frac{1}{s!}\left(\frac{2r}{a}\right)^s, \tag{79}$$

which converges to $e^{2r/a}$.

We must now examine how our solution χ_0 behaves for large values of r. We must distinguish between the two cases of H' positive and H' negative. For H' negative, a given by (75) will be real. Suppose we take the positive value for a. Then as $r \to \infty$ the sum of the series (77) will tend to infinity according to the same law as the sum of the series (79), i.e. the law $e^{2r/a}$. Thus, from (74), χ_0 will tend to infinity according to the law $e^{r/a}$ and will not represent a physically possible state. There is therefore in general no permissible solution of (73) for negative values of H'. An exception arises, however, whenever the series (77) terminates on the side of large s, in which case the boundary conditions are all satisfied. The condition for this termination of the series is that the coefficient of c_{s-1} in (78) shall vanish for some value of the suffix $s-1$ not less than its minimum value $n+1$, which is the same as the condition that

$$\frac{s}{a} - \frac{me^2}{\hbar^2} = 0$$

for some integer s not less than $n+1$. With the help of (75) this condition becomes

$$H' = -\frac{me^4}{2s^2\hbar^2}, \tag{80}$$

and is thus a condition for the energy level H'. Since s may be any positive integer, the formula (80) gives a discrete set of negative energy-levels for the hydrogen atom. These are in agreement with experiment. For each of them (except the lowest one $s = 1$) there are several independent states, as there are various possible values for n, namely any positive or zero integer less than s. This multiplicity of states belonging to an energy-level is in addition to that mentioned in the preceding section arising from the various possible values for a component of angular momentum, which latter multiplicity occurs with any central field of force. The n multiplicity occurs only with an inverse square law of force and even then is removed when one takes relativistic mechanics into account, as will be found in Chapter XI. The solution χ_0 of (73) when H' satisfies (80) tends to zero exponentially as $r \to \infty$ and thus represents a closed state (corresponding to an elliptic orbit in Bohr's theory).

For any positive values of H', a given by (75) will be pure imaginary. The series (77), which is like the series (79) for large r, will now have a sum that remains finite as $r \to \infty$. Thus χ_0 given by (74) will now remain finite as $r \to \infty$ and will therefore be a permissible solution of (73), giving a wave function ψ that tends to zero according to the law

r^{-1} as $r \to \infty$. Hence in addition to the discrete set of negative energy-levels (80), all positive energy-levels are allowed. The states of positive energy are not closed, since for them the integral to infinity $\int^{\infty} |\chi_0|^2 \, dr$ does not converge. (These states correspond to the hyperbolic orbits of Bohr's theory.)

40. Selection Rules

If a dynamical system is set up in a certain stationary state, it will remain in that stationary state so long as it is not acted upon by outside forces. Any atomic system in practice, however, frequently gets acted upon by external electromagnetic fields, under whose influence it is liable to cease to be in one stationary state and to make a transition to another. The theory of such transitions will be developed in §§ 44 and 45. A result of this theory is that, to a high degree of accuracy, transitions between two states cannot occur under the influence of electromagnetic radiation if, in a Heisenberg representation with these two stationary states as two of the basic states, the matrix element, referring to these two states, of the representative of the total electric displacement **D** of the system vanishes. Now it happens for many atomic systems that the great majority of the matrix elements of **D** in a Heisenberg representation do vanish, and hence there are severe limitations on the possibilities for transitions. The rules that express these limitations are called *selection rules*.

The idea of selection rules can be refined by a more detailed application of the theory of §§ 44 and 45, according to which the matrix elements of the different Cartesian components of the vector **D** are associated with different states of polarization of the electromagnetic radiation. The nature of this association is just what one would get if one considered the matrix elements, or rather their real parts, as the amplitudes of harmonic oscillators which interact with the field of radiation according to classical electrodynamics.

There is a general method for obtaining all selection rules, as follows. Let us call the constants of the motion which are diagonal in the Heisenberg representation α's and let D be one of the Cartesian components of **D**. We must obtain an algebraic equation connecting D and the α's which does not involve any dynamical variables other than D and the α's and which is linear in D. Such an equation will be of the form

$$\sum_r f_r D g_r = 0, \tag{81}$$

where the f_r's and g_r's are functions of the α's only. If this equation is expressed in terms of representatives, it gives us

$$\sum_r f_r(\alpha')\langle\alpha'|D|\alpha''\rangle g_r(\alpha'') = 0,$$

or

$$\langle\alpha'|D|\alpha''\rangle \sum_r f_r(\alpha') g_r(\alpha'') = 0,$$

which shows that $\langle\alpha'|D|\alpha''\rangle = 0$ unless

$$\sum_r f_r(\alpha') g_r(\alpha'') = 0. \tag{82}$$

This last equation, giving the connexion which must exist between α' and α'' in order that $\langle\alpha'|D|\alpha''\rangle$ may not vanish, constitutes the selection rule, so far as the component D of **D** is concerned.

Our work on the harmonic oscillator in § 34 provides an example of a selection rule. Equation (8) is of the form (81) with $\bar{\eta}$ for D and H playing the part of the α's, and it shows that the matrix elements $\langle H'|\bar{\eta}|H''\rangle$ of $\bar{\eta}$ all vanish except those for which $H'' - H' = \hbar\omega$. The conjugate complex of this result is that the matrix elements $\langle H'|\eta|H''\rangle$ of η all vanish except those for which $H'' - H' = -\hbar\omega$. Since q is a numerical multiple of $\eta - \bar{\eta}$, its matrix elements $\langle H'|q|H''\rangle$ all vanish except those for which $H'' - H' = \pm\hbar\omega$. If the harmonic oscillator carries an electric charge, its electric displacement D will be proportional to q. The selection rule is then that only those transitions can take place in which the energy H changes by a single quantum $\hbar\omega$.

We shall now obtain the selection rules for m_z and k for an electron moving in a central field of force. The components of electric displacement are here proportional to the Cartesian coordinates x, y, z. Taking first m_z, we have that m_z commutes with z, or that

$$m_z z - z m_z = 0.$$

This is an equation of the required type (81), giving us the selection rule

$$m'_z - m''_z = 0$$

for the z-component of the displacement. Again, from equations (23) we have

$$[m_z, [m_z, x]] = [m_z, y] = -x$$

or

$$m_z^2 x - 2 m_z x m_z + x m_z^2 - \hbar^2 x = 0,$$

which is also of the type (81) and gives us the selection rule

$$m_z'^2 - 2m_z'm_z'' + m_z''^2 - \hbar^2 = 0$$

or

$$(m_z' - m_z'' - \hbar)(m_z' - m_z'' + \hbar) = 0$$

for the x-component of the displacement. The selection rule for the y-component is the same. Thus our selection rules for m_z are that *in transitions associated with radiation with a polarization corresponding to an electric dipole in the z-direction, m_z' cannot change, while in transitions associated with a polarization corresponding to an electric dipole in the x-direction or y-direction, m_z' must change by $\pm\hbar$.*

We can determine more accurately the state of polarization of the radiation associated with a transition in which m_z' changes by $\pm\hbar$, by considering the condition for the non-vanishing of matrix elements of $x + iy$ and $x - iy$. We have

$$[m_z, x + iy] = y - ix = -i(x + iy)$$

or

$$m_z(x + iy) - (x + iy)(m_z + \hbar) = 0,$$

which is again of the type (81). It gives

$$m_z' - m_z'' - \hbar = 0$$

as the condition that $\langle m_z'|x + iy|m_z''\rangle$ shall not vanish. Similarly,

$$m_z' - m_z'' + \hbar = 0$$

is the condition that $\langle m_z'|x - iy|m_z''\rangle$ shall not vanish. Hence

$$\langle m_z'|x - iy|m_z' - \hbar\rangle = 0$$

or

$$\langle m_z'|x|m_z' - \hbar\rangle = i\langle m_z'|y|m_z' - \hbar\rangle = (a + ib)e^{i\omega t}$$

say, a, b, and ω being real. The conjugate complex of this is

$$\langle m_z' - \hbar|x|m_z'\rangle = -i\langle m_z' - \hbar|y|m_z'\rangle = (a - ib)e^{-i\omega t}.$$

Thus the vector $(1/2)(\langle m_z'|\mathbf{D}|m_z' - \hbar\rangle + \langle m_z' - \hbar|\mathbf{D}|m_z'\rangle)$, which determines the state of polarization of the radiation associated with transitions for

40. SELECTION RULES

which $m''_z = m'_z - \hbar$, has the following three components

$$\frac{1}{2}(\langle m'_z|x|m'_z - \hbar\rangle + \langle m'_z - \hbar|x|m'_z\rangle)$$
$$= \frac{1}{2}[(a+ib)e^{i\omega t} + (a-ib)e^{-i\omega t}] = a\cos\omega t - b\sin\omega t,$$
$$\frac{1}{2}(\langle m'_z|y|m'_z - \hbar\rangle + \langle m'_z - \hbar|y|m'_z\rangle) \tag{83}$$
$$= \frac{1}{2}i[-(a+ib)e^{i\omega t} + (a-ib)e^{-i\omega t}] = a\sin\omega t + b\cos\omega t,$$
$$\frac{1}{2}(\langle m'_z|z|m'_z - \hbar\rangle + \langle m'_z - \hbar|z|m'_z\rangle) = 0.$$

From the form of these components we see that the associated radiation moving in the z-direction will be circularly polarized, that moving in any direction in the xy-plane will be linearly polarized in this plane, and that moving in intermediate directions will be elliptically polarized. The direction of circular polarization for radiation moving in the z-direction will depend on whether ω is positive or negative, and this will depend on which of the two states m'_z or $m''_z = m'_z - \hbar$ has the greater energy.

We shall now determine the selection rule for k. We have

$$[k(k+\hbar), z] = [m_x^2, z] + [m_y^2, z] = -ym_x - m_x y + xm_y + m_y x$$
$$= 2(m_y x - m_x y + i\hbar z) - 2(m_y x - ym_x)$$
$$= 2(xm_y - m_x y).$$

Similarly,
$$[k(k+\hbar), x] = 2(ym_z - m_y z)$$
and
$$[k(k+\hbar), y] = 2(m_x z - xm_z).$$

Hence

$$[k(k+\hbar), [k(k+\hbar), z]]$$
$$= 2[k(k+\hbar), m_y x - m_x y + i\hbar z]$$
$$= 2m_y[k(k+\hbar), x] - 2m_x[k(k+\hbar), y] + 2i\hbar[k(k+\hbar), z]$$
$$= 4m_y(ym_z - m_y z) - 4m_x(m_x z - xm_z) + 2[k(k+\hbar)z - zk(k+\hbar)]$$
$$= 4(m_x x + m_y y + m_z z)m_z - 4(m_x^2 + m_y^2 + m_z^2)z$$
$$\quad + 2[k(k+\hbar)z - zk(k+\hbar)].$$

From (22)
$$m_x x + m_y y + m_z z = 0 \tag{84}$$

and hence
$$[k(k+\hbar), [k(k+\hbar), z]] = -2[k(k+\hbar)z + zk(k+\hbar)],$$
which gives
$$k^2(k+\hbar)^2 z - 2k(k+\hbar)zk(k+\hbar) + zk^2(k+\hbar)^2$$
$$- 2\hbar^2[k(k+\hbar)z + zk(k+\hbar)] = 0. \tag{85}$$
Similar equations hold for x and y. These equations are of the required type (81), and give us the selection rule
$$k'^2(k'+\hbar)^2 - 2k'(k'+\hbar)k''(k''+\hbar) + k''^2(k''+\hbar)^2$$
$$- 2\hbar^2 k'(k'+\hbar) - 2\hbar^2 k''(k''+\hbar) = 0,$$
which reduces to
$$(k' + k'' + 2\hbar)(k' + k'')(k' - k'' + \hbar)(k' - k'' - \hbar) = 0.$$
A transition can take place between two states k' and k'' only if one of these four factors vanishes.

Now the first of the factors, $(k' + k'' + 2\hbar)$, can never vanish, since the eigenvalues of k are all positive or zero. The second, $(k' + k'')$, can vanish only if $k' = 0$ and $k'' = 0$. But transitions between two states with these values for k cannot occur on account of other selection rules, as may be seen from the following argument. If two states (labelled respectively with a single prime and a double prime) are such that $k' = 0$ and $k'' = 0$, then from (41) and the corresponding results for m_x and m_y, $m'_x = m'_y = m'_z = 0$ and $m''_x = m''_y = m''_z = 0$. The selection rule for m_z now shows that the matrix elements of x and y referring to the two states must vanish, as the value of m_z does not change during the transition, and the similar selection rule for m_x or m_y shows that the matrix element of z also vanishes. Thus transitions between the two states cannot occur. Our selection rule for k now reduces to
$$(k' - k'' + \hbar)(k' - k'' - \hbar) = 0,$$
showing that k *must change by* $\pm\hbar$. This selection rule may be written
$$k'^2 - 2k'k'' + k''^2 - \hbar^2 = 0,$$
and since this is the condition that a matrix element $\langle k'|z|k''\rangle$ shall not vanish, we get the equation
$$k^2 z - 2kzk + zk^2 - \hbar^2 z = 0$$
or
$$[k, [k, z]] = -z, \tag{86}$$

a result which could not easily be obtained in a more direct way.

As a final example we shall obtain the selection rule for the magnitude K of the total angular momentum $\mathbf{M}$ of a general atomic system. Let x, y, z be the coordinates of one of the electrons. We must obtain the condition that the (K', K'') matrix element of x, y, or z shall not vanish. This is evidently the same as the condition that the (K', K'') matrix element of λ_1, λ_2, or λ_3 shall not vanish, where λ_1, λ_2, and λ_3 are any three independent linear functions of x, y, and z with numerical coefficients, or more generally with any coefficients that commute with K and are thus represented by matrices which are diagonal with respect to K. Let

$$\lambda_0 = M_x x + M_y y + M_z z, \quad \lambda_x = M_y z - M_z y - i\hbar x,$$
$$\lambda_y = M_z x - M_x z - i\hbar y, \quad \lambda_z = M_x y - M_y x - i\hbar z.$$

We have

$$M_x \lambda_x + M_y \lambda_y + M_z \lambda_z = \sum_{xyz}(M_x M_y z - M_x M_z y - i\hbar M_x x)$$
$$= \sum_{xyz}(M_x M_y - M_y M_x - i\hbar M_z)z = 0 \quad (87)$$

from (29). Thus λ_x, λ_y, and λ_z are not linearly independent functions of x, y, and z. Any two of them, however, together with λ_0 are three linearly independent functions of x, y, and z and may be taken as the above $\lambda_1, \lambda_2, \lambda_3$, since the coefficients M_x, M_y, M_z all commute with K. Our problem thus reduces to finding the condition that the (K', K'') matrix elements of λ_0, λ_x, λ_y, and λ_z shall not vanish. The physical meanings of these λ's are that λ_0 is proportional to the component of the vector (x, y, z) in the direction of the vector $\mathbf{M}$, and $\lambda_x, \lambda_y, \lambda_z$ are proportional to the Cartesian components of the component of (x, y, z) perpendicular to $\mathbf{M}$.

Since λ_0 is a scalar it must commute with K. It follows that only the diagonal elements $\langle K' | \lambda_0 | K' \rangle$ of λ_0 can differ from zero, so the selection rule is that K cannot change so far as λ_0 is concerned. Applying (30) to the vector $\lambda_x, \lambda_y, \lambda_z$, we have

$$[M_z, \lambda_x] = \lambda_y, \quad [M_z, \lambda_y] = -\lambda_x, \quad [M_z, \lambda_z] = 0.$$

These relations between M_z and $\lambda_x, \lambda_y, \lambda_z$ are of exactly the same form as the relations (23), (24) between m_z and x, y, z, and also (87) is of the same form as (84). The dynamical variables $\lambda_x, \lambda_y, \lambda_z$ thus have the same properties relative to the angular momentum $\mathbf{M}$ as x, y, z have relative to

m. The deduction of the selection rule for k when the electric displacement is proportional to (x, y, z) can therefore be taken over and applied to the selection rule for K when the electric displacement is proportional to $(\lambda_x, \lambda_y, \lambda_z)$. We find in this way that, so far as $\lambda_x, \lambda_y, \lambda_z$ are concerned, the selection rule for K is that it must change by $\pm\hbar$.

Collecting results, we have as the selection rule for K that it must change by 0 or $\pm\hbar$. We have considered the electric displacement produced by only one of the electrons, but the same selection rule must hold for each electron and thus also for the total electric displacement.

41. The Zeeman Effect for the Hydrogen Atom

We shall now consider the system of a hydrogen atom in a uniform magnetic field. The Hamiltonian (57) with $V = -e^2/r$, which describes the hydrogen atom in no external field, gets modified by the magnetic field, the modification, according to classical mechanics, consisting in the replacement of the components of momentum, p_x, p_y, p_z, by $p_x + e/c \cdot A_x, p_y + e/c \cdot A_y, p_z + e/c \cdot A_z$, where A_x, A_y, A_z are the components of the vector potential describing the field. For a uniform field of magnitude $\mathcal{H}$ in the direction of the z-axis we may take $A_x = -(1/2)\mathcal{H}y, A_y = (1/2)\mathcal{H}x, A_z = 0$. The classical Hamiltonian will then be

$$H = \frac{1}{2m}\left[\left(p_x - \frac{1}{2}\frac{e}{c}\mathcal{H}y\right)^2 + \left(p_y + \frac{1}{2}\frac{e}{c}\mathcal{H}x\right)^2 + p_z^2\right] - \frac{e^2}{r}.$$

This classical Hamiltonian may be taken over into the quantum theory if we add on to it a term giving the effect of the spin of the electron. According to experimental evidence and according to the theory of Chapter XI, the electron has a magnetic moment $-e\hbar/2mc \cdot \boldsymbol{\sigma}$, where $\boldsymbol{\sigma}$ is the spin vector of § 37. The energy of this magnetic moment in the magnetic field will be $e\hbar\mathcal{H}/2mc \cdot \sigma_z$. Thus the total quantum Hamiltonian will be

$$H = \frac{1}{2m}\left[\left(p_x - \frac{1}{2}\frac{e}{c}\mathcal{H}y\right)^2 + \left(p_y + \frac{1}{2}\frac{e}{c}\mathcal{H}x\right)^2 + p_z^2\right] - \frac{e^2}{r} + \frac{e\hbar\mathcal{H}}{2mc}\sigma_z. \quad (88)$$

There ought strictly to be other terms in this Hamiltonian giving the interaction of the magnetic moment of the electron with the electric field of the nucleus of the atom, but this effect is small, of the same order of magnitude as the correction one gets by taking relativistic mechanics into account, and will be neglected here. It will be taken into account in the relativistic theory of the electron given in Chapter XI.

41. THE ZEEMAN EFFECT FOR THE HYDROGEN ATOM

If the magnetic field is not too large, we can neglect terms involving $\mathcal{H}^2$, so that the Hamiltonian (88) reduces to

$$H = \frac{1}{2m}(p_x^2 + p_y^2 + p_z^2) - \frac{e^2}{r} + \frac{e\mathcal{H}}{2mc}(xp_y - yp_x) + \frac{e\hbar\mathcal{H}}{2mc}\sigma_z$$

$$= \frac{1}{2m}(p_x^2 + p_y^2 + p_z^2) - \frac{e^2}{r} + \frac{e\mathcal{H}}{2mc}(m_z + \hbar\sigma_z). \quad (89)$$

The extra terms due to the magnetic field are now $e\mathcal{H}/2mc \cdot (m_z + \hbar\sigma_z)$. But these extra terms commute with the total Hamiltonian and are thus constants of the motion. This makes the problem very easy. The stationary states of the system, i.e. the eigenstates of the Hamiltonian (89), will be those eigenstates of the Hamiltonian for no field that are simultaneously eigenstates of the observables m_z and σ_z, or at least of the one observable $m_z + \hbar\sigma_z$, and the energy-levels of the system will be those for the system with no field, given by (80) if one considers only closed states, increased by an eigenvalue of $e\mathcal{H}/2mc \cdot (m_z + \hbar\sigma_z)$. Thus stationary states of the system with no field for which m_z has the numerical value m'_z, an integral multiple of $\hbar$, and for which also σ_z has the numerical value $\sigma'_z = \pm 1$, will still be stationary states when the field is applied. Their energy will be increased by an amount consisting of the sum of two parts, a part $e\mathcal{H}/2mc \cdot m'_z$ arising from the orbital motion, which part may be considered as due to an orbital magnetic moment $-em'_z/2mc$, and a part $e\mathcal{H}/2mc \cdot \hbar\sigma'_z$ arising from the spin. The ratio of the orbital magnetic moment to the orbital angular momentum m'_z is $-e/2mc$, which is half the ratio of the spin magnetic moment to the spin angular momentum. This fact is sometimes referred to as the magnetic anomaly of the spin.

Since the energy-levels now involve m_z, the selection rule for m_z obtained in the preceding section becomes capable of direct comparison with experiment. We take a Heisenberg representation in which, among other constants of the motion, m_z and σ_z are diagonal. The selection rule for m_z now requires m_z to change by $\hbar$, 0, or $-\hbar$, while σ_z, since it commutes with the electric displacement, will not change at all. Thus the energy difference between the two states taking part in the transition process will differ by an amount $e\hbar\mathcal{H}/2mc$, 0, or $-e\hbar\mathcal{H}/2mc$ from its value for no magnetic field. Hence, from Bohr's frequency condition, the frequency of the associated electromagnetic radiation will differ by $e\mathcal{H}/4\pi mc$, 0, or $-e\mathcal{H}/4\pi mc$ from that for no magnetic field. This means that each spectral line for no magnetic field gets split up by the field into three components. If one considers radiation moving in the z-direction, then from (83) the two outer components will be circularly polarized, while the central undisplaced one will be of zero intensity. These results are in agreement with experiment and also with the classical theory of the Zeeman effect.

CHAPTER VII

Perturbation Theory

42. General Remarks

In the preceding chapter exact treatments were given of some simple dynamical systems in the quantum theory. Most quantum problems, however, cannot be solved exactly with the present resources of mathematics, as they lead to equations whose solutions cannot be expressed in finite terms with the help of the ordinary functions of analysis. For such problems one can often use a perturbation method. This consists in splitting up the Hamiltonian into two parts, one of which must be simple and the other small. The first part may then be considered as the Hamiltonian of a simplified or unperturbed system, which can be dealt with exactly, and the addition of the second will then require small corrections, of the nature of a perturbation, in the solution for the unperturbed system. The requirement that the first part shall be simple requires in practice that it shall not involve the time explicitly. If the second part contains a small numerical factor ϵ, we can obtain the solution of our equations for the perturbed system in the form of a power series in ϵ, which, provided it converges, will give the answer to our problem with any desired accuracy. Even when the series does not converge, the first approximation obtained by means of it is usually fairly accurate.

There are two distinct methods in perturbation theory. In one of these the perturbation is considered as causing *a modification of the states of motion* of the unperturbed system. In the other we do not consider any modification to be made in the states of the unperturbed system, but we suppose that the perturbed system, instead of remaining permanently in *one* of these states, is continually changing from one to another, or *making transitions*, under the influence of the perturbation. Which method is to be used in any particular case depends on the nature of the problem to be solved. The first method is useful usually only when the perturbing energy (the correction in the Hamiltonian for the undisturbed system) does not involve the time explicitly, and is then applied to the stationary states. It can be used for calculating things that do not refer to any definite time, such as the energy-levels of the stationary states of the perturbed system, or, in the case of collision problems, the probability of scattering through a given angle. The second method must, on the other hand, be used for solving all problems involving a consideration of time, such as those about the transient phenomena that occur when the perturbation is suddenly applied, or more generally problems in which the perturbation varies with the time in any way (i.e. in which the perturbing energy involves the time explicitly). Again, this second method must be used in collision problems, even though the perturbing energy does not here involve the time explicitly, if one wishes to calculate absorption and emission probabilities, since these probabilities, unlike a scattering probability, cannot be defined without reference to a state of affairs that varies with the time.

One can summarize the distinctive features of the two methods by saying that, with the first method, one compares the stationary states of the perturbed system with those of the unperturbed system; with the second method one takes a stationary state of the unperturbed system and sees how it varies with time under the influence of the perturbation.

43. The Change in the Energy-levels Caused by a Perturbation

The first of the above-mentioned methods will now be applied to the calculation of the changes in the energy-levels of a system caused by a perturbation. We assume the perturbing energy, like the Hamiltonian for the unperturbed system, not to involve the time explicitly. Our problem has a meaning, of course, only provided the energy-levels of the unperturbed system are discrete and the differences between them are large compared with the changes in them caused by the perturbation. This circumstance results in the treatment of perturbation problems by the first method having some different features according to whether the energy-levels of the unperturbed system are discrete or continuous.

Let the Hamiltonian of the perturbed system be

$$H = E + V, \qquad (1)$$

E being the Hamiltonian of the unperturbed system and V the small perturbing energy. By hypothesis each eigenvalue H' of H lies very close to one and only one eigenvalue E' of E. We shall use the same number of primes to specify any eigenvalue of H and the eigenvalue of E to which it lies very close. Thus we shall have H'' differing from E'' by a small quantity of order V and differing from E' by a quantity that is not small unless $E' = E''$. We must now take care always to use different numbers

43. THE CHANGE IN THE ENERGY-LEVELS CAUSED BY A PERTURBATION

of primes to specify eigenvalues of H and E which we do not want to lie very close together.

To obtain the eigenvalues of H, we have to solve the equation

$$H|H'\rangle = H'|H'\rangle$$

or

$$(H' - E)|H'\rangle = V|H'\rangle. \qquad (2)$$

Let $|0\rangle$ be an eigenket of E belonging to the eigenvalue E' and suppose the $|H'\rangle$ and H' that satisfy (2) to differ from $|0\rangle$ and E' only by small quantities and to be expressed as

$$|H'\rangle = |0\rangle + |1\rangle + |2\rangle + \cdots,$$
$$H' = E' + a_1 + a_2 + \cdots, \qquad (3)$$

where $|1\rangle$ and a_1 are of the first order of smallness (i.e. the same order as V), $|2\rangle$ and a_2 are of the second order, and so on. Substituting these expressions in (2), we obtain

$$(E' - E + a_1 + a_2 + \cdots)(|0\rangle + |1\rangle + |2\rangle + \cdots) = V(|0\rangle + |1\rangle + |2\rangle + \cdots).$$

If we now separate the terms of zero order, of the first order, of the second order, and so on, we get the following set of equations,

$$(E' - E)|0\rangle = 0,$$
$$(E' - E)|1\rangle + a_1|0\rangle = V|0\rangle,$$
$$(E' - E)|2\rangle + a_1|1\rangle + a_2|0\rangle = V|1\rangle, \qquad (4)$$
$$\cdots\cdots\cdots\cdots\cdots\cdots\cdots\cdots\cdots$$

The first of these equations tells us, what we have already assumed, that $|0\rangle$ is an eigenket of E belonging to the eigenvalue E'. The others enable us to calculate the various corrections $|1\rangle, |2\rangle, \ldots, a_1, a_2, \ldots$.

For the further discussion of these equations it is convenient to introduce a representation in which E is diagonal, i.e. a Heisenberg representation for the unperturbed system, and to take E itself as one of the observables whose eigenvalues label the representatives. Let the others, in the event of others being necessary, as is the case when there is more than one eigenstate of E belonging to any eigenvalue, be called β's. A basic bra is then $\langle E''\beta''|$. Since $|0\rangle$ is an eigenket of E belonging to the eigenvalue E', we have

$$\langle E''\beta''|0\rangle = \delta_{E''E'} f(\beta''), \qquad (5)$$

where $f(\beta'')$ is some function of the variables β''. With the help of this result the second of equations (4), written in terms of representatives, becomes

$$(E' - E'')\langle E''\beta''|1\rangle + a_1 \delta_{E''E'} f(\beta'') = \sum_{\beta'} \langle E''\beta''|V|E'\beta'\rangle f(\beta'). \quad (6)$$

Putting $E'' = E'$ here, we get

$$a_1 f(\beta'') = \sum_{\beta'} \langle E'\beta''|V|E'\beta'\rangle f(\beta'). \quad (7)$$

Equation (7) is of the form of the standard equation in the theory of eigenvalues, so far as the variables β' are concerned. It shows that the various possible values for a_1 are the eigenvalues of the matrix $\langle E'\beta''|V|E'\beta'\rangle$. This matrix is a part of the representative of the perturbing energy in the Heisenberg representation for the unperturbed system, namely, the part consisting of those elements that refer to the same unperturbed energy-level E' for their row and column. Each of these values for a_1 gives, to the first order, an energy level of the perturbed system lying close to the energy-level E' of the unperturbed system.[1] There may thus be several energy-levels of the perturbed system lying close to the one energy-level E' of the unperturbed system, their number being anything not exceeding the number of independent states of the unperturbed system belonging to the energy-level E'. In this way the perturbation may cause a separation or partial separation of the energy-levels that coincide at E' for the unperturbed system.

Equation (7) also determines, to the zero order, the representatives $\langle E''\beta''|0\rangle$ of the stationary states of the perturbed system belonging to energy-levels lying close to E', any solution $f(\beta')$ of (7) substituted in (5) giving one such representative. Each of these stationary states of the perturbed system approximates to one of the stationary states of the unperturbed system, but the converse, that each stationary state of the unperturbed system approximates to one of the stationary states of the perturbed system, is not true, since the general stationary state of the unperturbed system belonging to the energy level E' is represented by the right-hand side of (5) with an arbitrary function $f(\beta'')$. The problem of

---

[1]To distinguish these energy-levels one from another we should require some more elaborate notation, since according to the present notation they must all be specified by the same number of primes, namely by the number of primes specifying the energy-level of the unperturbed system from which they arise. For our present purposes, however, this more elaborate notation is not required.

43. THE CHANGE IN THE ENERGY-LEVELS CAUSED BY A PERTURBATION

finding which stationary states of the unperturbed system approximate to stationary states of the perturbed system, i.e. the problem of finding the solutions $f(\beta')$ of (7), corresponds to the problem of 'secular perturbations' in classical mechanics. It should be noted that the above results are independent of the values of all those matrix elements of the perturbing energy which refer to two different energy-levels of the unperturbed system.

Let us see what the above results become in the specially simple case when there is only one stationary state of the unperturbed system belonging to each energy-level.[1] In this case E alone fixes the representation, no β's being required. The sum in (7) now reduces to a single term and we get

$$a_1 = \langle E'|V|E'\rangle. \tag{8}$$

There is only one energy-level of the perturbed system lying close to any energy-level of the unperturbed system and *the change in energy is equal, in the first order, to the corresponding diagonal element of the perturbing energy in the Heisenberg representation for the unperturbed system, or to the average value of the perturbing energy for the corresponding unperturbed state.* The latter formulation of the result is the same as in classical mechanics when the unperturbed system is multiply periodic.

We shall proceed to calculate the second-order correction a_2 in the energy-level for the case when the unperturbed system is non-degenerate. Equation (5) for this case reads

$$\langle E''|0\rangle = \delta_{E''E'},$$

with neglect of an unimportant numerical factor, and equation (6) reads

$$(E' - E'')\langle E''|1\rangle + a_1 \delta_{E''E'} = \langle E''|V|E'\rangle.$$

This gives us the value of $\langle E''|1\rangle$ when $E'' \neq E'$, namely

$$\langle E''|1\rangle = \frac{\langle E''|V|E'\rangle}{E' - E''}. \tag{9}$$

The third of equations (4), written in terms of representatives, becomes

$$(E' - E'')\langle E''|2\rangle + a_1\langle E''|1\rangle + a_2 \delta_{E''E'} = \sum_{E'''} \langle E''|V|E'''\rangle\langle E'''|1\rangle.$$

---

[1] A system with only one stationary state belonging to each energy-level is often called *non-degenerate* and one with two or more stationary states belonging to an energy-level is called *degenerate*, although these words are not very appropriate from the modern point of view.

Putting $E'' = E'$ here, we get
$$a_1\langle E'|1\rangle + a_2 = \sum_{E'''}\langle E'|V|E'''\rangle\langle E'''|1\rangle,$$
which reduces, with the help of (8), to
$$a_2 = \sum_{E''\neq E'}\langle E'|V|E''\rangle\langle E''|1\rangle.$$
Substituting for $\langle E''|1\rangle$ from (9), we obtain finally
$$a_2 = \sum_{E''\neq E'}\frac{\langle E'|V|E''\rangle\langle E''|V|E'\rangle}{E' - E''},$$
giving for the total energy change to the second order
$$a_1 + a_2 = \langle E'|V|E'\rangle + \sum_{E''\neq E'}\frac{\langle E'|V|E''\rangle\langle E''|V|E'\rangle}{E' - E''}. \tag{10}$$

The method may be developed for the calculation of the higher approximations if required. General recurrence formulas giving the nth order corrections in terms of those of lower order have been obtained by Born, Heisenberg, and Jordan.[1]

44. The Perturbation Considered as Causing Transitions

We shall now consider the second of the two perturbation methods mentioned in § 42. We suppose again that we have an unperturbed system governed by a Hamiltonian E which does not involve the time explicitly, and a perturbing energy V which can now be an arbitrary function of the time. The Hamiltonian for the perturbed system is again $H = E + V$. For the present method it does not make any essential difference whether the energy-levels of the unperturbed system, i.e. the eigenvalues of E, form a discrete or continuous set. We shall, however, take the discrete case, for definiteness. We shall again work with a Heisenberg representation for the unperturbed system, but as there will now be no advantage in taking E itself as one of the observables whose eigenvalues label the representatives, we shall suppose we have a general set of α's to label the representatives.

Let us suppose that at the initial time t_0 the system is in a state for which the α's certainly have the values α'. The ket corresponding to this state is the basic ket $|\alpha'\rangle$. If there were no perturbation, i.e. if the Hamiltonian were E, this state would be stationary. The perturbation

---
[1] *Z.f. Physik*, **35** (1925), 565.

44. THE PERTURBATION CONSIDERED AS CAUSING TRANSITIONS

causes the state to change. At time t the ket corresponding to the state in Schrödinger's picture will be $T|\alpha'\rangle$, according to equation (1) of § 27. The probability of the α's then having the values α'' is

$$P(\alpha'\alpha'') = |\langle\alpha''|T|\alpha'\rangle|^2. \tag{11}$$

For $\alpha'' \neq \alpha'$, $P(\alpha'\alpha'')$ is the probability of a transition taking place from state α' to state α'' during the time interval $t_0 \to t$, while $P(\alpha'\alpha')$ is the probability of no transition taking place at all. The sum of $P(\alpha'\alpha'')$ for all α'' is, of course, unity.

Let us now suppose that initially the system, instead of being certainly in the state α', is in one or other of various states α' with the probability $P_{\alpha'}$ for each. The Gibbs density corresponding to this distribution is, according to (68) of § 33

$$\rho = \sum_{\alpha'} |\alpha'\rangle P_{\alpha'} \langle\alpha'|. \tag{12}$$

At time t, each ket $|\alpha'\rangle$ will have changed to $T|\alpha'\rangle$ and each bra $\langle\alpha'|$ to $\langle\alpha'|\overline{T}$, so ρ will have changed to

$$\rho_t = \sum_{\alpha'} T|\alpha'\rangle P_{\alpha'} \langle\alpha'|\overline{T}. \tag{13}$$

The probability of the α's then having the values α'' will be, from (73) of § 33,

$$\langle\alpha''|\rho_t|\alpha''\rangle = \sum_{\alpha'} \langle\alpha''|T|\alpha'\rangle P_{\alpha'} \langle\alpha'|\overline{T}|\alpha''\rangle$$
$$= \sum_{\alpha'} P_{\alpha'} P(\alpha'\alpha'') \tag{14}$$

with the help of (11). This result expresses that the probability of the system being in the state α'' at time t is the sum of the probabilities of the system being initially in any state $\alpha' \neq \alpha''$, and making a transition from state α' to state α'' and the probability of its being initially in the state α'' and making no transition. Thus the various transition probabilities act independently of one another, according to the ordinary laws of probability.

The whole problem of calculating transitions thus reduces to the determination of the probability amplitudes $\langle\alpha''|T|\alpha'\rangle$. These can be worked out from the differential equation for T, equation (6) of § 27, or

$$i\hbar\frac{dT}{dt} = HT = (E + V)T. \tag{15}$$

The calculation can be simplified by working with
$$T^* = e^{iE(t-t_0)/\hbar} T. \tag{16}$$

We have
$$\begin{aligned} i\hbar \frac{dT^*}{dt} &= e^{iE(t-t_0)/\hbar}\left(-ET + i\hbar \frac{dT}{dt}\right) \\ &= e^{iE(t-t_0)/\hbar} VT = V^*T^*, \end{aligned} \tag{17}$$

where
$$V^* = e^{iE(t-t_0)/\hbar} V e^{-iE(t-t_0)/\hbar}, \tag{18}$$

i.e. V^* is the result of applying a certain unitary transformation to V. Equation (17) is of a more convenient form than (15), because (17) makes the change in T^* depend entirely on the perturbation V, and for $V = 0$ it would make T^* equal its initial value, namely unity. We have from (16)
$$\langle \alpha''|T^*|\alpha'\rangle = e^{iE''(t-t_0)/\hbar} \langle \alpha''|T|\alpha'\rangle,$$

so that
$$P(\alpha'\alpha'') = |\langle \alpha''|T^*|\alpha'\rangle|^2, \tag{19}$$

showing that T^* and T are equally good for determining transition probabilities.

Our work up to the present has been exact. We now assume V is a small quantity of the first order and express T^* in the form
$$T^* = 1 + T_1^* + T_2^* + \cdots, \tag{20}$$

where T_1^* is of the first order, T_2^* is of the second, and so on. Substituting (20) into (17) and equating terms of equal order, we get
$$\begin{aligned} i\hbar \frac{dT_1^*}{dt} &= V^*, \\ i\hbar \frac{dT_2^*}{dt} &= V^*T_1^*, \end{aligned} \tag{21}$$
$$\cdots\cdots\cdots\cdots\cdots$$

From the first of these equations we obtain
$$T_1^* = -i\hbar^{-1} \int_{t_0}^{t} V^*(t')\,dt', \tag{22}$$

from the second we obtain
$$T_2^* = -\hbar^{-2} \int_{t_0}^{t} V^*(t')\,dt' \int_{t_0}^{t'} V^*(t'')\,dt'', \tag{23}$$

44. THE PERTURBATION CONSIDERED AS CAUSING TRANSITIONS

and so on. For many practical problems it is sufficiently accurate to retain only the term T_1^*, which gives for the transition probability $P(\alpha'\alpha'')$ with $\alpha'' \neq \alpha'$

$$P(\alpha'\alpha'') = \hbar^{-2}\left|\langle\alpha''|\int_{t_0}^{t} V^*(t')\,dt'|\alpha'\rangle\right|^2$$
$$= \hbar^{-2}\left|\int_{t_0}^{t} \langle\alpha''|V^*(t')|\alpha'\rangle\,dt'\right|^2. \qquad (24)$$

We obtain in this way the transition probability to the second order of accuracy. The result depends only on the matrix element $\langle\alpha''|V^*(t')|\alpha'\rangle$ of $V^*(t')$ referring to the two states concerned, with t' going from t_0 to t. Since V^* is real, like V,

$$\langle\alpha''|V^*(t')|\alpha'\rangle = \overline{\langle\alpha'|V^*(t')|\alpha''\rangle}$$

and hence

$$P(\alpha'\alpha'') = P(\alpha''\alpha') \qquad (25)$$

to the second order of accuracy.

Sometimes one is interested in a transition $\alpha' \to \alpha''$ such that the matrix element $\langle\alpha''|V^*|\alpha'\rangle$ vanishes, or is small compared with other matrix elements of V^*. It is then necessary to work to a higher accuracy. If we retain only the terms T_1^* and T_2^*, we get, for $\alpha'' \neq \alpha'$,

$$P(\alpha'\alpha'') = \hbar^{-2}\Bigg|\int_{t_0}^{t} \langle\alpha''|V^*(t')|\alpha'\rangle\,dt'$$
$$- i\hbar^{-1}\sum_{\alpha'''\neq\alpha',\alpha''}\int_{t_0}^{t} \langle\alpha''|V^*(t')|\alpha'''\rangle\,dt' \int_{t_0}^{t'} \langle\alpha'''|V^*(t'')|\alpha'\rangle\,dt''\Bigg|^2. \qquad (26)$$

The terms $\alpha''' = \alpha'$ and $\alpha''' = \alpha''$ are omitted from the sum since they are small compared with other terms of the sum, on account of the smallness of $\langle\alpha''|V^*|\alpha'\rangle$. To interpret the result (26), we may suppose that the term

$$\int_{t_0}^{t} \langle\alpha''|V^*(t')|\alpha'\rangle\,dt' \qquad (27)$$

gives rise to a transition directly from state α' to state α'', while the term

$$- i\hbar^{-1}\int_{t_0}^{t} \langle\alpha''|V^*(t')|\alpha'''\rangle\,dt' \int_{t_0}^{t'} \langle\alpha'''|V^*(t'')|\alpha'\rangle\,dt'' \qquad (28)$$

gives rise to a transition from state α' to state α''', followed by a transition from state α''' to state α''. The state α''' is called an *intermediate state* in this interpretation. We must add the term (27) to the various terms (28) corresponding to different intermediate states and then take the square of the modulus of the sum, which means that there is interference between the different transition processes—the direct one and those involving intermediate states—and one cannot give a meaning to the probability for one of these processes by itself. For each of these processes, however, there is a probability amplitude. If one carries out the perturbation method to a higher degree of accuracy, one obtains a result which can be interpreted similarly, with the help of more complicated transition processes involving a succession of intermediate states.

45. Application to Radiation

In the preceding section a general theory of the perturbation of an atomic system was developed, in which the perturbing energy could vary with the time in an arbitrary way. A perturbation of this kind can be realized in practice by allowing incident electromagnetic radiation to fall on the system. Let us see what our result (24) reduces to in this case.

If we neglect the effects of the magnetic field of the incident radiation, and if we further assume that the wave-lengths of the harmonic components of this radiation are all large compared with the dimensions of the atomic system, then the perturbing energy is simply the scalar product

$$V = (\mathbf{D}, \mathcal{E}), \tag{29}$$

where $\mathbf{D}$ is the total electric displacement of the system and $\mathcal{E}$ is the electric force of the incident radiation. We suppose $\mathcal{E}$ to be a given function of the time. If we take for simplicity the case when the incident radiation is plane polarized with its electric vector in a certain direction and let D denote the Cartesian component of $\mathbf{D}$ in this direction, the expression (29) for V reduces to the ordinary product

$$V = D\mathcal{E},$$

where $\mathcal{E}$ is the magnitude of the vector $\mathcal{E}$. The matrix elements of V are

$$\langle \alpha''|V|\alpha'\rangle = \langle \alpha''|D|\alpha'\rangle \mathcal{E},$$

since $\mathcal{E}$ is a number. The matrix element $\langle \alpha''|D|\alpha'\rangle$ is independent of t. From (18)

$$\langle \alpha''|V^*(t)|\alpha'\rangle = \langle \alpha''|D|\alpha'\rangle e^{i(E''-E')(t-t_0)/\hbar} \mathcal{E}(t),$$

and hence the expression (24) for the transition probability becomes

$$P(\alpha'\alpha'') = \hbar^{-2}|\langle\alpha''|D|\alpha'\rangle|^2 \left|\int_{t_0}^{t} e^{i(E''-E')(t'-t_0)/\hbar} \mathcal{E}(t')\,dt'\right|^2. \quad (30)$$

If the incident radiation during the time interval t_0 to t is resolved into its Fourier components, the energy crossing unit area per unit frequency range about the frequency v will be, according to classical electrodynamics,

$$E_v = \frac{c}{2\pi}\left|\int_{t_0}^{t} e^{2\pi i v(t'-t_0)} \mathcal{E}(t')\,dt'\right|^2. \quad (31)$$

Comparing this with (30), we obtain

$$P(\alpha'\alpha'') = 2\pi c^{-1}\hbar^{-2}|\langle\alpha''|D|\alpha'\rangle|^2 E_v, \quad (32)$$

where

$$v = \frac{|E'' - E'|}{h}. \quad (33)$$

From this result we see in the first place that the transition probability depends only on that Fourier component of the incident radiation whose frequency v is connected with the change of energy by (33). This gives us *Bohr's frequency condition* and shows how the ideas of Bohr's atomic theory, which was the forerunner of quantum mechanics, can be fitted in with quantum mechanics.

The present elementary theory does not tell us anything about the energy of the field of radiation. It would be reasonable to assume, though, that the energy absorbed or liberated by the atomic system in the transition process comes from or goes into the component of the radiation with frequency v given by (33). This assumption will be justified by the more complete theory of radiation given in Chapter X. The result (32) is then to be interpreted as the probability of the system, if initially in the state of lower energy, absorbing radiation and being carried to the upper state, and if initially in the upper state, being *stimulated* by the incident radiation to emit and fall to the lower state. The present theory does not account for the experimental fact that the system, if in the upper state with no incident radiation, can emit spontaneously and fall to the lower state, but this also will be accounted for by the more complete theory of Chapter X.

The existence of the phenomenon of stimulated emission was inferred by Einstein,[1] long before the discovery of quantum mechanics,

---

[1] Einstein, *Phys. Zeits.* **18** (1917), 121.

from a consideration of statistical equilibrium between atoms and a field of black-body radiation satisfying Planck's law. Einstein showed that the transition probability for stimulated emission must equal that for absorption between the same pair of states, in agreement with the present quantum theory, and deduced also a relation connecting this transition probability with that for spontaneous emission, which relation is in agreement with the theory of Chapter X.

The matrix element $\langle \alpha''|D|\alpha'\rangle$ in (32) plays the part of the amplitude of one of the Fourier components of D in the classical theory of a multiply-periodic system interacting with radiation. In fact it was the idea of replacing classical Fourier components by matrix elements which led Heisenberg to the discovery of quantum mechanics in 1925. Heisenberg assumed that the formulas describing the interaction with radiation of a system in the quantum theory can be obtained from the classical formulas by substituting for the Fourier components of the total electric displacement of the system the corresponding matrix elements. According to this assumption applied to spontaneous emission, a system having an electric moment $\mathbf{D}$ will, when in the state α', spontaneously emit radiation of frequency $v = (E' - E'')/h$, where E'' is an energy-level, less than E', of some state α'', at the rate

$$\frac{4}{3}\frac{(2\pi v)^4}{c^3}|\langle \alpha''|\mathbf{D}|\alpha'\rangle|^2. \tag{34}$$

The distribution of this radiation over the different directions of emission and its state of polarization for each direction will be the same as that for a classical electric dipole of moment equal to the real part of $\langle \alpha''|\mathbf{D}|\alpha'\rangle$. To interpret this rate of emission of radiant energy as a transition probability, we must divide it by the quantum of energy of this frequency, namely hv, and call it the probability per unit time of this quantum being spontaneously emitted, with the atomic system simultaneously dropping to the state α'' of lower energy. These assumptions of Heisenberg are justified by the present radiation theory, supplemented by the spontaneous transition theory of Chapter X.

46. Transitions Caused by a Perturbation Independent of the Time

The perturbation method of § 44 is still valid when the perturbing energy V does not involve the time t explicitly. Since the total Hamiltonian H in this case does not involve t explicitly, we could now, if desired, deal with the system by the perturbation method of § 43 and find its stationary states. Whether this method would be convenient or not would depend on what we want to find out about the system. If what we have to calculate makes an explicit reference to the time, e.g. if we have to calculate the probability of the system being in a certain state at one time when we are given that it is in a certain state at another time, the method of § 44 would be the more convenient one.

Let us see what the result (24) for the transition probability becomes when V does not involve t explicitly and let us take $t_0 = 0$ to simplify the writing. The matrix element $\langle \alpha''|V|\alpha'\rangle$ is now independent of t, and from (18)

$$\langle \alpha''|V^*(t')|\alpha'\rangle = \langle \alpha''|V|\alpha'\rangle e^{i(E''-E')t'/\hbar} \qquad (35)$$

so

$$\int_0^t \langle \alpha''|V(t')|\alpha'\rangle\, dt' = \langle \alpha''|V|\alpha'\rangle \frac{e^{i(E''-E')t/\hbar} - 1}{i(E''-E')/\hbar},$$

provided $E'' \neq E'$. Thus the transition probability (24) becomes

$$P(\alpha'\alpha'') = |\langle \alpha''|V|\alpha'\rangle|^2 \frac{[e^{i(E''-E')t/\hbar} - 1][e^{-i(E''-E')t/\hbar} - 1]}{(E'' - E')^2}$$

$$= 2|\langle \alpha''|V|\alpha'\rangle|^2 \frac{1 - \cos[(E''-E')t/\hbar]}{(E''-E')^2}. \qquad (36)$$

If E'' differs appreciably from E' this transition probability is small and remains so for all values of t. This result is required by the law of the conservation of energy. The total energy H is constant and hence the proper-energy E (i.e. the energy with neglect of the part V due to the perturbation), being approximately equal to H, must be approximately constant. This means that if E initially has the numerical value E', at any later time there must be only a small probability of its having a numerical value differing considerably from E'.

On the other hand, when the initial state α' is such that there exists another state α'' having the same or very nearly the same proper-energy E, the probability of a transition to the final state α'' may be quite large. The case of physical interest now is that in which there is a continuous range of final states α'' having a continuous range of proper-energy levels E'' passing through the value E' of the proper-energy of the initial state. The initial state must not be one of the continuous range of final states, but may be either a separate discrete state or one of another continuous range of states. We shall now have, remembering the rules of § 18 for the interpretation of probability amplitudes with continuous ranges of states, that, with $P(\alpha'\alpha'')$ having the value (36), the probability of a transition to a final state within the small range α'' to $\alpha'' + d\alpha''$ will be $P(\alpha'\alpha'')\, d\alpha''$

if the initial state α' is discrete and will be proportional to this quantity if α' is one of a continuous range.

We may suppose that the α's describing the final state consist of E together with a number of other dynamical variables β, so that we have a representation like that of § 43 for the degenerate case. (The β's, however, need have no meaning for the initial state α'.) We shall suppose for definiteness that the β's have only discrete eigenvalues. The total probability of a transition to a final state α'' for which the β's have the values β'' and E has any value (there will be a strong probability of its having a value near the initial value E') will now be (or be proportional to)

$$\int P(\alpha'\alpha'') dE''$$
$$= 2 \int_{-\infty}^{\infty} |\langle E''\beta''|V|\alpha'\rangle|^2 \frac{1 - \cos[(E'' - E')t/\hbar]}{(E'' - E')^2} dE'' \quad (37)$$
$$= 2t\hbar^{-1} \int_{-\infty}^{\infty} |\langle E' + \frac{\hbar x}{t}, \beta''|V|\alpha'\rangle|^2 \frac{1 - \cos x}{x^2} dx$$

if one makes the substitution $(E'' - E')t/\hbar = x$. For large values of t this reduces to

$$2t\hbar^{-1}|\langle E'\beta''|V|\alpha'\rangle|^2 \int_{-\infty}^{\infty} \frac{1 - \cos x}{x^2} dx$$
$$= 2\pi t\hbar^{-1}|\langle E'\beta''|V|\alpha'\rangle|^2. \quad (38)$$

Thus the total probability up to time t of a transition to a final state for which the β's have the values β'' is proportional to t. There is therefore a definite *probability coefficient*, or probability per unit time, for the transition process under consideration, having the value

$$2\pi\hbar^{-1}|\langle E'\beta''|V|\alpha'\rangle|^2. \quad (39)$$

It is proportional to the square of the modulus of the matrix element, associated with this transition, of the perturbing energy.

If the matrix element $\langle E'\beta''|V|\alpha'\rangle$ is small compared with other matrix elements of V, we must work with the more accurate formula (26). We have from (35)

$$\int_0^t \langle \alpha''|V^*(t')|\alpha'''\rangle dt' \int_0^{t'} \langle \alpha'''|V^*(t'')|\alpha'\rangle dt''$$
$$= \langle \alpha''|V|\alpha'''\rangle\langle \alpha'''|V|\alpha'\rangle \int_0^t e^{i(E''-E''')t'/\hbar} dt' \int_0^{t'} e^{i(E'''-E')t'/\hbar} dt''$$

$$= \frac{\langle \alpha^* | V | \alpha''' \rangle \langle \alpha''' | V | \alpha' \rangle}{i(E''' - E')/\hbar} \int_0^t [e^{i(E''-E')t'/\hbar} - e^{i(E''-E''')t'/\hbar}] \, dt'.$$

For E'' close to E', only the first term in the integrand here gives rise to a transition probability of physical importance and the second term may be discarded. Using this result in (26) we get

$$P(\alpha'\alpha'') = 2 \left| \langle \alpha'' | V | \alpha' \rangle - \sum_{\alpha''' \neq \alpha', \alpha''} \frac{\langle \alpha'' | V | \alpha''' \rangle \langle \alpha''' | V | \alpha' \rangle}{E''' - E'} \right|^2$$

$$\frac{1 - \cos[(E'' - E')t/\hbar]}{(E'' - E')^2},$$

which replaces (36). Proceeding as before, we obtain for the transition probability per unit time to a final state for which the β's have the values β'' and E has a value close to its initial value E'

$$\frac{2\pi}{\hbar} \left| \langle E'\beta'' | V | \alpha' \rangle - \sum_{\alpha''' \neq \alpha', \alpha''} \frac{\langle E'\beta'' | V | \alpha''' \rangle \langle \alpha''' | V | \alpha' \rangle}{E''' - E'} \right|^2. \quad (40)$$

This formula shows how intermediate states, differing from the initial state and final state, play a role in the determination of a probability coefficient.

In order that the approximations used in deriving (39) and (40) may be valid, the time t must be not too small and not too large. It must be large compared with the periods of the atomic system in order that the approximate evaluation of the integral (37) leading to the result (38) may be valid, while it must not be excessively large or else the general formula (24) or (26) will break down. In fact one could make the probability (38) greater than unity by taking t large enough. The upper limit to t is fixed by the condition that the probability (24) or (26), or t times (39) or (40), must be small compared with unity. There is no difficulty in t satisfying both these conditions simultaneously provided the perturbing energy V is sufficiently small.

47. The Anomalous Zeeman Effect

One of the simplest examples of the perturbation method of § 43 is the calculation of the first-order change in the energy-levels of an atom caused by a uniform magnetic field. The problem of a hydrogen atom in a uniform magnetic field has already been dealt with in § 41 and was so simple that perturbation theory was unnecessary. The case of a general atom is not much more complicated when we make a few approximations such that we can set up a simple model for the atom.

We first of all consider the atom in the absence of the magnetic field and look for constants of the motion or quantities that are approximately constants of the motion. The total angular momentum of the atom, the vector **j** say, is certainly a constant of the motion. This angular momentum may be regarded as the sum of two parts, the total orbital angular momentum of all the electrons, **l** say, and the total spin angular momentum, **s** say. Thus we have **j** = **l** + **s**. Now the effect of the spin magnetic moments on the motion of the electrons is small compared with the effect of the Coulomb forces and may be neglected as a first approximation. With this approximation the spin angular momentum of each electron is a constant of the motion, there being no forces tending to change its orientation. Thus **s**, and hence also **l**, will be constants of the motion. The magnitudes, l, s, and j say, of **l**, **s**, and **j** will be given by

$$l + \frac{1}{2}\hbar = \left(l_x^2 + l_y^2 + l_z^2 + \frac{1}{4}\hbar^2\right)^{1/2},$$

$$s + \frac{1}{2}\hbar = \left(s_x^2 + s_y^2 + s_z^2 + \frac{1}{4}\hbar^2\right)^{1/2},$$

$$j + \frac{1}{2}\hbar = \left(j_x^2 + j_y^2 + j_z^2 + \frac{1}{4}\hbar^2\right)^{1/2},$$

corresponding to equation (39) of § 36. They commute with each other, and from (47) of § 36 we see that with given numerical values for l and s the possible numerical values for j are

$$l + s, \quad l + s - \hbar, \quad l + s - 2\hbar, \quad \ldots, \quad |l - s|.$$

Let us consider a stationary state for which l, s, and j have definite numerical values in agreement with the above scheme. The energy of this state will depend on l, but one might think that with neglect of the spin magnetic moments it would be independent of s, and also of the direction of the vector **s** relative to **l**, and thus of j. It will be found in Chapter IX, however, that the energy depends very much on the magnitude s of the vector **s**, although independent of its direction when one neglects the spin magnetic moments, on account of certain phenomena arising from the fact that the electrons are indistinguishable one from another. There are thus different energy-levels of the system for each different value of l and s. This means that l and s are functions of the energy, according to the general definition of a function given in § 11, since the l and s of a stationary state are fixed when the energy of that state is fixed.

47. THE ANOMALOUS ZEEMAN EFFECT

We can now take into account the effect of the spin magnetic moments, treating it as a small perturbation according to the method of § 43. The energy of the unperturbed system will still be approximately a constant of the motion and hence l and s, being functions of this energy, will still be approximately constants of the motion. The directions of the vectors **l** and **s**, however, not being functions of the unperturbed energy, need not now be approximately constants of the motion and may undergo large secular variations. Since the vector **j** is constant, the only possible variation of **l** and **s** is a precession about the vector **j**. We thus have an approximate model of the atom consisting of the two vectors **l** and **s** of constant lengths precessing about their sum **j**, which is a fixed vector. The energy is determined mainly by the magnitudes of **l** and **s** and depends only slightly on their relative directions, specified by j. Thus states with the same l and s and different j will have only slightly different energy-levels, forming what is called a *multiplet* term.

Let us now take this atomic model as our unperturbed system and suppose it to be subjected to a uniform magnetic field of magnitude $\mathcal{H}$ in the direction of the z-axis. The extra energy due to this magnetic field will consist of a term

$$\frac{e\mathcal{H}}{2mc}(m_z + \hbar\sigma_z), \tag{41}$$

like the last term in equation (89) of § 41, contributed by each electron, and will thus be altogether

$$\frac{e\mathcal{H}}{2mc}\sum(m_z + \hbar\sigma_z) = \frac{e\mathcal{H}}{2mc}(l_z + 2s_z) = \frac{e\mathcal{H}}{2mc}(j_z + s_z). \tag{42}$$

This is our perturbing energy V. We shall now use the method of § 43 to determine the changes in the energy-levels caused by this V. The method will be legitimate only provided the field is so weak that V is small compared with the energy differences within a multiplet.

Our unperturbed system is degenerate, on account of the direction of the vector **j** being undetermined. We must therefore take, from the representative of V in a Heisenberg representation for the unperturbed system, those matrix elements that refer to one particular energy-level for their row and column, and obtain the eigenvalues of the matrix thus formed. We can do this best by first splitting up V into two parts, one of which is a constant of the unperturbed motion, so that its representative contains only matrix elements referring to the same unperturbed energy-level for their row and column, while the representative of the other contains only matrix elements referring to two different unperturbed energy-levels for their row and column, so that this second part does not affect the first-order perturbation. The term involving j_z in (42) is a constant of the unperturbed motion and thus belongs entirely to the first part. For the term involving s_z we have

$$s_z(j_x^2+j_y^2+j_z^2) = j_z(s_xj_x+s_yj_y+s_zj_z)+(s_zj_x-j_zs_x)j_x+(s_zj_y-j_zs_y)j_y$$

or

$$s_z = \frac{j_z}{j(j+\hbar)}\frac{1}{2}[j(j+\hbar)-l(l+\hbar)+s(s+\hbar)]-(\gamma_yj_x-\gamma_xj_y)\frac{1}{j(j+\hbar)}, \quad (43)$$

where

$$\gamma_x = s_zj_y - j_zs_y = s_zl_y - l_zs_y = l_ys_z - l_zs_y,$$
$$\gamma_y = j_zs_x - s_zj_x = l_zs_x - s_zl_x = l_zs_x - l_xs_z, \quad (44)$$

The first term in this expression for s_z is a constant of the unperturbed motion and thus belongs entirely to the first part, while the second term, as we shall now see, belongs entirely to the second part.

Corresponding to (44) we can introduce

$$\gamma_z = l_xs_y - l_ys_x.$$

It can now easily be verified that

$$j_x\gamma_x + j_y\gamma_y + j_z\gamma_z = 0$$

and from (30) of § 35

$$[j_z,\gamma_x] = \gamma_y, \quad [j_z,\gamma_y] = -\gamma_x, \quad [j_z,\gamma_z] = 0.$$

These relations connecting j_x, j_y, j_z and $\gamma_x, \gamma_y, \gamma_z$ are of the same form as the relations connecting m_x, m_y, m_z and x, y, z in the calculation in § 40 of the selection rule for the matrix elements of z in a representation with k diagonal. From the result there obtained that all matrix elements of z vanish except those referring to two k values differing by $\pm\hbar$, we can infer that all matrix elements of γ_z, and similarly of γ_x and γ_y, in a representation with j diagonal, vanish except those referring to two j values differing by $\pm\hbar$. The coefficients of γ_x and γ_y in the second term on the right-hand side of (43) commute with j, so the representative of the whole of this term will contain only matrix elements referring to two j values differing by $\pm\hbar$, and thus referring to two different energy-levels of the unperturbed system.

Hence the perturbing energy V becomes, when we neglect that part of it whose representative consists of matrix elements referring to two different unperturbed energy-levels,

$$\frac{e\mathcal{H}}{2mc}j_z\left[1 + \frac{j(j+\hbar) - l(l+\hbar) + s(s+\hbar)}{2j(j+\hbar)}\right]. \tag{45}$$

The eigenvalues of this give the first-order changes in the energy-levels. We can make the representative of this expression diagonal by choosing our representation such that j_z is diagonal, and it then gives us directly the first-order changes in the energy-levels caused by the magnetic field. This expression is known as Landé's formula.

The result (45) holds only provided the perturbing energy V is small compared with the energy differences within a multiplet. For larger values of V a more complicated theory is required. For very strong fields, however, for which V is large compared with the energy differences within a multiplet, the theory is again very simple. We may now neglect altogether the energy of the spin magnetic moments for the atom with no external field, so that for our unperturbed system the vectors **l** and **s** themselves are constants of the motion, and not merely their magnitudes l and s. Our perturbing energy V, which is still $e\mathcal{H}/2mc \cdot (j_z + s_z)$, is now a constant of the motion for the unperturbed system, so that its eigenvalues give directly the changes in the energy-levels. These eigenvalues are integral or half-odd integral multiples of $e\mathcal{H}\hbar/2mc$ according to whether the number of electrons in the atom is even or odd.

CHAPTER VIII

Collision Problems

48. General Remarks

In this chapter we shall investigate problems connected with a particle which, coming from infinity, encounters or 'collides with' some atomic system and, after being scattered through a certain angle, goes off to infinity again. The atomic system which does the scattering we shall call, for brevity, the *scatterer*. We thus have a dynamical system composed of an incident particle and a scatterer interacting with each other, which we must deal with according to the laws of quantum mechanics, and for which we must, in particular, calculate the probability of scattering through any given angle. The scatterer is usually assumed to be of infinite mass and to be at rest throughout the scattering process. The problem was first solved by Born by a method substantially equivalent to that of the next section. We must take into account the possibility that the scatterer, considered as a system by itself, may have a number of different stationary states and that if it is initially in one of these states when the particle arrives from infinity, it may be left in a different one when the particle goes off to infinity again. The colliding particle may thus induce transitions in the scatterer.

The Hamiltonian for the whole system of scatterer plus particle will not involve the time explicitly, so that this whole system will have stationary states represented by periodic solutions of Schrödinger's wave equation. The meaning of these stationary states requires a little care to be properly understood. It is evident that for *any* state of motion of the system the particle will spend nearly all its time at infinity, so that the time average of the probability of the particle being in any finite volume will be zero. Now for a *stationary* state the probability of the particle being in a given finite volume, like any other result of observation, must be independent of the time, and hence this probability will equal its time average, which we have seen is zero. Thus only the relative probabilities of the particle being in different finite volumes will be physically significant, their absolute values being all zero. The total energy of the system has a continuous range of eigenvalues, since the initial energy of the particle can be anything. Thus a ket, $|s\rangle$ say, corresponding to a stationary state, being an eigenket of the total energy, must be of infinite length. We can see a physical reason for this, since if $|s\rangle$ were normalized and if Q denotes that observable—a certain function of the position of the particle—that is equal to unity if the particle is in a given finite volume and zero otherwise, then $\langle s|Q|s\rangle$ would be zero, meaning that the average value of Q, i.e. the probability of the particle being in the given volume, is zero. Such a ket $|s\rangle$ would not be a convenient one to work with. However, with $|s\rangle$ of infinite length, $\langle s|Q|s\rangle$ can be finite and would then give the relative probability of the particle being in the given volume.

In picturing a state of a system corresponding to a ket $|x\rangle$ which is not normalized, but for which $\langle x|x\rangle = n$ say, it may be convenient to suppose that we have n similar systems all occupying the same space but with no interaction between them, so that each one follows out its own motion independently of the others, as we had in the theory of the Gibbs ensemble in § 33. We can then interpret $\langle x|\alpha|x\rangle$, where α is any observable, directly as the total α for all the n systems. In applying these ideas to the above-mentioned $|s\rangle$ of infinite length, corresponding to a stationary state of the system of scatterer plus colliding particle, we should picture an infinite number of such systems with the scatterers all located at the same point and the particles distributed continuously throughout space. The number of particles in a given finite volume would be pictured as $\langle s|Q|s\rangle$, Q being the observable defined above, which has the value unity when the particle is in the given volume and zero otherwise. If the ket is represented by a Schrödinger wave function involving the Cartesian coordinates of the particle, then the square of the modulus of the wave function could be interpreted directly as the density of particles in the picture. One must remember, however, that *each of these particles has its own individual scatterer*. Different particles may belong to scatterers in different states. There will thus be one particle density for each state of the scatterer, namely the density of those particles belonging to scatterers in that state. This is taken account of by the wave function involving variables describing the state of the scatterer in addition to those describing the position of the particle.

For determining scattering coefficients we have to investigate *stationary states* of the whole system of scatterer plus particle. For instance, if we want to determine the probability of scattering in various directions when the scatterer is initially in a given stationary state and the incident particle has initially a given velocity in a given direction, we must investigate that stationary state of the whole system whose picture, according to the above method, contains at great distances from the point of location of the scatterers only particles moving with the given initial velocity and direction and belonging each to a scatterer in the given initial stationary state, together with particles moving *outward* from the point of location of the scatterers and belonging possibly to scatterers in various stationary states. This picture corresponds closely to the actual state of affairs in an experimental determination of scattering coefficients, with the difference that the picture really describes only one *actual* system of scatterer plus particle. The distribution of outward moving particles at infinity in the picture gives us immediately all the information about scattering coefficients that could be obtained by experiment. For practical calculations about the stationary state described by this picture one may use a perturbation method somewhat like that of § 43, taking as unperturbed system, for example, that for which there is no interaction between the scatterer and particle.

In dealing with collision problems, a further possibility to be taken into consideration is that the scatterer may perhaps be capable of absorbing and re-emitting the particle. This possibility arises when there exists one or more *states of absorption* of the whole system, a state of absorption being an *approximately* stationary state which is closed in the sense mentioned at the end of § 38 (i.e. for which the probability of the particle being at a greater distance than r from the scatterer tends to zero as $r \to \infty$). Since a state of absorption is only approximately stationary, its property of being closed will be only a transient one, and after a sufficient lapse of time there will be a finite probability of the particle being on its way to infinity. Physically this means there is a finite probability of spontaneous emission of the particle. The fact that we had to use the word 'approximately' in stating the conditions required for the phenomena of emission and absorption to be able to occur shows that these conditions are not expressible in exact mathematical language. One can give a meaning to these phenomena only with reference to a perturbation method. They occur when the unperturbed system (of scatterer plus particle) has stationary states that are closed. The introduction of the perturbation spoils the stationary property of these states and gives rise to spontaneous emission and its converse absorption.

For calculating absorption and emission probabilities it is necessary to deal with *non-stationary states* of the system, in contradistinction to the case for scattering coefficients, so that the perturbation method of § 44 must be used. Thus for calculating an emission coefficient we must consider the non-stationary states of absorption described above. Again, since an absorption is always followed by a re-emission, it cannot be distinguished from a scattering in any experiment involving a steady state of affairs, corresponding to a stationary state of the system. The distinction can be made only by reference to a non-steady state of affairs, e.g. by use of a stream of incident particles that has a sharp beginning, so that the scattered particles will appear immediately after the incident particles meet the scatterers, while those that have been absorbed and re-emitted will begin to appear only some time later. This stream of particles would be the picture of a certain ket of infinite length, which could be used for calculating the absorption coefficient.

49. The Scattering Coefficient

We shall now consider the calculation of scattering coefficients, taking first the case when there is no absorption and emission, which means that our unperturbed system has no closed stationary states. We may conveniently take this unperturbed system to be that for which there is no interaction between the scatterer and particle. Its Hamiltonian will thus be of the form

$$E = H_s + W, \qquad (1)$$

where H_s is that for the scatterer alone and W that for the particle alone, namely, with neglect of relativistic mechanics,

$$W = \frac{1}{2m}(p_x^2 + p_y^2 + p_z^2). \qquad (2)$$

The perturbing energy V, assumed small, will now be a function of the Cartesian coordinates of the particle x, y, z, and also, perhaps, of its momenta p_x, p_y, p_z, together with dynamical variables describing the scatterer.

Since we are now interested only in stationary states of the whole system, we use a perturbation method like that of § 43. Our unperturbed system now necessarily has a continuous range of energy-levels, since it contains a free particle, and this gives rise to certain modifications in the perturbation method. The question of the change in the energy-levels caused by the perturbation, which was the main question of § 43, no longer has a meaning, and the convention in § 43 of using the same number of primes to denote nearly equal eigenvalues of E and H now drops out. Again, the splitting of energy-levels which we had in § 43 when the unperturbed system is degenerate cannot now arise, since if the unperturbed system is degenerate the perturbed one, which must also have a continuous range of energy-levels, will also be degenerate to exactly the same extent.

We again use the general scheme of equations developed at the beginning of § 43, equations (1) to (4) there, but we now take our unperturbed stationary state forming the zero-order approximation to belong to an energy-level E' just equal to the energy-level H' of our perturbed stationary state. Thus the a's introduced in the second of equations (3) § 43 are now all zero and the second of equations (4) there now reads

$$(E' - E)|1\rangle = V|0\rangle. \tag{3}$$

Similarly, the third of equations (4) § 43 now reads

$$(E' - E)|2\rangle = V|1\rangle. \tag{4}$$

We shall proceed to solve equation (3) and to obtain the scattering coefficient to the first order. We shall need equation (4) in § 51.

Let α denote a complete set of commuting observables describing the scatterer, which are constants of the motion when the scatterer is alone and may thus be used for labelling the stationary states of the scatterer. This requires that H_s shall commute with the α's and be a function of them. We can now take a representation of the whole system in which the α's and x, y, z, the coordinates of the particle, are diagonal. This will make H_s diagonal. Let $|0\rangle$ be represented by $\langle \mathbf{x}\alpha'|0\rangle$ and $|1\rangle$ by $\langle \mathbf{x}\alpha'|1\rangle$, the single variable $\mathbf{x}$ being written to denote x, y, z and the prime being omitted from $\mathbf{x}$ for brevity. Also the single differential d^3x will be written to denote the product $dx\,dy\,dz$. Equation (3), written in terms of representatives, becomes, with the help of (1) and (2),

$$\left[E' - H_s(\alpha') + \frac{\hbar^2}{2m}\nabla^2 \right] \langle \mathbf{x}\alpha'|1\rangle = \sum_{\alpha''} \int \langle \mathbf{x}\alpha'|V|\mathbf{x}''\alpha''\rangle \, d^3x'' \, \langle \mathbf{x}''\alpha''|0\rangle. \tag{5}$$

Suppose that the incident particle has the momentum $\mathbf{p}^0$ and that the initial stationary state of the scatterer is α^0. The stationary state of our unperturbed system is now the one for which $\mathbf{p} = \mathbf{p}^0$ and $\alpha = \alpha^0$, and hence its representative is

$$\langle \mathbf{x}\alpha'|0\rangle = \delta_{\alpha'\alpha^0} e^{i(\mathbf{p}^0, \mathbf{x})/\hbar}. \tag{6}$$

This makes equation (5) reduce to

$$\left[E' - H_s(\alpha') + \frac{\hbar^2}{2m}\nabla^2 \right] \langle \mathbf{x}\alpha'|1\rangle = \int \langle \mathbf{x}\alpha'|V|\mathbf{x}^0\alpha^0\rangle \, d^3x^0 \, e^{i(\mathbf{p}^0, \mathbf{x}^0)/\hbar}$$

or
$$(k^2 + \nabla^2)\langle \mathbf{x}\alpha'|1\rangle = F, \tag{7}$$
where
$$k^2 = 2m\hbar^{-2}[E' - H_s(\alpha')] \tag{8}$$
and
$$F = 2m\hbar^{-2}\int \langle \mathbf{x}\alpha'|V|\mathbf{x}^0\alpha^0\rangle\, d^3x^0\, e^{i(\mathbf{p}^0,\mathbf{x}^0)/\hbar}, \tag{9}$$
a definite function of x, y, z, and α'. We must also have
$$E' = H_s(\alpha^0) + \frac{\mathbf{p}^{02}}{2m}. \tag{10}$$

Our problem now is to obtain a solution $\langle \mathbf{x}\alpha'|1\rangle$ of (7) which, for values of x, y, z denoting points far from the scatterer, represents only outward moving particles. The square of its modulus, $|\langle \mathbf{x}\alpha'|1\rangle|^2$, will then give the density of scattered particles belonging to scatterers in the state α' when the density of the incident particles is $|\langle \mathbf{x}\alpha^0|0\rangle|^2$, which is unity. If we transform to polar coordinates r, θ, ϕ, equation (7) becomes

$$\left[k^2 + \frac{\partial^2}{\partial r^2} + \frac{2}{r}\frac{\partial}{\partial r} + \frac{1}{r^2\sin\theta}\frac{\partial}{\partial \theta}\sin\theta\frac{\partial}{\partial \theta} + \frac{1}{r^2\sin^2\theta}\frac{\partial^2}{\partial \phi^2}\right]\langle r\theta\phi\alpha'|1\rangle = F. \tag{11}$$

Now F must tend to zero as $r \to \infty$, on account of the physical requirement that the interaction energy between the scatterer and particle must tend to zero as the distance between them tends to infinity. If we neglect F in (11) altogether, an approximate solution for large r is

$$\langle r\theta\phi\alpha'|1\rangle = u(\theta\phi\alpha')r^{-1}e^{ikr}, \tag{12}$$

where u is an arbitrary function of θ, ϕ, and α', since this expression substituted in the left-hand side of (11) gives a result of order r^{-3}. When we do not neglect F, the solution of (11) will still be of the form (12) for large r, provided F tends to zero sufficiently rapidly as $r \to \infty$, but the function u will now be definite and determined by the solution for smaller values of r.

For values α' of the α's such that k^2, defined by (8), is positive, the k in (12) must be chosen to be the positive square root of k^2, in order that (12) may represent only outward moving particles, i.e. particles for which the radial component of momentum, which from § 38 equals $p_r - i\hbar r^{-1}$ or $-i\hbar(\partial/\partial r + r^{-1})$, has a positive value. We now have that the density of scattered particles belonging to scatterers in state α', equal to the square of the modulus of (12), falls off with increasing r according to the inverse square law, as is physically necessary, and their angular distribution is given by $|u(\theta\phi\alpha')|^2$. Further, the magnitude, P' say, of the momentum of these scattered particles must equal $k\hbar$, the momentum being radial for large r, so that their energy is equal to

$$\frac{P'^2}{2m} = \frac{k^2\hbar^2}{2m} = E' - H_s(\alpha') = H_s(\alpha^0) - H_s(\alpha') + \frac{\mathbf{p}^{02}}{2m},$$

with the help of (8) and (10). This is just the energy of an incident particle, namely $\mathbf{p}^{02}/2m$, reduced by the increase in energy of the scatterer, namely $H_s(\alpha') - H_s(\alpha^0)$, in agreement with the law of conservation of energy. For values α' of the α's such that k^2 is negative there are no scattered particles, the total initial energy being insufficient for the scatterer to be left in the state α'.

We must now evaluate $u(\theta\phi\alpha')$ for a set of values α' for the α's such that k^2 is positive, and obtain the angular distribution of the scattered particles belonging to scatterers in state α'. It is sufficient to evaluate u for the direction $\theta = 0$ of the pole of the polar coordinates, since this direction is arbitrary. We make use of Green's theorem, which states that for any two functions of position A and B the volume integral $\int (A\nabla^2 B - B\nabla^2 A)\, d^3x$ taken over any volume equals the surface integral $\int (A\partial B/\partial n - B\partial A/\partial n)\, dS$ taken over the boundary of the volume, $\partial/\partial n$ denoting differentiation along the normal to the surface. We take

$$A = e^{-ikr\cos\theta}, \quad B = \langle r\theta\phi\alpha'|1\rangle$$

and apply the theorem to a large sphere with the origin as centre. The volume integrand is thus

$$e^{-ikr\cos\theta}\nabla^2\langle r\theta\phi\alpha'|1\rangle - \langle r\theta\phi\alpha'|1\rangle\nabla^2 e^{-ikr\cos\theta}$$
$$= e^{-ikr\cos\theta}(\nabla^2 + k^2)\langle r\theta\phi\alpha'|1\rangle = e^{-ikr\cos\theta}F$$

from (7) or (11), while the surface integrand is, with the help of (12),

$$e^{-ikr\cos\theta}\frac{\partial}{\partial r}\langle r\theta\phi\alpha'|1\rangle - \langle r\theta\phi\alpha'|1\rangle\frac{\partial}{\partial r}e^{-ikr\cos\theta}$$
$$= e^{-ikr\cos\theta}u\left(-\frac{1}{r^2} + \frac{ik}{r}\right)e^{ikr} + i\frac{u}{r}e^{ikr}k\cos\theta e^{-ikr\cos\theta}$$
$$= ikur^{-1}(1+\cos\theta)e^{ikr(1-\cos\theta)}$$

with neglect of r^{-2}. Hence we get

$$\int e^{-ikr\cos\theta} F\, d^3x$$

$$= \int_0^{2\pi} d\phi \int_0^\pi r^2 \sin\theta \, d\theta \, [ikur^{-1}(1+\cos\theta)e^{ikr(1-\cos\theta)}].$$

the volume integral on the left being taken over the whole of space. The right-hand side becomes, on being integrated by parts with respect to θ,

$$\int_0^{2\pi} d\phi \left\{ [u(1+\cos\theta)e^{ikr(1-\cos\theta)}]_{\theta=0}^{\theta=\pi} \right.$$
$$\left. - \int_0^\pi e^{ikr(1-\cos\theta)} \frac{\partial}{\partial\theta} [u(1+\cos\theta)] \, d\theta \right\}.$$

The second term in the { } brackets is of the order of magnitude of r^{-1}, as would be revealed by further partial integrations, and may therefore be neglected. We are thus left with

$$\int e^{-ikr\cos\theta} F \, d^3x = -2 \int_0^{2\pi} d\phi \, u(0\phi\alpha') = -4\pi \, u(0\phi\alpha'),$$

giving the value of $u(\theta\phi\alpha')$ for the direction $\theta = 0$.

This result may be written

$$u(0\phi\alpha') = -(4\pi)^{-1} \int e^{-iP'r\cos\theta/\hbar} F \, d^3x, \tag{13}$$

since $P' = k\hbar$. If the vector $\mathbf{p}'$ denotes the momentum of the scattered electrons coming off in a certain direction (and is thus of magnitude P'), the value of u for this direction will be

$$u(\theta'\phi'\alpha') = -(4\pi)^{-1} \int e^{-i(\mathbf{p}',\mathbf{x})/\hbar} F \, d^3x,$$

as follows from (13) if one takes this direction to be the pole of the polar coordinates. This becomes, with the help of (9),

$$u(\theta'\phi'\alpha')$$
$$= -(2\pi)^{-1} m\hbar^{-2} \iint e^{-i(\mathbf{p}',\mathbf{x}/\hbar)} d^3x \, \langle \mathbf{x}\alpha' | V | \mathbf{x}^0 \alpha^0 \rangle \, d^3x^0 \, e^{i(\mathbf{p}^0,\mathbf{x}^0)/\hbar}$$
$$= -2\pi m\hbar \langle \mathbf{p}'\alpha' | V | \mathbf{p}^0 \alpha^0 \rangle, \tag{14}$$

when one makes a transformation from the coordinates $\mathbf{x}$ to the momenta $\mathbf{p}$ of the particle, using the transformation function (54) of § 23. The single letter $\mathbf{p}$ is here used as a label for the three components of momentum.

The density of scattered particles belonging to scatterers in state α' is now given by $|u(\theta'\phi'\alpha')|^2/r^2$. Since their velocity is P'/m, the rate at which these particles appear per unit solid angle about the direction of the vector $\mathbf{p}'$ will be $P'/m \cdot |u(\theta'\phi'\alpha')|^2$. The density of the incident particles is, as we have seen, unity, so that the number of incident particles crossing unit area per unit time is equal to their velocity P^0/m, where P^0 is the magnitude of $\mathbf{p}^0$. Hence the effective area that must be hit by an incident particle in order to be scattered in a unit solid angle about the direction $\mathbf{p}'$ and then belong to a scatterer in state α' will be

$$\frac{P'}{P^0}|u(\theta'\phi'\alpha')|^2 = 4\pi^2 m^2 h^2 \frac{P'}{P^0} |\langle \mathbf{p}'\alpha'|V|\mathbf{p}^0\alpha^0\rangle|^2. \tag{15}$$

This is the scattering coefficient for transitions $\alpha^0 \to \alpha'$ of the scatterer. It depends on that matrix element $\langle \mathbf{p}'\alpha'|V|\mathbf{p}^0\alpha^0\rangle$ of the perturbing energy V whose column $\mathbf{p}^0\alpha^0$ and whose row $\mathbf{p}'\alpha'$ refer respectively to the initial and final states of the unperturbed system, between which the scattering transition process takes place. The result (15) is thus in some ways analogous to the result (24) of § 44, although the numerical coefficients are different in the two cases, corresponding to the different natures of the two transition processes.

50. Solution with the Momentum Representation

The result (15) for the scattering coefficient makes a reference only to that representation in which the momentum $\mathbf{p}$ is diagonal. One would thus expect to be able to get a more direct proof of the result by working all the time in the $\mathbf{p}$-representation, instead of working in the $\mathbf{x}$-representation and transforming at the end to the $\mathbf{p}$-representation, as was done in § 49. This would not at first sight appear to be a great improvement, as the lack of directness of the $\mathbf{x}$-representation method is offset by more direct applicability, it being possible to picture the square of the modulus of the $\mathbf{x}$-representative of a state as the density of a stream of particles in process of being scattered. The $\mathbf{x}$-representation method has, however, other more serious disadvantages. One of the main applications of the theory of collisions is to the case of photons as incident particles. Now a photon is not a simple particle but has a polarization. It is evident from classical electromagnetic theory that a photon with a definite momentum, i.e. one moving in a definite direction with a definite frequency, may have a definite state of polarization (linear, circular, etc.), while a photon with a definite position, which is to be pictured as an electromagnetic disturbance confined to a very small volume, cannot have any definite polarization. These facts mean that the polarization observable of a photon commutes with its momentum but not with its position. This results in the $\mathbf{p}$-representation method being immediately applicable to the case of photons, it being only necessary to introduce the polarizing variable into the representatives and treat it along with the α's describing the scatterer, while the **x**-representation method is not applicable. Further, in dealing with photons, it is necessary to take relativistic mechanics into account. This can easily be done in the **p**-representation method, but not so easily in the **x**-representation method.

Equation (3) still holds with relativistic mechanics, but W is now given by

$$\frac{W^2}{c^2} = m^2c^2 + P^2 = m^2c^2 + p_x^2 + p_y^2 + p_z^2 \tag{16}$$

instead of by (2). Written in terms of **p**-representatives, equation (3) gives

$$[E' - H_s(\alpha') - W]\langle \mathbf{p}\alpha'|1\rangle = \langle \mathbf{p}\alpha'|V|0\rangle,$$

p being written instead of $\mathbf{p}'$ for brevity and W being understood as a definite function of p_x, p_y, p_z given by (16). This may be written

$$(W' - W)\langle \mathbf{p}\alpha'|1\rangle = \langle \mathbf{p}\alpha'|V|0\rangle, \tag{17}$$

where

$$W' = E' - H_s(\alpha') \tag{18}$$

and is the energy required by the law of conservation of energy for a scattered particle belonging to a scatterer in state α'. The ket $|0\rangle$ is represented by (6) in the **x**-representation and the basic ket $|\mathbf{p}^0\alpha^0\rangle$ is represented by

$$\langle \mathbf{x}\alpha'|\mathbf{p}^0\alpha^0\rangle = \delta_{\alpha'\alpha^0}\langle \mathbf{x}|\mathbf{p}^0\rangle = \delta_{\alpha'\alpha^0} h^{-3/2} e^{i(\mathbf{p}^0,\mathbf{x})/\hbar},$$

from the transformation function (54) of § 23. Hence

$$|0\rangle = h^{3/2}|\mathbf{p}^0\alpha^0\rangle, \tag{19}$$

and equation (17) may be written

$$(W' - W)\langle \mathbf{p}\alpha'|1\rangle = h^{3/2}\langle \mathbf{p}\alpha'|V|\mathbf{p}^0\alpha^0\rangle. \tag{20}$$

We now make a transformation from the Cartesian coordinates p_x, p_y, p_z of **p** to its polar coordinates P, ω, χ, given by

$$p_x = P\cos\omega, \quad p_y = P\sin\omega\cos\chi, \quad p_z = P\sin\omega\sin\chi.$$

If in the new representation we take the weight function $P^2 \sin\omega$, then the weight attached to any volume of **p**-space will be the same as in the previous **p**-representation, so that the transformation will mean simply a relabelling of the rows and columns of the matrices without any alteration of the matrix elements. Thus (20) will become in the new representation

$$(W' - W)\langle P\omega\chi\alpha'|1\rangle = h^{3/2}\langle P\omega\chi\alpha'|V|P^0\omega^0\chi^0\alpha^0\rangle, \tag{21}$$

W being now a function of the single variable P.

50. SOLUTION WITH THE MOMENTUM REPRESENTATION

The coefficient of $\langle P\omega\chi\alpha'|1\rangle$, namely $W' - W$, is now simply a multiplying factor and not a differential operator as it was with the x-representation method. We can therefore divide out by this factor and obtain an explicit expression for $\langle P\omega\chi\alpha'|1\rangle$. When, however, α' is such that W', defined by (18), is greater than mc^2, this factor will have the value zero for a certain point in the domain of the variable P, namely the point $P = P'$, given in terms of W' by (16). The function $\langle P\omega\chi\alpha'|1\rangle$ will then have a singularity at this point. This singularity shows that $\langle P\omega\chi\alpha'|1\rangle$ represents an infinite number of particles moving about at great distances from the scatterers with energies indefinitely close to W' and it is therefore this singularity that we have to study to get the angular distribution of the particles at infinity.

The result of dividing out (21) by the factor $W' - W$ is, according to (13) of § 15,

$$\langle P\omega\chi\alpha'|1\rangle = \frac{h^{3/2}\langle P\omega\chi\alpha'|V|P^0\omega^0\chi^0\alpha^0\rangle}{W' - W} + \lambda(\omega\chi\alpha')\delta(W' - W), \tag{22}$$

where λ is an arbitrary function of ω, χ, and α'. To give a meaning to the first term on the right-hand side of (22), we make the convention that its integral with respect to P over a range that includes the value P' is the limit when $\epsilon \to 0$ of the integral when the small domain $P' - \epsilon$ to $P' + \epsilon$ is excluded from the range of integration. This is sufficient to make the meaning of (22) precise, since we are interested effectively only in the integrals of the representatives of states when the representation has continuous ranges of rows and columns. We see that equation (21) is inadequate to determine the representative $\langle P\omega\chi\alpha'|1\rangle$ completely, on account of the arbitrary function λ occurring in (22). We must choose this λ such that $\langle P\omega\chi\alpha'|1\rangle$ represents only outward moving particles, since we want the only inward moving particles to be those corresponding to $|0\rangle$.

Let us take first the general case when the representative $\langle P\omega\chi|\rangle$ of a state of the particle satisfies an equation of the type

$$(W' - W)\langle P\omega\chi|\rangle = f(P\omega\chi), \tag{23}$$

where $f(P\omega\chi)$ is any function of P, ω, and χ, and W' is a number greater than mc^2, so that $\langle P\omega\chi|\rangle$ is of the form

$$\langle P\omega\chi|\rangle = \frac{f(P\omega\chi)}{W' - W} + \lambda(\omega\chi)\delta(W' - W), \tag{24}$$

and let us determine now what λ must be in order that $\langle P\omega\chi|\rangle$ may represent only outward moving particles. We can do this by transforming $\langle P\omega\chi|\rangle$ to the **x**-representation, or rather the $(r\theta\phi)$-representation, and comparing it with (12) for large values of r. The transformation function is

$$\langle r\theta\phi|P\omega\chi\rangle = h^{-3/2}e^{i(\mathbf{p},\mathbf{x})/\hbar} = h^{-3/2}e^{iPr\{\cos\omega\cos\theta+\sin\omega\sin\theta\cos(\chi-\phi)\}/\hbar}.$$

For the direction $\theta = 0$ we find

$$\langle r0\phi|\rangle = h^{-3/2}\int_0^\infty P^2\,dP\int_0^{2\pi}d\chi\int_0^\pi \sin\omega\,d\omega\,e^{iPr\cos\omega/\hbar}\langle P\omega\chi|\rangle$$

$$= h^{-3/2}\int_0^\infty P^2\,dP\int_0^{2\pi}d\chi\left(-\left[\frac{e^{iPr\cos\omega/\hbar}}{iPr/\hbar}\langle P\omega\chi|\rangle\right]_{\omega=0}^{\omega=\pi}\right.$$

$$\left.+\int_0^\pi d\omega\,\frac{e^{iPr\cos\omega/\hbar}}{iPr/\hbar}\frac{\partial}{\partial\omega}\langle P\omega\chi|\rangle\right).$$

The second term in the ( ) parentheses is of order r^{-2}, as may be verified by further partial integrations with respect to ω, and can therefore be neglected. We are left with

$$\langle r0\phi|\rangle = ih^{-1/2}(2\pi r)^{-1}\int_0^\infty P\,dP\int_0^{2\pi}d\chi\,(e^{-iPr/\hbar}\langle P\pi\chi|\rangle$$

$$- e^{iPr/\hbar}\langle P0\chi|\rangle)$$

$$= ih^{-1/2}r^{-1}\int_0^\infty P\,dP\,(e^{-iPr/\hbar}\langle P\pi\chi|\rangle - e^{iPr/\hbar}\langle P0\chi|\rangle). \quad (25)$$

When we substitute for $\langle P\omega\chi|\rangle$ its value given by (24), the first term in the integrand in (25) gives

$$ih^{-1/2}r^{-1}\int_0^\infty P\,dP\,e^{-iPr/\hbar}\left[\frac{f(P\pi\chi)}{W'-W} + \lambda(\pi\chi)\delta(W'-W)\right]. \quad (26)$$

The term involving $\delta(W'-W)$ here may be integrated immediately and gives, when one uses the relation $P\,dP = W\,dW/c^2$, which follows from (16),

$$ih^{-1/2}c^{-2}r^{-1}\int_{mc^2}^\infty W\,dW\,e^{-iPr/\hbar}\lambda(\pi\chi)\delta(W'-W)$$

$$= ih^{-1/2}c^{-2}r^{-1}W'\lambda(\pi\chi)e^{-iP'r/\hbar}. \quad (27)$$

To integrate the other term in (26) we use the formula

$$\int_0^\infty g(P)\frac{e^{-iPr/\hbar}}{P'-P}\,dP = g(P')\int_0^\infty \frac{e^{-iPr/\hbar}}{P'-P}\,dP, \quad (28)$$

50. SOLUTION WITH THE MOMENTUM REPRESENTATION

with neglect of terms involving r^{-1}, for any continuous function $g(P)$, which formula holds since $\int_0^\infty K(P) e^{-iPr/\hbar}\, dP$ is of order r^{-1} for any continuous function $K(P)$ ate since the difference

$$\frac{g(P)}{P'-P} - \frac{g(P')}{P'-P}$$

is continuous. The right-hand side of (28), when evaluated with neglect of terms involving r^{-1}, and also with neglect of the small domain $P' - \epsilon$ to $P' + \epsilon$ in the domain of integration, gives

$$g(P') \int_{-\infty}^{\infty} \frac{e^{-iPr/\hbar}}{P'-P}\, dP = g(P') e^{-iP'r/\hbar} \int_{-\infty}^{\infty} \frac{e^{i(P'-P)r/\hbar}}{P'-P}\, dP$$

$$= i g(P') e^{-iP'r/\hbar} \int_{-\infty}^{\infty} \frac{\sin(P'-P)r/\hbar}{P'-P}\, dP = i\pi\, g(P') e^{iP'r/\hbar}. \quad (29)$$

In our present example $g(P)$ is

$$g(P) = \frac{i h^{-1/2} r^{-1} P\, f(P\pi\chi)(P'-P)}{W' - W},$$

which has the limiting value when $P = P'$,

$$g(P') = \frac{i h^{-1/2} r^{-1} P'\, f(P'\pi\chi) W'}{P'c^2} = i h^{-1/2} c^{-2} r^{-1} W'\, f(P'\pi\chi).$$

Substituting this in (29) and adding on the expression (27), we obtain the following value for the integral (26)

$$h^{-1/2} c^{-2} r^{-1} W'[-\pi f(P'\pi\chi) + i\lambda(\pi\chi)] e^{-iP'r/\hbar}. \quad (30)$$

Similarly the second term in the integrand in (25) gives

$$h^{-1/2} c^{-2} r^{-1} W'[-\pi f(P'0\chi) - i\lambda(0\chi)] e^{iP'r/\hbar}. \quad (31)$$

The sum of these two expressions is the value of $\langle r0\phi|\rangle$ when r is large.

We require that $\langle r0\phi|\rangle$ shall represent only outward moving particles, and hence it must be of the form of a multiple of $e^{iP'r/\hbar}$. Thus (30) must vanish, so that

$$\lambda(\pi\chi) = -i\pi f(P'\pi\chi). \quad (32)$$

We see in this way that the condition that $\langle r\theta\phi|\rangle$ shall represent only outward moving particles in the direction $\theta = 0$ fixes the value of λ for the opposite direction $\theta = \pi$. Since the direction $\theta = 0$ or $\omega = 0$ of the pole of our polar coordinates is not in any way singular, we can generalize (32) to

$$\lambda(\omega\chi) = -i\pi f(P'\omega\chi), \quad (33)$$

which gives the value of λ for an arbitrary direction. This value substituted in (24) gives a result that may be written

$$\langle P\omega\chi| \rangle = f(P\omega\chi)\left[\frac{1}{W'-W} - i\pi\,\delta(W'-W)\right], \quad (34)$$

since one can substitute P' for P in the coefficient of a term involving $\delta(W'-W)$ as a factor without changing the value of the term. *The condition that $\langle P\omega\chi|\rangle$ shall represent only outward moving particles is thus that it shall contain the factor*

$$\left[\frac{1}{W'-W} - i\pi\,\delta(W'-W)\right]. \quad (35)$$

It is interesting to note that this factor is of the form of the right-hand side of equation (15) of § 15.

With λ given by (33), expression (30) vanishes and the value of $\langle r0\phi|\rangle$ for large r is given by expression (31) alone, thus

$$\langle r0\phi|\rangle = -2\pi h^{-1/2} c^{-2} r^{-1} W' f(P'0\chi) e^{iP'r/\hbar}.$$

This may be generalized to

$$\langle r\theta\phi|\rangle = -2\pi h^{-1/2} c^{-2} r^{-1} W' f(P'\omega\chi) e^{iP'r/\hbar},$$

giving the value of $\langle r\theta\phi|\rangle$ for any direction θ, ϕ in terms of $f(P'\omega\chi)$ for the same direction labelled by ω, χ. This is of the form (12) with

$$u(\theta\phi) = -2\pi h^{-1/2} c^{-2} W' f(P'\omega\chi)$$

and thus represents a distribution of outward moving particles of momentum P' whose number is

$$\frac{c^2 P'}{W'} |u|^2 = \frac{4\pi^2 W' P'}{hc^2} |f(P'\omega\chi)|^2 \quad (36)$$

per unit solid angle per unit time. This distribution is the one represented by the $\langle P\omega\chi|\rangle$ of (34).

From this general result we can infer that, whenever we have a representative $\langle P\omega\chi|\rangle$ representing only outward moving particles and satisfying an equation of the type (23), the number per unit solid angle per unit time of these particles is given by (36). If this $\langle P\omega\chi|\rangle$ occurs in a problem in which the number of incident particles is one per unit volume, it will correspond to a scattering coefficient of amount

$$\frac{4\pi^2 W^0 W' P'}{hc^4 P^0} |f(P'\omega\chi)|^2. \quad (37)$$

It is only the value of the function $f(P\omega\chi)$ for the point $P = P'$ that is of importance.

If we now apply this general theory to our equations (21) and (22), we have
$$f(P\omega\chi) = h^{3/2}\langle P\omega\chi\alpha'|V|P^0\omega^0\chi^0\alpha^0\rangle.$$
Hence from (37) the scattering coefficient is
$$\frac{4\pi^2 h^2 W^0 W' P'}{c^4 P^0}|\langle P'\omega\chi\alpha'|V|P^0\omega^0\chi^0\alpha^0\rangle|^2. \tag{38}$$
If one neglects relativity and puts $W^0 W'/c^4 = m^2$, this result reduces to the result (15) obtained in the preceding section by means of Green's theorem.

51. Dispersive Scattering

We shall now determine the scattering when the incident particle is capable of being absorbed, that is, when our unperturbed system of scatterer plus particle has closed stationary states with the particle absorbed. The existence of these closed states for the unperturbed system will be found to have a considerable effect on the scattering for the perturbed system, and indeed an effect that depends very much on the energy of the incident particle, giving rise to the phenomenon of dispersion in optics when the incident particle is taken to be a photon.

We use a representation for which the basic kets correspond to the stationary states of the unperturbed system, as was the case with the p-representation of the preceding section. We take these stationary states to be the states $(\mathbf{p}'\alpha')$ for which the particle has a definite momentum $\mathbf{p}'$ and the scatterer is in a definite state α', together with the closed states, k say, which form a separate discrete set, and assume that these states are all independent and orthogonal. This assumption is not accurate when the particle is an electron or atomic nucleus, since in this case for an absorbed state k the particle will still certainly be somewhere, so that one would expect to be able to expand $|k\rangle$ in terms of the eigenkets $|\mathbf{x}'\alpha'\rangle$ of x, y, z, and the α's, and hence also in terms of the $|\mathbf{p}'\alpha'\rangle$'s. On the other hand, when the particle is a photon it will no longer exist for the absorbed states, which are then certainly independent of and orthogonal to the states $(\mathbf{p}'\alpha')$ for which the particle does exist. Thus the assumption is valid in this case, which is an important practical one.

Since we are concerned with scattering, we must still deal with *stationary* states of the whole system. We shall now, however, have to work

VIII. COLLISION PROBLEMS

to the second order of accuracy, so that we cannot use merely the first-order equation (3), but must use also (4). Equation (3) becomes, when written in terms of representatives in our present representation,

$$(W' - W)\langle \mathbf{p}\alpha'|1\rangle = \langle \mathbf{p}\alpha'|V|0\rangle,$$
$$(E' - E_k)\langle k|1\rangle = \langle k|V|0\rangle, \tag{39}$$

where W' is the function of E' and the α''s given by (18) and E_k is the energy of the stationary state k of the unperturbed system. Similarly, equation (4) becomes

$$(W' - W)\langle \mathbf{p}\alpha'|2\rangle = \langle \mathbf{p}\alpha'|V|1\rangle,$$
$$(E' - E_k)\langle k|2\rangle = \langle k|V|1\rangle. \tag{40}$$

Expanding the right-hand sides by matrix multiplication, we get

$$(W' - W)\langle \mathbf{p}\alpha'|2\rangle$$
$$= \sum_{\alpha''} \int \langle \mathbf{p}\alpha'|V|\mathbf{p}''\alpha''\rangle \, d^3p'' \, \langle \mathbf{p}''\alpha''|1\rangle + \sum_{k''} \langle \mathbf{p}\alpha'|V|k''\rangle \langle k''|1\rangle,$$
$$(E' - E_k)\langle k|2\rangle \tag{41}$$
$$= \sum_{\alpha''} \int \langle k|V|\mathbf{p}''\alpha''\rangle \, d^3p'' \, \langle \mathbf{p}''\alpha''|1\rangle + \sum_{k''} \langle k|V|k''\rangle \langle k''|1\rangle.$$

The ket $|0\rangle$ is still given by (19), so (39) may be written

$$(W' - W)\langle \mathbf{p}\alpha'|1\rangle = h^{3/2}\langle \mathbf{p}\alpha'|V|\mathbf{p}^0\alpha^0\rangle, \tag{42}$$
$$(E' - E_k)\langle k|1\rangle = h^{3/2}\langle k|V|\mathbf{p}^0\alpha^0\rangle. \tag{43}$$

We may assume that the matrix elements $\langle k'|V|k''\rangle$ of V vanish, since these matrix elements are not essential to the phenomena under investigation, and if they did not vanish it would mean simply that the absorbed states k had not been suitably chosen. We shall further assume that the matrix elements $\langle \mathbf{p}'\alpha'|V|\mathbf{p}''\alpha''\rangle$ are of the second order of smallness when the matrix elements $\langle k'|V|\mathbf{p}''\alpha''\rangle$, $\langle \mathbf{p}'\alpha'|V|k''\rangle$ are taken to be of the first order of smallness. This assumption will be justified for the case of photons in § 64. We now have from (43) and (42) that $\langle k|1\rangle$ is of the first order of smallness, provided E' does not lie near one of the discrete set of energy-levels E_k, and $\langle \mathbf{p}\alpha'|1\rangle$ is of the second order. The value of $\langle \mathbf{p}\alpha'|2\rangle$ to the second order will thus be given, from the first of equations (41), by

$$(W' - W)\langle \mathbf{p}\alpha'|2\rangle = h^{3/2} \sum_{k''} \frac{\langle \mathbf{p}\alpha'|V|k''\rangle \langle k''|V|\mathbf{p}^0\alpha^0\rangle}{(E' - E_{k''})}.$$

The total correction in the wave function to the second order, namely ⟨p**α**′|1⟩ plus ⟨p**α**′|2⟩, therefore satisfies

$$(W' - W)(\langle \mathbf{p}\alpha'|1\rangle + \langle \mathbf{p}\alpha'|2\rangle)$$
$$= h^{3/2}\left(\langle \mathbf{p}\alpha'|V|\mathbf{p}^0\alpha^0\rangle + \sum_k \frac{\langle \mathbf{p}\alpha'|V|k\rangle\langle k|V|\mathbf{p}^0\alpha^0\rangle}{E' - E_k}\right).$$

This equation is of the type (23), provided α' is such that $W' > mc^2$, which means that α' as a final state for the scatterer is not inconsistent with the law of conservation of energy. We can therefore infer from the general result (37) that the scattering coefficient is

$$\frac{4\pi^2 h^2 W^0 W' P'}{c^4 P^0}\left|\langle \mathbf{p}\alpha'|V|\mathbf{p}^0\alpha^0\rangle + \sum_k \frac{\langle \mathbf{p}'\alpha'|V|k\rangle\langle k|V|\mathbf{p}^0\alpha^0\rangle}{E' - E_k}\right|^2. \quad (44)$$

The scattering may now be considered as composed of two parts, a part that arises from the matrix element $\langle \mathbf{p}'\alpha'|V|\mathbf{p}^0\alpha^0\rangle$ of the perturbing energy and a part that arises from the matrix elements $\langle \mathbf{p}'\alpha'|V|k\rangle$ and $\langle k|V|\mathbf{p}^0\alpha^0\rangle$. The first part, which is the same as our previously obtained result (38), may be called the direct scattering. The second part may be considered as arising from an absorption of the incident particle into some state k, followed immediately by a re-emission in a different direction, and is like the transitions through an intermediate state considered in § 44. The fact that we have to add the two terms before taking the square of the modulus denotes interference between the two kinds of scattering. There is no experimental way of separating the two kinds, the distinction between them being only mathematical.

52. Resonance Scattering

Suppose the energy of the incident particle to be varied continuously while the initial state α^0 of the scatterer is kept fixed, so that the total energy E' or H' varies continuously. The formula (44) now shows that as E' approaches one of the discrete set of energy-levels E_k, the scattering becomes very large. In fact, according to formula (44) the scattering should be infinite when E' is exactly equal to an E_k. An infinite scattering coefficient is, of course, physically impossible, so that we can infer that the approximations used in deriving (44) are no longer legitimate when E' is close to an E_k. To investigate the scattering in this case we must therefore go back to the exact equation

$$(E' - E)|H'\rangle = V|H'\rangle,$$

equation (2) of § 43 with E' written for H', and use a different method of approximating to its solution. This exact equation, written in terms of representatives like (41), becomes

$$(W' - W)\langle \mathbf{p}\alpha' | H' \rangle$$
$$= \sum_{\alpha''} \int \langle \mathbf{p}\alpha' | V | \mathbf{p}''\alpha'' \rangle d^3 p'' \langle \mathbf{p}''\alpha'' | H' \rangle$$
$$+ \sum_{k''} \langle \mathbf{p}\alpha' | V | k'' \rangle \langle k'' | H' \rangle,$$
$$(E' - E_k)\langle k | H' \rangle \qquad (45)$$
$$= \sum_{\alpha''} \int \langle k | V | \mathbf{p}''\alpha'' \rangle d^3 p'' \langle \mathbf{p}''\alpha'' | H' \rangle$$
$$+ \sum_{k''} \langle k | V | k'' \rangle \langle k'' | H' \rangle.$$

Let us take one particular E_k and consider the case when E' is close to it. The large term in the scattering coefficient (44) now arises from those elements of the matrix representing V that lie in row k or in column k, i.e. those of the type $\langle k | V | \mathbf{p}\alpha' \rangle$ or $\langle \mathbf{p}\alpha' | V | k \rangle$. The scattering arising from the other matrix elements of V is of a smaller order of magnitude. This suggests that in our exact equations (45) we should make the approximation of neglecting all the matrix elements of V except the important ones, which are those of the type $\langle \mathbf{p}\alpha' | V | k \rangle$ or $\langle k | V | \mathbf{p}\alpha' \rangle$, where α' is a state of the scatterer that has not too much energy to be disallowed as a final state by the law of conservation of energy. These equations then reduce to

$$(W' - W)\langle \mathbf{p}\alpha' | H' \rangle = \langle \mathbf{p}\alpha' | V | k \rangle \langle k | H' \rangle,$$
$$(E' - E_k)\langle k | H' \rangle = \sum_{\alpha'} \int \langle k | V | \mathbf{p}\alpha' \rangle d^3 p \, \langle \mathbf{p}\alpha' | H' \rangle, \qquad (46)$$

the α' summation being over those values of α' for which W' given by (18) is $> mc^2$. These equations are now sufficiently simple for us to be able to solve exactly without further approximation.

From the first of equations (46) we obtain by division

$$\langle \mathbf{p}\alpha' | H' \rangle = \frac{\langle \mathbf{p}\alpha' | V | k \rangle \langle k | H' \rangle}{W' - W} + \lambda \delta(W' - W). \qquad (47)$$

We must choose λ, which may be any function of the momentum $\mathbf{p}$ and α', such that (47) represents the incident particles corresponding to $|0\rangle$ or

52. RESONANCE SCATTERING

$h^{3/2}|\mathbf{p}^0\alpha^0\rangle$ together with only outward moving particles. [The representative of $h^{3/2}|\mathbf{p}^0\alpha^0\rangle$ is actually of the form $\lambda\,\delta(W'-W)$, since the conditions $\alpha' = \alpha^0$ and $\mathbf{p} = \mathbf{p}^0$ for it not to vanish lead to $W' = E' - H_s(\alpha') = E' - H_s(\alpha^0) = W^0 = W$.] Thus (47) must be

$$\langle \mathbf{p}\alpha'|H'\rangle = h^{3/2}\langle \mathbf{p}\alpha'|\mathbf{p}^0\alpha^0\rangle$$
$$+ \langle \mathbf{p}\alpha'|V|k\rangle\langle k|H'\rangle\left[\frac{1}{W'-W} - i\pi\,\delta(W'-W)\right], \tag{48}$$

and from the general formula (37) the scattering coefficient will be

$$\frac{4\pi^2 W^0 W' P'}{hc^4 P^0}|\langle \mathbf{p}'\alpha'|V|k\rangle|^2|\langle k|H'\rangle|^2. \tag{49}$$

It remains for us to determine the value of $\langle k|H'\rangle$. We can do this by substituting for $\langle \mathbf{p}\alpha'|H'\rangle$ in the second of equations (46) its value given by (48). This gives

$$(E' - E_k)\langle k|H'\rangle$$
$$= h^{3/2}\langle k|V|\mathbf{p}^0\alpha^0\rangle$$
$$+ \langle k|H'\rangle\sum_{\alpha'}\int|\langle k|V|\mathbf{p}\alpha'\rangle|^2\left[\frac{1}{W-W'} - i\pi\,\delta(W'-W)\right]d^3p$$
$$= h^{3/2}\langle k|V|\mathbf{p}^0\alpha^0\rangle + \langle k|H'\rangle(a - ib)$$

where

$$a = \sum_{\alpha'}\int\frac{|\langle k|V|\mathbf{p}\alpha'\rangle|^2}{W'-W}d^3p \tag{50}$$

and

$$b = \pi\sum_{\alpha'}\int|\langle k|V|\mathbf{p}\alpha'\rangle|^2\,\delta(W'-W)\,d^3p$$
$$= \pi\sum_{\alpha'}\iiint|\langle k|V|P\omega\chi\alpha'\rangle|^2\,\delta(W'-W)P^2\,dP\,\sin\omega\,d\omega\,d\chi$$
$$= \pi\sum_{\alpha'}P'W'c^{-2}\iint|\langle k|V|P'\omega\chi\alpha'\rangle|^2\,\sin\omega\,d\omega\,d\chi. \tag{51}$$

Thus

$$\langle k|H'\rangle = h^{3/2}\frac{\langle k|V|\mathbf{p}^0\alpha^0\rangle}{E' - E_k - a + ib}. \tag{52}$$

Note that a and b are real and that b is positive.

This value for $\langle k|H'\rangle$ substituted in (49) gives for the scattering coefficient

$$\frac{4\pi^2 h^2 W^0 W' P'}{c^4 P^0} \frac{|\langle \mathbf{p}'\alpha'|V|k\rangle|^2 |\langle k|V|\mathbf{p}^0\alpha^0\rangle|^2}{(E' - E_k - a)^2 + b^2}. \tag{53}$$

One can obtain the total effective area that the incident particle must hit in order to be scattered anywhere by integrating (53) over all directions of scattering, i.e. by integrating over all directions of the vector $\mathbf{p}'$ with its magnitude kept fixed at P', and then summing over all α' that are to be taken into consideration, i.e. for which $W' > mc^2$. This gives, with the help of (51), the result

$$\frac{4\pi h^2 W^0}{c^2 P^0} \frac{b |\langle k|V|\mathbf{p}^0\alpha^0\rangle|^2}{(E' - E_k - a)^2 + b^2}. \tag{54}$$

If we suppose E' to vary continuously through the value E_k, the main variation of (53) or (54) will be due to the small denominator $(E' - E_k - a)^2 + b^2$. If we neglect the dependence of the other factors in (53) and (54) on E', then the maximum scattering will occur when E' has the value $E_k + a$ and the scattering will be half its maximum when E differs from this value by an amount b. The large amount of scattering that occurs for values of the energy of the incident particle that make E' nearly equal to E_k give rise to the phenomenon of an absorption line. The centre of the line is displaced by an amount a from the resonance energy of the incident particle, i.e. the energy which would make the total energy just E_k, while the quantity b is what is sometimes called the half-width of the line.

53. Emission and Absorption

For studying emission and absorption we must consider non-stationary states of the system and must use the perturbation method of § 44. To determine the coefficient of spontaneous emission we must take an initial state for which the particle is absorbed, corresponding to a ket $|k\rangle$, and determine the probability that at some later time the particle shall be on its way to infinity with a definite momentum. The method of § 46 can now be applied. From the result (39) of that section we see that the probability per unit time per unit range of ω and χ, of the particle being emitted in any direction ω', χ' with the scatterer being left in state α' is

$$2\pi\hbar^{-1} |\langle W'\omega'\chi'\alpha'|V|k\rangle|^2, \tag{55}$$

provided, of course, that α' is such that the energy W', given by (18), of the particle is greater than mc^2. For values of α' that do not satisfy this condition there is no emission possible. The matrix element $\langle W'\omega'\chi'\alpha'|V|k\rangle$ here must refer to a representation in which W, ω, χ, and α are diagonal with the weight function unity. The matrix elements of V appearing in the three preceding sections refer to a representation in which p_x, p_y, p_z are diagonal with the weight function unity, or P, ω, χ are diagonal with the weight function $P^2 \sin\omega$. They would thus refer to a representation in which W, ω, χ are diagonal with the weight function $dP/dW \cdot P^2 \sin\omega = WP/c^2 \cdot \sin\omega$. Thus the matrix element $\langle W'\omega'\chi'\alpha'|V|k\rangle$ in (55) is equal to $(W'P'/c^2 \cdot \sin\omega')^{1/2}$ times our previous matrix element $\langle W'\omega'\chi'\alpha'|V|k\rangle$ or $\langle \mathbf{p}'\alpha'|V|k\rangle$, so that (55) is equal to

$$\frac{2\pi}{\hbar} \frac{W'P'}{c^2} \sin\omega' |\langle \mathbf{p}'\alpha'|V|k\rangle|^2.$$

The probability of emission per unit solid angle per unit time, with the scatterer simultaneously dropping to state α', is thus

$$\frac{2\pi}{\hbar} \frac{W'P'}{c^2} |\langle \mathbf{p}'\alpha'|V|k\rangle|^2. \tag{56}$$

To obtain the total probability per unit time of the particle being emitted in any direction, with any final state for the scatterer, we must integrate (56) over all angles ω', χ' and sum over all states α' whose energy $H_s(\alpha')$ is such that $H_s(\alpha') + mc^2 < E_k$. The result is just $2b/\hbar$, where b is defined by (51). *There is thus this simple relation between the total emission coefficient and the half-width b of the absorption line.*

Let us now consider absorption. This requires that we shall take an initial state for which the particle is certainly not absorbed but is incident with a definite momentum. Thus the ket corresponding to the initial state must be of the form (19). We must now determine the probability of the particle being absorbed after time t. Since our final state k is not one of a continuous range, we cannot use directly the result (39) of § 46. If, however, we take

$$|0\rangle = |\mathbf{p}^0 \alpha^0\rangle, \tag{57}$$

as the ket corresponding to the initial state, the analysis of §§ 44 and 46 is still applicable as far as equation (36) and shows us that the probability of the particle being absorbed into state k after time t is

$$\frac{2|\langle k|V|\mathbf{p}^0\alpha^0\rangle|^2 \{1 - \cos[(E_k - E')t/\hbar]\}}{(E_k - E')^2}.$$

This corresponds to a distribution of incident particles of density h^{-3}, owing to the omission of the factor $h^{3/2}$ from (57), as compared with (19). The probability of there being an absorption after time t when there is one incident particle crossing unit area per unit time is therefore

$$\frac{2h^3 W^0}{c^2 P^0} |\langle k|V|\mathbf{p}^0\alpha^0\rangle|^2 \frac{\{1-\cos[(E_k - E')t/\hbar]\}}{(E_k - E')^2}. \tag{58}$$

To obtain the absorption coefficient we must consider the incident particles not all to have exactly the same energy $W^0 = E' - H_s(\alpha^0)$, but to have a distribution of energy values about the correct value $E_k - H_s(\alpha^0)$ required for absorption. If we take a beam of incident particles consisting of one crossing unit area per unit time per unit energy range, the probability of there being an absorption after time t will be given by the integral of (58) with respect to E'. This integral may be evaluated in the same way as (37) of § 46 and is equal to

$$\frac{4\pi^2 h^2 W^0 t}{c^2 P^0} |\langle k|V|\mathbf{p}^0\alpha^0\rangle|^2.$$

The probability per unit time of an absorption taking place with an incident beam of one particle per unit area per unit time per unit energy range is therefore

$$\frac{4\pi^2 h^2 W^0}{c^2 P^0} |\langle k|V|\mathbf{p}^0\alpha^0\rangle|^2, \tag{59}$$

which is the absorption coefficient.

The connexion between the absorption and emission coefficients (59) and (56) and the resonance scattering coefficients calculated in the preceding section should be noted. When the incident beam does not consist of particles all with the same energy, but consists of a unit distribution of particles per unit energy range crossing unit area per unit time, the total number of incident particles with energies near an absorption line that get scattered will be given by the integral of (54) with respect to E'. If one neglects the dependence of the numerator of (54) on E', this integral will, since

$$\int_{-\infty}^{\infty} \frac{b}{(E' - E_k - a)^2 + b^2} dE' = \pi,$$

have just the value (59). Thus *the total number of scattered particles in the neighbourhood of an absorption line is equal to the total number absorbed*. We can therefore regard all these scattered particles as absorbed particles that are subsequently re-emitted in a different direction. Further, the number of particles in the neighbourhood of the absorption line that

53. EMISSION AND ABSORPTION

get scattered per unit solid angle about a given direction specified by $\mathbf{p}'$ and then belong to scatterers in state α' will be given by the integral with respect to E' of (53), which integral has in the same way the value

$$\frac{4\pi^2 h^2 W^0 W' P'}{c^4 P^0} \frac{\pi}{b} |\langle \mathbf{p}'\alpha'|V|k\rangle|^2 |\langle k|V|\mathbf{p}^0 \alpha^0\rangle|^2.$$

This is just equal to the absorption coefficient (59) multiplied by the emission coefficient (56) divided by $2b/\hbar$, the total emission coefficient. This is in agreement with the point of view of regarding the resonance scattered particles as those that are absorbed and then re-emitted, with the absorption and emission processes governed independently each by its own probability law, since this point of view would make the fraction of the total number of absorbed particles that are re-emitted in a unit solid angle about a given direction just the emission coefficient for this direction divided by the total emission coefficient.

CHAPTER IX

Systems Containing Several Similar Particles

54. Symmetrical and Antisymmetrical States

If a system in atomic physics contains a number of particles of the same kind, e.g. a number of electrons, the particles are absolutely indistinguishable one from another. No observable change is made when two of them are interchanged. This circumstance gives rise to some curious phenomena in quantum mechanics having no analogue in the classical theory, which arise from the fact that in quantum mechanics a transition may occur resulting in merely the interchange of two similar particles, which transition then could not be detected by any observational means. A satisfactory theory ought, of course, to count two observationally indistinguishable states as the same state and to deny that any transition does occur when two similar particles exchange places. We shall find that it is possible to reformulate the theory so that this is so.

Suppose we have a system containing n similar particles. We may take as our dynamical variables a set of variables ξ_1 describing the first particle, the corresponding set ξ_2 describing the second particle, and so on up to the set ξ_n describing the nth particle. We shall then have the ξ_r's commuting with the ξ_s's for $r \neq s$. (We may require certain extra variables, describing what the system consists of in addition to the n similar particles, but it is not necessary to mention these explicitly in the present chapter.) The Hamiltonian describing the motion of the system will now be expressible as a function of the $\xi_1, \xi_2, \ldots, \xi_n$. The fact that the particles are similar requires that *the Hamiltonian shall be a symmetrical function of the* $\xi_1, \xi_2, \ldots, \xi_n$, i.e. it shall remain unchanged when the sets of variables ξ_r are interchanged or permuted in any way. This condition must hold, no matter what perturbations are applied to the system. In fact, any quantity of physical significance must be a symmetrical function of the ξ's.

Let $|a_1\rangle, |b_1\rangle, |c_1\rangle, \ldots$ be kets for the first particle considered as a dynamical system by itself. There will be corresponding kets $|a_2\rangle, |b_2\rangle, |c_2\rangle, \ldots$ for the second particle by itself, and so on. We can get a ket for the assembly by taking the product of kets for each particle by itself, for example

$$|a_1\rangle|b_2\rangle|c_3\rangle \cdots |g_n\rangle = |a_1 b_2 c_3 \cdots g_n\rangle \qquad (1)$$

say, according to the notation of (65) of § 20. The ket (1) corresponds to a special kind of state for the assembly, which may be described by saying that each particle is in its own state, corresponding to its own factor on the left-hand side of (1). The general ket for the assembly is of the form of a sum or integral of kets like (1), and corresponds to a state for the assembly for which one cannot say that each particle is in its own state, but only that each particle is partly in several states, in a way which is correlated with the other particles being partly in several states. If the kets $|a_1\rangle, |b_1\rangle, |c_1\rangle, \ldots$ are a set of basic kets for the first particle by itself, the kets $|a_2\rangle, |b_2\rangle, |c_2\rangle, \ldots$ will be a set of basic kets for the second particle by itself, and so on, and the kets (1) will be a set of basic kets for the assembly. We call the representation provided by such basic kets for the assembly a *symmetrical representation*, as it treats all the particles on the same footing.

In (1) we may interchange the kets for the first two particles and get another ket for the assembly, namely

$$|b_1\rangle|a_2\rangle|c_3\rangle \cdots |g_n\rangle = |b_1 a_2 c_3 \cdots g_n\rangle.$$

More generally, we may interchange the role of the first two particles in any ket for the assembly and get another ket for the assembly. The process of interchanging the first two particles is an operator which can be applied to kets for the assembly, and is evidently a linear operator, of the type dealt with in § 7. Similarly, the process of interchanging any pair of particles is a linear operator, and by repeated applications of such interchanges we get any permutation of the particles appearing as a linear operator which can be applied to kets for the assembly. A permutation is called an *even permutation* or an *odd permutation* according to whether it can be built up from an even or an odd number of interchanges.

A ket for the assembly $|X\rangle$ is called *symmetrical* if it is unchanged by any permutation, i.e. if

$$P|X\rangle = |X\rangle \qquad (2)$$

for any permutation P. It is called *antisymmetrical* if it is unchanged by any even permutation and has its sign changed by any odd permutation, i.e. if

$$P|X\rangle = \pm|X\rangle, \qquad (3)$$

the + or − sign being taken according to whether P is even or odd. The state corresponding to a symmetrical ket is called a *symmetrical state*, and the state corresponding to an antisymmetrical ket is called an *antisymmetrical state*. In a symmetrical representation, the representative of a symmetrical ket is a symmetrical function of the variables referring to the various particles and the representative of an antisymmetrical ket is an antisymmetrical function.

In the Schrödinger picture, the ket corresponding to a state of the assembly will vary with time according to Schrödinger's equation of motion. If it is initially symmetrical it must always remain symmetrical, since, owing to the Hamiltonian being symmetrical, there is nothing to disturb the symmetry. Similarly if the ket is initially antisymmetrical it must always remain antisymmetrical. Thus *a state which is initially symmetrical always remains symmetrical and a state which is initially antisymmetrical always remains antisymmetrical*. In consequence, it may be that for a particular kind of particle only symmetrical states occur in nature, or only antisymmetrical states occur in nature. If either of these possibilities held, it would lead to certain special phenomena for the particles in question.

Let us suppose first that only antisymmetrical states occur in nature. The ket (1) is not antisymmetrical and so does not correspond to a state occurring in nature. From (1) we can in general form an antisymmetrical ket by applying all possible permutations to it and adding the results, with the coefficient -1 inserted before those terms arising from an odd permutation, so as to get

$$\sum_P \pm P |a_1 b_2 c_3 \cdots g_n\rangle, \qquad (4)$$

the + or − sign being taken according to whether P is even or odd. The ket (4) may be written as a determinant

$$\begin{vmatrix} |a_1\rangle & |a_2\rangle & |a_3\rangle & \cdots & |a_n\rangle \\ |b_1\rangle & |b_2\rangle & |b_3\rangle & \cdots & |b_n\rangle \\ |c_1\rangle & |c_2\rangle & |c_3\rangle & \cdots & |c_n\rangle \\ \cdots & \cdots & \cdots & \cdots & \cdots \\ |g_1\rangle & |g_2\rangle & |g_3\rangle & \cdots & |g_n\rangle \end{vmatrix} \qquad (5)$$

and its representative in a symmetrical representation is a determinant. The ket (4) or (5) is not the general antisymmetrical ket, but is a specially simple one. It corresponds to a state for the assembly for which one can say that certain particle-states, namely the states $a, b, c, \ldots, g$, are occupied, but one cannot say which particle is in which state, each particle being equally likely to be in any state. If two of the particle-states $a, b, c, \ldots, g$ are the same, the ket (4) or (5) vanishes and does not correspond to any state for the assembly. Thus *two particles cannot occupy the same state*. More generally, *the occupied states must be all independent*, otherwise (4) or (5) vanishes. This is an important characteristic of particles for which only antisymmetrical states occur in nature. It leads to a special statistics, which was first studied by Fermi, so we shall call particles for which only antisymmetrical states occur in nature fermions.

Let us suppose now that only symmetrical states occur in nature. The ket (1) is not symmetrical, except in the special case when all the particle-states $a, b, c, \ldots, g$ are the same, but we can always obtain a symmetrical ket from it by applying all possible permutations to it and adding the results, so as to get

$$\sum_P P|a_1 b_2 c_3 \cdots g_n\rangle. \tag{6}$$

The ket (6) is not the general symmetrical ket, but is a specially simple one. It corresponds to a state for the assembly for which one can say that certain particle-states are occupied, namely the states $a, b, c, \ldots, g$, without being able to say which particle is in which state. It is now possible for two or more of the states $a, b, c, \ldots, g$ to be the same, so that two or more particles can be in the same state. In spite of this, the statistics of the particles is not the same as the usual statistics of the classical theory. The new statistics was first studied by Bose, so we shall call particles for which only symmetrical states occur in nature *bosons*.

We can see the difference of Bose statistics from the usual statistics by considering a special case—that of only two particles and only two independent states a and b for a particle. According to classical mechanics, if the assembly of two particles is in thermodynamic equilibrium at a high temperature, each particle will be equally likely to be in either state. There is thus a probability $1/4$ of both particles being in state a, a probability $1/4$ of both particles being in state b, and a probability $1/2$ of one particle being in each state. In the quantum theory there are three independent symmetrical states for the pair of particles, corresponding to the symmetrical kets $|a_1\rangle|a_2\rangle$, $|b_1\rangle|b_2\rangle$, and $|a_1\rangle|b_2\rangle + |a_2\rangle|b_1\rangle$, and describable as both particles in state a, both particles in state b, and one particle in each state respectively. For thermodynamic equilibrium at a high temperature these three states are equally probable, as was shown in § 33, so that there is a probability $1/3$ of both particles being in state a, a probability $1/3$ of both particles being in state b, and a probability $1/3$

of one particle being in each state. Thus *with Bose statistics the probability of two particles being in the same state is greater than with classical statistics*. Bose statistics differ from classical statistics in the opposite direction to Fermi statistics, for which the probability of two particles being in the same state is zero.

In building up a theory of atoms on the lines mentioned at the beginning of § 38, to get agreement with experiment one must assume that two electrons are never in the same state. This rule is known as *Pauli's exclusion principle*. It shows us that *electrons are fermions*. Planck's law of radiation shows us that *photons are bosons*, as only the Bose statistics for photons will lead to Planck's law. Similarly, for each of the other kinds of particle known in physics, there is experimental evidence to show either that they are fermions, or that they are bosons. Protons, neutrons, positrons are fermions, α-particles are bosons. It appears that all particles occurring in nature are either fermions or bosons, and thus only antisymmetrical or symmetrical states for an assembly of similar particles are met with in practice. Other more complicated kinds of symmetry are possible mathematically, but do not apply to any known particles. With a theory which allows only antisymmetrical or only symmetrical states for a particular kind of particle, one cannot make a distinction between two states which differ only through a permutation of the particles, so that the transitions mentioned at the beginning of this section disappear.

55. Permutations as Dynamical Variables

We shall now build up a general theory for a system containing n similar particles when states with any kind of symmetry properties are allowed, i.e. when there is no restriction to only symmetrical or only antisymmetrical states. The general state now will not be symmetrical or antisymmetrical, nor will it be expressible linearly in terms of symmetrical and antisymmetrical states when $n > 2$. This theory will not apply directly to any particles occurring in nature, but all the same it is useful for setting up an approximate treatment for an assembly of electrons, as will be shown in § 58.

We have seen that each permutation P of the n particles is a linear operator which can be applied to any ket for the assembly. Hence we can regard P as a dynamical variable in our system of n particles. There are $n!$ permutations, each of which can be regarded as a dynamical variable. One of them, P_1 say, is the identical permutation, which is equal to unity. The product of any two permutations is a third permutation and hence any

function of the permutations is reducible to a linear function of them. Any permutation P has a reciprocal P^{-1} satisfying
$$PP^{-1} = P^{-1}P = P_1 = 1.$$

A permutation P can be applied to a bra $\langle X|$ for the assembly, to give another bra, which we shall denote for the present by $P\langle X|$. If P is applied to both factors of the product $\langle X|Y\rangle$, the product must be unchanged, since it is just a number, independent of any order of the particles. Thus
$$(P\langle X|)P|Y\rangle = \langle X|Y\rangle$$
showing that
$$P\langle X| = \langle X|P^{-1} \tag{7}$$
Now $P\langle X|$ is the conjugate imaginary of $P|X\rangle$ and is thus equal to $\langle X|\overline{P}$, and hence from (7)
$$\overline{P} = P^{-1}. \tag{8}$$
Thus a permutation is not in general a real dynamical variable, its conjugate complex being equal to its reciprocal.

Any permutation of the numbers $1, 2, \ldots, n$ may be expressed in the cyclic notation, e.g. with $n = 8$
$$P_a = (143)(27)(58)(6), \tag{9}$$
in which each number is to be replaced by the succeeding number in a bracket, unless it is the last in a bracket, when it is to be replaced by the first in that bracket. Thus P_a changes the numbers 12345678 into 47138625. The type of any permutation is specified by the partition of the number n which is provided by the number of numbers in each of the brackets. Thus the type of P_a is specified by the partition $8 = 3+2+2+1$. Permutations of the same type, i.e. corresponding to the same partition, we shall call *similar*. Thus, for example, P_a in (9) is similar to
$$P_b = (871)(35)(46)(2). \tag{10}$$
The whole of the $n!$ possible permutations may be divided into sets of similar permutations, each such set being called a *class*. The permutation $P_1 = 1$ forms a class by itself. Any permutation is similar to its reciprocal.

When two permutations P_a and P_b are similar, either of them P_b may be obtained by making a certain permutation P_x in the other P_a. Thus, in our example (9), (10) we can take P_x to be the permutation that changes 14327586 into 87135462, i.e. the permutation
$$P_x = (18623)(475).$$

Different ways of writing P_a and P_b in the cyclic notation would lead to different P_x's. Any of these P_x's applied to the product $P_a|X\rangle$ would change it into $P_b P_x |X\rangle$, i.e.

$$P_x P_a |X\rangle = P_b P_x |X\rangle.$$

Hence
$$P_b = P_x P_a P_x^{-1}, \tag{11}$$

which expresses the condition for P_a and P_b to be similar as an algebraic equation. The existence of any P_x satisfying (11) is sufficient to show that P_a and P_b are similar.

56. Permutations as Constants of the Motion

Any symmetrical function V of the dynamical variables of all the particles is unchanged by the application of any permutation P, so P applied to the product $V|X\rangle$ affects only the factor $|X\rangle$, thus

$$PV|X\rangle = VP|X\rangle.$$

Hence
$$PV = VP, \tag{12}$$

showing that *a symmetrical function of the dynamical variables commutes with every permutation*. The Hamiltonian is a symmetrical function of the dynamical variables and thus commutes with every permutation. It follows that *each permutation is a constant of the motion*. This holds even if the Hamiltonian is not constant. If $|Xt\rangle$ is any solution of Schrödinger's equation of motion, $P|Xt\rangle$ is another.

In dealing with any system in quantum mechanics, when we have found a constant of the motion α, we know that if for any state of motion, α initially has the numerical value α', then it always has this value, so that we can assign different numbers α' to the different states and so obtain a classification of the states. The procedure is not so straightforward, however, when we have several constants of the motion α which do not commute (as is the case with our permutations P), since we cannot in general assign numerical values for all the α's simultaneously to any state. Let us first take the case of a system whose Hamiltonian does not involve the time explicitly. The existence of constants of the motion α which do not commute is then a sign that the system is degenerate. This is because, for a non-degenerate system, the Hamiltonian H by itself forms a complete set of commuting observables and hence, from Theorem 2 of § 19, each of the α's is a function of H and therefore commutes with any other α.

We must now look for a function β of the α's which has one and the same numerical value β' for all those states belonging to one energy-level H', so that we can use β for classifying the energy-levels of the system. We can express the condition for β by saying that it must be a function of H and must therefore commute with every dynamical variable that commutes with H, i.e. with every constant of the motion. If the α's are the only constants of the motion, or if they are a set that commute with all other independent constants of the motion, our problem reduces to finding a function β of the α's which commutes with all the α's. We can then assign a numerical value β' for β to each energy-level of the system. If we can find several such functions β, they must all commute with each other, so that we can give them all numerical values simultaneously. We obtain thus a classification of the energy-levels. When the Hamiltonian involves the time explicitly one cannot talk about energy-levels, but the β's will still give a useful classification of the states.

We follow this method in dealing with our permutations P. We must find a function χ of the P's such that $P\chi P^{-1} = \chi$ for every P. It is evident that a possible χ is $\sum P_c$, the sum of all the permutations in a certain class c, i.e. the sum of a set of similar permutations, since $\sum PP_cP^{-1}$ must consist of the same permutations summed in a different order. There will be one such χ for each class. Further, there can be no other independent χ, since an arbitrary function of the P's can be expressed as a linear function of them with numerical coefficients, and it will not then commute with every P unless the coefficients of similar P's are always the same. We thus obtain all the χ's that can be used for classifying the states. It is convenient to define each χ as an average instead of a sum, thus

$$\chi_c = n_c^{-1} \sum P_c,$$

where n_c is the number of P's in the class c. An alternative expression for χ_c is

$$\chi_c = n!^{-1} \sum_P PP_cP^{-1}, \qquad (13)$$

the sum being extended over all the $n!$ permutations P, it being easy to verify that this sum contains each member of the class c the same number of times. For each permutation P there is one χ, $\chi(P)$ say, equal to the average of all permutations similar to P. One of the χ's is $\chi(P_1) = 1$.

The constants of the motion $\chi_1, \chi_2, \ldots, \chi_m$ obtained in this way will each have a definite numerical value for every stationary state of the system, in the case when the Hamiltonian does not involve the time explicitly, and also in the general case can be used for classifying the states, there being one set of states for every permissible set of numerical values $\chi'_1, \chi'_2, \ldots, \chi'_m$ for the χ's. Since the χ's are always constants of the motion, these sets of states will be *exclusive*, i.e. transitions will never take place from a state in one set to a state in another.

The permissible sets of values χ' that one can give to the χ's are limited by the fact that there exist algebraic relations between the χ's. The product of any two χ's, $\chi_p \chi_q$, is of course expressible as a linear function of the P's, and since it commutes with every P it must be expressible as a linear function of the χ's, thus

$$\chi_p \chi_q = a_1 \chi_1 + a_2 \chi_2 + \cdots + a_m \chi_m, \tag{14}$$

where the a's are numbers. Any numerical values χ' that one gives to the χ's must be eigenvalues of the χ's and must satisfy these same algebraic equations. For every solution χ' of these equations there is one exclusive set of states. One solution is evidently $\chi'_p = 1$ for every χ_p, giving the set of symmetrical states. A second obvious solution, giving the set of antisymmetrical states, is $\chi'_p = \pm 1$, the + or − sign being taken according to whether the permutations in the class p are even or odd. The other solutions may be worked out in any special case by ordinary algebraic methods, as the coefficients a in (14) may be obtained directly by a consideration of the types of permutation to which the χ's concerned refer. Any solution is, apart from a certain factor, what is called in group theory a *character* of the group of permutations. The χ's are all real dynamical variables, since each P and its conjugate complex P^{-1} are similar and will occur added together in the definition of any χ, so that the χ''s must be all real numbers.

The number of possible solutions of the equations (14) may easily be determined, since it must equal the number of different eigenvalues of an arbitrary function B of the χ's. We can express B as a linear function of the χ's with the help of equations (14); thus

$$B = b_1 \chi_1 + b_2 \chi_2 + \cdots + b_m \chi_m. \tag{15}$$

Similarly, we can express each of the quantities $B^2, B^3, \ldots, B^m$ as a linear function of the χ's. From the m equations thus obtained, together with the equation $\chi(P_1) = 1$, we can eliminate the m unknowns $\chi_1, \chi_2, \ldots, \chi_m$, obtaining as result an algebraic equation of degree m for B,

$$B^m + c_1 B^{m-1} + c_2 B^{m-2} + \cdots + c_m = 0.$$

The m solutions of this equation give the m possible eigenvalues for B, each of which will, according to (15), be a linear function of $b_1, b_2, \ldots, b_m$

Whose coefficients are a permissible set of values $\chi'_1, \chi'_2, \ldots, \chi'_m$. The sets of values χ' thus obtained must be all different, since if there were fewer than m different permissible sets of values χ' for the χ's, there would exist a linear function of the χ's every one of whose eigenvalues vanishes, which would mean that the linear function itself vanishes and the χ's are not linearly independent. Thus the number of permissible sets of numerical values for the χ's is just equal to m, which is the number of classes of permutations or the number of partitions of n. This number is therefore the number of exclusive sets of states.

All dynamical variables of physical importance and all observable quantities are symmetrical between the particles and thus commute with all the P's. Thus the only functions of the P's of physical importance are the χ's. The states corresponding to $|\chi'\rangle$ and to $f(P)|\chi'\rangle$, where $|\chi'\rangle$ is any eigenket of the χ's belonging to the eigenvalues χ' and $f(P)$ is any function of the P's such that $f(P)|\chi'\rangle \neq 0$, are observationally indistinguishable and are thus physically equivalent. There is a definite number, $n(\chi')$ say, of independent kets which can be formed by multiplying $|\chi'\rangle$ by functions of the P's, which number depends only on the χ''s. It is the number of rows and columns in a matrix representation of the P's in which each χ is equal to χ'. If $|\chi'\rangle$ corresponds to a stationary state, $n(\chi')$ will be its degree of degeneracy (so far as concerns degeneracy caused by the symmetry between the particles). This degeneracy cannot be removed by any perturbation that is symmetrical between the particles.

57. Determination of the Energy-levels

Let us apply the perturbation method of § 43 and make a first-order calculation of the energy-levels in the case when the Hamiltonian does not involve the time explicitly. We suppose that for our unperturbed stationary states of the assembly each of the similar particles has its own individual state. With n particles, we shall have n of these states, corresponding to kets $|\alpha^1\rangle, |\alpha^2\rangle, \ldots, |\alpha^n\rangle$ say, which we assume for the present to be all orthogonal. The ket for the assembly is then

$$|X\rangle = |\alpha_1^1\rangle|\alpha_2^2\rangle \cdots |\alpha_n^n\rangle, \tag{16}$$

like (1) with $\alpha^1, \alpha^2, \ldots$ instead of $a, b, \ldots$. If we apply any permutation P to it we get another ket

$$P|X\rangle = |\alpha_r^1\rangle|\alpha_s^2\rangle \cdots |\alpha_z^n\rangle \tag{17}$$

say, $r, s, \ldots, z$ being some permutation of the numbers $1, 2, \ldots, n$, corresponding to another stationary state of the assembly with the same energy.

57. DETERMINATION OF THE ENERGY-LEVELS

There are thus altogether $n!$ unperturbed states with this energy, if we assume there are no other causes of degeneracy. According to the method of § 43 when the unperturbed system is degenerate, we must consider those elements of the matrix representing the perturbing energy V that refer to two states with the same energy, i.e. those of the type $\langle X|P_a V P_b|X\rangle$. These will form a matrix with $n!$ rows and columns, whose eigenvalues are the first-order corrections in the energy-levels.

We must now introduce another kind of permutation operator which can be applied to kets of the form (17), namely a permutation which acts on the indices of the α's. We denote such a permutation operator by P^α. The essential difference between the P's and the P^α's may be seen in the following way. Let us consider a permutation in the general sense, say that consisting of the interchange of 2 and 3. This may be interpreted either as the interchange of the objects 2 and 3 or as the interchange of the objects in the places 2 and 3, these two operations producing in general quite different results. The first of these interpretations is the one that gives the operators P, the objects concerned being the similar particles. A permutation P can be applied to an arbitrary ket for the assembly. A permutation with the second interpretation has a meaning, however, only when applied to a ket of the form (17), for which each of the particles is in a 'place' specified by an α, or to a sum of kets of the form (17). A permutation P may be considered as an ordinary dynamical variable. A permutation P^α may be considered as a dynamical variable in a restricted sense, valid when one is dealing only with states obtainable by superposition of the various states (17). This is the case for our present perturbation problem.

We can form algebraic functions of the P^α which will be other operators applicable to kets of the form (17). In particular we can form $\chi(P_c^\alpha)$, the average of all P^α's in a certain class c. This must equal $\chi(P_c)$, the average of the permutation operators P in the same class, since the total set of all permutations in a given class must evidently be the same whether the permutations are applied to the particles or to the places the particles are in. Any P commutes with any P^α, i.e.

$$P_a P_b^\alpha = P_b^\alpha P_a. \tag{18}$$

By labelling the α's by the same numbers $1, 2, \ldots, n$ which label the particles, we set up a one-one correspondence between the α's and the particles, so that given any permutation P_a applying to the particles, we can give a meaning to the same permutation P_a^α applying to the α's. This meaning is such that, for the ket $|X\rangle$ given by (16),

$$P_a^\alpha P_a |X\rangle = |X\rangle. \tag{19}$$

Since the various kets $|\alpha^1\rangle, |\alpha^2\rangle, \ldots$ are orthogonal, $|X\rangle$ and $P|X\rangle$ are orthogonal unless $P = 1$. It follows that, for any coefficients c_P,

$$\sum_P c_P \langle X | P^\alpha P_a | X \rangle = c_{P_a}, \tag{20}$$

provided $|X\rangle$ is normalized, the summation being over all the $n!$ permutations P or P^α, with P_a fixed. Now define V_P by

$$V_P = \langle X | V P | X \rangle. \tag{21}$$

We then have, for any two permutations P_x and P_y,

$$\langle X | P_x V P_y | X \rangle = \langle X | V P_x P_y | X \rangle = V_{P_x P_y}$$
$$= \sum_P V_P \langle X | P^\alpha P_x P_y | X \rangle$$

with the help of (20). From (18) this gives

$$\langle X | P_x V P_y | X \rangle = \sum_P V_P \langle X | P_x P^\alpha P_y | X \rangle. \tag{22}$$

We may write this result as

$$V \approx \sum_P V_P P^\alpha, \tag{23}$$

where the sign $\approx$ means an equation in a restricted sense, the operators on the two sides being equal so long as they are used only with kets of the form $P|X\rangle$ and their conjugate imaginary bras.

The formula (23) shows that the perturbing energy V is equal, in the restricted sense, to a linear function of the permutation operators P^α with coefficients V_P given by (21). The restricted sense is adequate for the calculation of the first-order correction in the energy-levels, as this calculation involves only those matrix elements of V given by (22). The formula (23) is a very convenient one because the expression on its right-hand side is easily handled.

As an example of an application of (23) we shall determine the average energy of all those states, arising from the unperturbed state (16), that belong to one exclusive set. This requires us to calculate the average eigenvalue of V for those states (17) for which the χ's have specified numerical values χ'. Now the average eigenvalue of P_a^α for any of these states equals that of $P^\alpha P_a^\alpha (P^\alpha)^{-1}$ for arbitrary P^α and thus equals that of $n!^{-1} \sum_{P^a} P^a P_a^a (P^a)^{-1}$, which is $\chi'(P_a^a)$ or $\chi'(P_a)$. Hence the average eigenvalue of V is $\sum_P V_P \chi'(P)$. A similar method could be used for calculating the average eigenvalue of any function of V, it being necessary only to replace each P^a by $\chi'(P)$ to perform the averaging.

The number of energy-levels in an exclusive set $\chi = \chi'$ that arise from a given state of the unperturbed system is equal to the number of eigenvalues of the right-hand side of (23) that are consistent with the equations $\chi = \chi'$. This number is the number $n(\chi')$ introduced at the end of the preceding section, and is thus just the degree of degeneracy of the states in this set.

We have assumed that the individual kets $|\alpha^1\rangle, |\alpha^2\rangle, \ldots$ which determine the unperturbed state according to (16) are all orthogonal. The theory can easily be extended to the case when some of these kets are equal, any two that are not equal being still restricted to be orthogonal. We now have some permutations P^a such that $P^a |X\rangle = |X\rangle$, namely those permutations which involve only interchanges of equal α's. Equation (20) will now hold if the summation is extended only over those P's which make $P^a |X\rangle$ different. With this change in the meaning of $\sum_P$, all the previous equations still hold, including the result (23). For the present $|X\rangle$ there will be restrictions on the possible numerical values of the χ's, e.g. they cannot have those values corresponding to $|X\rangle$ being antisymmetrical.

58. Application to Electrons

Let us consider the case when the similar particles are electrons. This requires, according to Pauli's exclusion principle discussed in § 54, that we take into account only the antisymmetrical states. It is now necessary to make explicit reference to the fact that electrons have spins, which show themselves through an angular momentum and a magnetic moment. The effect of the spin on the motion of an electron in an electromagnetic field is not very great. There are additional forces on the electron due to its magnetic moment, requiring additional terms in the Hamiltonian. The spin angular momentum does not have any direct action on the motion, but it comes into play when there are forces tending to rotate the magnetic moment, since the magnetic moment and angular momentum are constrained to be always in the same direction. In the absence of a strong magnetic field these effects are all small, of the same order of magnitude as the corrections required by relativistic mechanics, and there would be no point in taking them into account in a non-relativistic theory. The importance of the spin lies not in these small effects on the motion of the electron, but in the fact that it gives two internal states to the electron, corresponding to the two possible values of the spin component in any assigned direction, which causes a doubling in the number of independent states of an electron. This fact has far-reaching consequences when combined with Pauli's exclusion principle.

In dealing with an assembly of electrons we have two kinds of dynamical variables. The first kind, which we may call the *orbital variables*, consists of the coordinates x, y, z of all the electrons and their conjugate momenta p_x, p_y, p_z. The second kind consists of the spin variables, the variables $\sigma_x, \sigma_y, \sigma_z$, as introduced in § 37, for all the electrons. These two kinds of variables belong to different degrees of freedom. According to §§ 20 and 21, a ket fixing the state of the whole system may be of the form $|A\rangle|B\rangle$, where $|A\rangle$ is a ket referring to the orbital variables alone and $|B\rangle$ is a ket referring to the spin variables alone, and the general ket fixing a state of the whole system is a sum or integral of kets of this form. This way of looking at things enables us to introduce two kinds of permutation operators, the first kind, P^x say, applying to the orbital variables only and operating only on the factor $|A\rangle$ and the second kind, P^σ say, applying only to the spin variables and operating only on the factor $|B\rangle$. The P^x's and P^σ's can each be applied to any ket for the whole system, not merely to certain special kets, like the P^α's of the preceding section. The permutations P that we have had up to the present apply to all the dynamical variables of the particles concerned, so for electrons they will apply to both the orbital and the spin variables. This means that each P_a equals the product

$$P_a = P_a^x P_a^\sigma. \tag{24}$$

We can now see the need for taking the spin variables into account when applying Pauli's exclusion principle, even if we neglect the spin forces in the Hamiltonian. For any state occurring in nature each P_a must have the value ± 1, according to whether it is an even or an odd permutation, so from (24)

$$P_a^x P_a^\sigma = \pm 1. \tag{25}$$

The theory of the three preceding sections would become trivial if applied directly to electrons, for which each $P_a = \pm 1$. We may, however, apply it to the P^x permutations of electrons. The P^σ's are constants of the motion if we neglect the terms in the Hamiltonian that arise from the spin forces, since this neglect results in the Hamiltonian not involving the spin dynamical variables σ at all. The P^x's must then also be constants of the motion. We can now introduce new χ's, equal to the average of all of the P^x's in each class, and assert that for any permissible set of numerical values χ' for these χ's there will be one exclusive set of states. Thus there exist exclusive sets of states for systems containing many electrons even when we restrict ourselves to a consideration of only those states that satisfy Pauli's principle. The exclusiveness of the sets of states is now, of course, only approximate, since the χ's are constants only so long as we neglect the spin forces. There will actually be a small probability for a transition from a state in one set to a state in another.

Equation (25) gives us a simple connexion between the P^x's and P^σ's, which means that instead of studying the dynamical variables P^x we can get all the results we want, e.g. the characters χ', by studying the dynamical variables P^σ. The P^σ's are much easier to study on account of there being only two independent states of spin for each electron. This fact results in there being fewer characters χ' for the group of permutations of the σ-variables than for the group of general permutations, since it prevents a ket in the spin variables from being antisymmetrical in more than two of them.

The study of the P^σ's is made specially easy by the fact that we can express them as algebraic functions of the dynamical variables σ. Consider the quantity

$$O_{12} = \frac{1}{2}(1 + \sigma_{x1}\sigma_{x2} + \sigma_{y1}\sigma_{y2} + \sigma_{z1}\sigma_{z2}) = \frac{1}{2}[1 + (\sigma_1, \sigma_2)].$$

With the help of equations (50) and (51) of § 37 we find readily that

$$(\sigma_1, \sigma_2)^2 = (\sigma_{x1}\sigma_{x2} + \sigma_{y1}\sigma_{y2} + \sigma_{z1}\sigma_{z2})^2 = 3 - 2(\sigma_1, \sigma_2), \qquad (26)$$

and hence that

$$O_{12}^2 = \frac{1}{4}[1 + 2(\sigma_1, \sigma_2) + (\sigma_1, \sigma_2)^2] = 1. \qquad (27)$$

Again, we find

$$O_{12}\sigma_{x1} = \frac{1}{2}(\sigma_{x1} + \sigma_{x2} - i\sigma_{z1}\sigma_{y2} + i\sigma_{y1}\sigma_{z2}),$$

$$\sigma_{x2}O_{12} = \frac{1}{2}(\sigma_{x2} + \sigma_{x1} + i\sigma_{y1}\sigma_{z2} - i\sigma_{z1}\sigma_{y2})$$

and hence

$$O_{12}\sigma_{x1} = \sigma_{x2}O_{12}.$$

Similar relations hold for σ_{y1} and σ_{z1} so that we have

$$O_{12}\sigma_1 = \sigma_2 O_{12}$$

or
$$O_{12}\sigma_1 O_{12}^{-1} = \sigma_2.$$
From this we can obtain with the help of (27)
$$O_{12}\sigma_2 O_{12}^{-1} = \sigma_1.$$
These commutation relations for O_{12} with σ_1 and σ_2 are precisely the same as those for P_{12}^σ, the permutation consisting of the interchange of the spin variables of electrons 1 and 2. Thus we can put
$$O_{12} = c P_{12}^\sigma,$$
where c is a number. Equation (27) shows that $c = \pm 1$. To determine which of these values for c is the correct one, we observe that the eigenvalues of P_{12}^σ are $1, 1, 1, -1$, corresponding to the fact that there exist three independent symmetrical and one antisymmetrical state in the spin variables of two electrons, namely, with the notation of § 37, the states represented by the three symmetrical functions $f_\alpha(\sigma'_{z1}) f_\alpha(\sigma'_{z2})$, $f_\beta(\sigma'_{z1}) f_\beta(\sigma'_{z2})$, $f_\alpha(\sigma'_{z1}) f_\beta(\sigma'_{z2}) + f_\beta(\sigma'_{z1}) f_\alpha(\sigma'_{z2})$, and the one antisymmetrical function $f_\alpha(\sigma'_{z1}) f_\beta(\sigma'_{z2}) - f_\beta(\sigma'_{z1}) f_\alpha(\sigma'_{z2})$. Thus the mean of the eigenvalues of P_{12}^σ is $1/2$. Now the mean of the eigenvalues of (σ_1, σ_2) is evidently zero and hence the mean of the eigenvalues of O_{12} is $1/2$. Thus we must have $c = +1$, and so we can put

$$P_{12}^\sigma = \frac{1}{2}[1 + (\sigma_1, \sigma_2)]. \tag{28}$$

In this way any permutation P^σ consisting simply of an interchange can be expressed as an algebraic function of the σ's. Any other permutation P^σ can be expressed as a product of interchanges and can therefore also be expressed as a function of the σ's. With the help of (25) we can now express the P^x's as algebraic functions of the σ's and eliminate the P^σ's from the discussion. We have, since the $-$ sign must be taken in (25) when the permutations are interchanges and since the square of an interchange is unity,

$$P_{12}^x = -\frac{1}{2}[1 + (\sigma_1, \sigma_2)]. \tag{29}$$

The formula (29) may conveniently be used for the evaluation of the characters χ' which define the exclusive sets of states. We have, for example, for the permutations consisting of interchanges,

$$\chi_{12} = \chi(P_{12}^x) = -\frac{1}{2}\left\{1 + \frac{2}{n[n-1]} \sum_{r<t}(\sigma_r, \sigma_t)\right\}.$$

58. APPLICATION TO ELECTRONS

If we introduce the dynamical variable s to describe the magnitude of the total spin angular momentum, $(1/2) \sum_r \sigma_r$ in units of $\hbar$, through the formula

$$s(s+1) = \left(\frac{1}{2} \sum_r \sigma_r, \frac{1}{2} \sum_t \sigma_t \right),$$

in agreement with (39) of § 36, we have

$$2 \sum_{r<t} (\sigma_r, \sigma_t) = \left(\sum_r \sigma_r, \sum_t \sigma_t \right) - \sum_r (\sigma_r, \sigma_r)$$
$$= 4s(s+1) - 3n.$$

Hence

$$\chi_{12} = -\frac{1}{2} \left[1 + \frac{4s(s+1) - 3n}{n(n-1)} \right] = -\frac{n(n-4) + 4s(s+1)}{2n(n-1)}. \tag{30}$$

Thus χ_{12} is expressible as a function of the dynamical variable s and of n the number of electrons. Any of the other χ's could be evaluated on similar lines and would have to be a function of s and n only, since there are no other symmetrical functions of all the σ dynamical variables which could be involved. There is therefore one set of numerical values χ' for the χ's, and thus one exclusive set of states, for each eigenvalue s' of s. The eigenvalues of s are

$$\frac{1}{2}n, \quad \frac{1}{2}n - 1, \quad \frac{1}{2}n - 2, \quad \ldots,$$

the series terminating with 0 or $1/2$.

We see in this way that each of the stationary states of a system with several electrons is an eigenstate of s, the magnitude in units of $\hbar$ of the total spin angular momentum $(1/2) \sum_r \sigma_r$, belonging to a definite eigenvalue s'. For any given s' there will be $2s'+1$ possible values for a component of the total spin vector in any direction and these will correspond to $2s'+1$ independent stationary states with the same energy. When we do not neglect the forces due to the spin magnetic moments these $2s'+1$ states will in general be split up into $2s'+1$ states with slightly different energies, and will thus form a multiplet of multiplicity $2s'+1$. Transitions in which s' changes, i.e. transitions from one multiplicity to another, cannot occur when the spin forces are neglected and will have only a small probability of occurrence when the spin forces are not neglected.

We can determine the energy-levels of a system with several electrons to the first approximation by applying the theory of the preceding section with the kets $|\alpha^r\rangle$ referring only to the orbital variables and using formula (23). If we consider only the Coulomb forces between the electrons, then the interaction energy V will consist of a sum of parts each referring to only two electrons, which will result in all the matrix elements V_P vanishing except those for which P^x is the identical permutation or is simply an interchange of two electrons. Thus (23) will reduce to

$$V \approx V_1 + \sum_{r<s} V_{rs} P^\alpha_{rs}, \qquad (31)$$

V_{rs} being the matrix element referring to the interchange of electrons r and s. Since the P^α's have the same properties as the P^x's, any function of the P^α's will have the same eigenvalues as the corresponding function of the P^x's, so that the right-hand side of (31) will have the same eigenvalues as

$$V_1 + \sum_{r<s} V_{rs} P^x_{rs},$$

or

$$V_1 - \frac{1}{2} \sum_{r<s} V_{rs}[1 + (\sigma_r, \sigma_s)] \qquad (32)$$

from (29). The eigenvalues of (32) will give the first-order corrections in the energy-levels. The form of (32) shows that a model which assumes a coupling energy between the spins of the various electrons, of magnitude $-[1/2]V_{rs}(\sigma_r, \sigma_s)$ for the electrons in the r and s orbital states, would meet with a fair amount of success. This coupling energy is much greater than that of the spin magnetic moments. Such models of the atom were in use before the justification by quantum mechanics was obtained.

We may have two of the orbital states of the unperturbed system the same, i.e. the kets $|\alpha^r\rangle$ in the orbital variables for two electrons may be the same. Suppose $|\alpha^1\rangle$ and $|\alpha^2\rangle$ are the same. Then we must take only those eigenvalues of (31) that are consistent with $P^\alpha_{12} = 1$, or those eigenvalues of (32) that are consistent with $P^x_{12} = 1$ or $P^\sigma_{12} = -1$. From (28) this condition gives $(\sigma_1, \sigma_2) = -3$, so that $(\sigma_1 + \sigma_2)^2 = 0$. Thus the resultant of the two spins σ_1 and σ_2 is zero, which may be interpreted as the spins σ_1 and σ_2 being antiparallel. Thus we may say that two electrons in the same orbital state have their spins antiparallel. More than two electrons cannot be in the same orbital state.

CHAPTER X

Theory of Radiation

59. An Assembly of Bosons

We consider a dynamical system composed of u' similar particles. We set up a representation for one of the particles with discrete basic kets $|\alpha^{(1)}\rangle, |\alpha^{(2)}\rangle, |\alpha^{(3)}\rangle, \ldots$. Then, as explained in § 54, we get a symmetrical representation of the assembly of u' particles by taking as basic kets the products

$$|\alpha_1^a\rangle|\alpha_2^b\rangle|\alpha_3^c\rangle \cdots |\alpha_{u'}^g\rangle = |\alpha_1^a \alpha_2^b \alpha_3^c \cdots \alpha_{u'}^g\rangle \tag{1}$$

in which there is one factor for each particle, the suffixes $1, 2, 3, \ldots, u'$ of the α's being the labels of the particles and the indices $a, b, c, \ldots, g$ denoting indices $^{(1)}, ^{(2)}, ^{(3)}, \ldots$ in the basic kets for one particle. If the particles are bosons, so that only symmetrical states occur in nature, then we need to work with only the symmetrical kets that can be constructed from the kets (1). The states corresponding to these symmetrical kets will form a complete set of states for the assembly of bosons. We can build up a theory of them as follows.

We introduce the linear operator S defined by

$$S = u'!^{-1/2} \sum P, \tag{2}$$

the sum being taken over all the $u'!$ permutations of the u' particles. Then S applied to any ket for the assembly gives a symmetrical ket. We may therefore call S the *symmetrizing operator*. From (8) of § 55 it is real. Applied to the ket (1) it gives

$$u'!^{-1/2} \sum P|\alpha_1^a \alpha_2^b \alpha_3^c \cdots \alpha_{u'}^g\rangle = S|\alpha^a \alpha^b \alpha^c \cdots \alpha^g\rangle, \tag{3}$$

the labels of the particles being omitted on the right-hand side as they are no longer relevant. The ket (3) corresponds to a state for the assembly of u' bosons with a definite distribution of the bosons among the various boson states, without any particular boson being assigned to any particular state. The distribution of bosons is specified if we specify how many bosons are in each boson state. Let $n_1', n_2', n_3', \ldots$ be the numbers of bosons in the states $\alpha^{(1)}, \alpha^{(2)}, \alpha^{(3)}, \ldots$ respectively with this distribution. The n''s are defined algebraically by the equation

$$\alpha^a + \alpha^b + \alpha^c + \cdots + \alpha^g = n'_1 \alpha^{(1)} + n'_2 \alpha^{(2)} + n'_3 \alpha^{(3)} + \cdots . \qquad (4)$$

The sum of the n''s is of course u'. The number of n''s is equal to the number of basic kets $|\alpha^{(r)}\rangle$, which in most applications of the theory is very much greater than u', so most of the n''s will be zero. If $\alpha^a, \alpha^b, \alpha^c, \ldots, \alpha^g$ are all different, i.e. if the n''s are all 0 or 1, the ket (3) is normalized, since in this case the terms on the left-hand side of (3) are all orthogonal to one another and each contributes $u'!^{-1}$ to the squared length of the ket. However, if $\alpha^a, \alpha^b, \alpha^c, \ldots, \alpha^g$ are not all different, those terms on the left-hand side of (3) will be equal which arise from permutations P which merely interchange bosons in the same state. The number of equal terms will be $n'_1! n'_2! n'_3! \cdots$, so the squared length of the ket (3) will be

$$\langle \alpha^a \alpha^b \alpha^c \cdots \alpha^g | S^2 | \alpha^a \alpha^b \alpha^c \cdots \alpha^g \rangle = n'_1! n'_2! n'_3! \ldots . \qquad (5)$$

For dealing with a general state of the assembly we can introduce the numbers $n_1, n_2, n_3, \ldots$ of bosons in the states $\alpha^{(1)}, \alpha^{(2)}, \alpha^{(3)}, \ldots$ respectively and treat the n's as dynamical variables or as observables. They have the eigenvalues $0, 1, 2, \ldots, u'$. The ket (3) is a simultaneous eigenket of all the n's, belonging to the eigenvalues $n'_1, n'_2, n'_3, \ldots$. The various kets (3) form a complete set for the dynamical system consisting of u' bosons, so the n's all commute (see the converse to the theorem of § 13). Further, there is only one independent ket (3) belonging to any set of eigenvalues $n'_1, n'_2, n'_3, \ldots$. Hence the n's form a complete set of commuting observables. If we normalize the kets (3) and then label the resulting kets by the eigenvalues of the n's to which they belong, i.e. if we put

$$(n'_1! n'_2! n'_3! \cdots)^{-1/2} S |\alpha^a \alpha^b \alpha^c \cdots \alpha^g \rangle = |n'_1 n'_2 n'_3 \cdots \rangle, \qquad (6)$$

we get a set of kets $|n'_1 n'_2 n'_3 \cdots \rangle$, with the n''s taking on all non-negative integral values adding up to u', which kets will form the basic kets of a representation with the n's diagonal.

The n's can be expressed as functions of the observables $\alpha_1, \alpha_2, \alpha_3, \ldots, \alpha_{u'}$ which define the basic kets of the individual bosons by means of the equations

$$n_a = \sum_r \delta_{\alpha_r \alpha_r^a}, \qquad (7)$$

or the equations

$$\sum_a n_a f(\alpha^a) = \sum_r f(\alpha_r) \qquad (8)$$

holding for any function f.

Let us now suppose that the number of bosons in the assembly is not given, but is variable. This number is then a dynamical variable or observable u, with eigenvalues $0, 1, 2, \ldots$, and the ket (3) is an eigenket of u belonging to the eigenvalue u'. To get a complete set of kets for our dynamical system we must now take all the symmetrical kets (3) for all values of u'. We may arrange them in order thus

$$|\rangle, \quad |\alpha^a\rangle, \quad S|\alpha^a\alpha^b\rangle, \quad S|\alpha^a\alpha^b\alpha^c\rangle, \quad \ldots, \tag{9}$$

where first is written the ket, with no label, corresponding to the state with no bosons present, then come the kets corresponding to states with one boson present, then those corresponding to states with two bosons, and so on. A general state corresponds to a ket which is a sum of the various kets (9). The kets (9) are all orthogonal to one another, two kets referring to the same number of bosons being orthogonal as before, and two referring to different numbers of bosons being orthogonal since they are eigenkets of u belonging to different eigenvalues. By normalizing all the kets (9), we get a set of kets like (6) with no restriction on the n''s (i.e. each n' taking on all non-negative integral values) and these kets form the basic kets of a representation with the n's diagonal for the dynamical system consisting of a variable number of bosons.

If there is no interaction between the bosons and if the basic kets $|\alpha^{(1)}\rangle, |\alpha^{(2)}\rangle, |\alpha^3\rangle \ldots$ correspond to stationary states of a boson, the kets (9) will correspond to stationary states for the assembly of bosons. The number u of bosons is now constant in time, but it need not be a specified number, i.e. the general state is a superposition of states with various values for u. If the energy of one boson is $H(\alpha)$, the energy of the assembly will be

$$\sum_r H(\alpha_r) = \sum_a n_a H^a \tag{10}$$

from (8), H^a being short for the number $H(\alpha^a)$. This gives the Hamiltonian for the assembly as a function of the dynamical variables n.

60. The Connexion between Bosons and Oscillators

In § 34 we studied the harmonic oscillator, a dynamical system of one degree of freedom describable in terms of a canonical q and p, such that the Hamiltonian is a sum of squares of q and p, with numerical coefficients. We define a general oscillator mathematically as a system of one degree of freedom describable in terms of a canonical q and p, such that the Hamiltonian is a power series in q and p, and remains so if the system is perturbed in any way. We shall now study a dynamical system composed of several of these oscillators. We can describe each oscillator in terms of, instead of q and p, a complex dynamical variable η, like the η of § 34, and its conjugate complex $\bar{\eta}$, satisfying the commutation relation (7) of § 34. We attach labels $1, 2, 3, \ldots$ to the different oscillators, so that the whole set of oscillators is describable in terms of the dynamical variables $\eta_1, \eta_2, \eta_3, \ldots, \bar{\eta}_1, \bar{\eta}_2, \bar{\eta}_3, \ldots$ satisfying the commutation relations

$$\eta_a \eta_b - \eta_b \eta_a = 0,$$
$$\bar{\eta}_a \bar{\eta}_b - \bar{\eta}_b \bar{\eta}_a = 0, \tag{11}$$
$$\bar{\eta}_a \eta_b - \eta_b \bar{\eta}_a = \delta_{ab}.$$

Put

$$\eta_a \bar{\eta}_a = n_a, \tag{12}$$

so that

$$\bar{\eta}_a \eta_a = n_a + 1. \tag{13}$$

The n's are observables which commute with one another and the work of § 34 shows that each of them has as eigenvalues all nonnegative integers. For the ath oscillator there is a standard ket for the Fock representation, $|0_a\rangle$ say, which is a normalized eigenket of n_a belonging to the eigenvalue zero. By multiplying all these standard kets together we get a standard ket for the Fock representation for the set of oscillators,

$$|0_1\rangle |0_2\rangle |0_3\rangle \cdots, \tag{14}$$

which is a simultaneous eigenket of all the n's belonging to the eigenvalues zero. We shall denote it simply by $|0\rangle$. From (13) of § 34

$$\bar{\eta}_a |0\rangle = 0 \tag{15}$$

for any a. The work of § 34 also shows that, if $n'_1, n'_2, n'_3, \ldots$ are any nonnegative integers,

$$\eta_1^{n'_1} \eta_2^{n'_2} \eta_3^{n'_3} \cdots |0\rangle \tag{16}$$

is a simultaneous eigenket of all the n's belonging to the eigenvalues $n'_1, n'_2, n'_3, \ldots$ respectively. The various kets (16) obtained by taking different n''s form a complete set of kets all orthogonal to one another and the square of the length of one of them is, from (16) of § 34, $n'_1! n'_2! n'_3! \cdots$. From this we see, bearing in mind the result (5), that the kets (16) have just the same properties as the kets (9), so that we can equate each ket (16) to the ket (9) referring to the same n' values without getting any inconsistency. This involves putting

$$S|\alpha^a\alpha^b\alpha^c \cdots \alpha^g\rangle = \eta_a\eta_b\eta_c \cdots \eta_g|0\rangle. \tag{17}$$

The standard ket $|0\rangle$ becomes equal to the first of the kets (9), corresponding to no bosons present.

The effect of equation (17) is to identify the states of an assembly of bosons with the states of a set of oscillators. This means that *the dynamical system consisting of an assembly of similar bosons is equivalent to the dynamical system consisting of a set of oscillators—the two systems are just the same system looked at from two different points of view.* There is one oscillator associated with each independent boson state. We have here one of the most fundamental results of quantum mechanics, which enables a unification of the wave and corpuscular theories of light to be effected.

Our work in the preceding section was built up on a discrete set of basic kets $|\alpha^a\rangle$ for a boson. We could pass to a different discrete set of basic kets, $|\beta^A\rangle$ say, and build up a similar theory on them. The basic kets for the assembly would then be, instead of (9),

$$|\rangle, \quad |\beta^A\rangle, \quad S|\beta^A\beta^B\rangle, \quad S|\beta^A\beta^B\beta^C\rangle, \quad \ldots. \tag{18}$$

The first of the kets (18), referring to no bosons present, is the same as the first of the kets (9). Those kets (18) referring to one boson present are linear functions of those kets (9) referring to one boson present, namely

$$|\beta^A\rangle = \sum_a |\alpha^a\rangle\langle\alpha^a|\beta^A\rangle, \tag{19}$$

and generally those kets (18) referring to u' bosons present are linear functions of those kets (9) referring to u' bosons present. Associated with the new basic states $|\beta^A\rangle$ for a boson there will be a new set of oscillator variables η_A, and corresponding to (17) we shall have

$$S|\beta^A\beta^B\beta^C\cdots\rangle = \eta_A\eta_B\eta_C \cdots |0\rangle. \tag{20}$$

Thus a ket $\eta_A\eta_B\eta_C \cdots |0\rangle$ with u' factors $\eta_A, \eta_B, \eta_C, \ldots$ must be a linear function of kets $\eta_a\eta_b\eta_c \cdots |0\rangle$ with u' factors $\eta_a, \eta_b, \eta_c, \ldots$. It follows that each linear operator η_A must be a linear function of the η_a's. Equation (19) gives

$$\eta_A|0\rangle = \sum_a \eta_a|0\rangle\langle\alpha^a|\beta^A\rangle$$

and hence
$$\eta_A = \sum_a \eta_a \langle \alpha^a | \beta^A \rangle. \tag{21}$$

Thus *the η's transform according to the same law as the basic kets for a boson*. The transformed η's satisfy, with their conjugate complexes, the same commutation relations (11) as the original ones. The transformed η's are on just the same footing as the original ones and hence, when we look upon our dynamical system as a set of oscillators, the different degrees of freedom have no invariant significance.

The $\bar{\eta}$'s transform according to the same law as the basic bras for a boson, and thus the same law as the numbers $\langle \alpha^a | x \rangle$ forming the representative of a state x. This similarity people often describe by saying that the $\bar{\eta}_a$'s are given by a process of *second quantization* applied to $\langle \alpha^a | x \rangle$, meaning thereby that, after one has set up a quantum theory for a single particle and so introduced the numbers $\langle \alpha^a | x \rangle$ representing a state of the particle, one can make these numbers into linear operators satisfying with their conjugate complexes the correct commutation relations, like (11), and one then has the appropriate mathematical basis for dealing with an assembly of the particles, provided they are bosons. There is a corresponding procedure for fermions, which will be given in § 65.

Since an assembly of bosons is the same as a set of oscillators, it must be possible to express any symmetrical function of the boson variables in terms of the oscillator variables η and $\bar{\eta}$. An example of this is provided by equation (10) with $\eta_a \bar{\eta}_a$ substituted for n_a. Let us see how it goes in general. Take first the case of a function of the boson variables of the form
$$U_T = \sum_r U_r, \tag{22}$$

where each U_r is a function only of the dynamical variables of the rth boson, so that it has a representative $\langle \alpha_r^a | U_r | \alpha_r^b \rangle$ referring to the basic kets $|\alpha_r^a\rangle$ of the rth boson. In order that U_T may be symmetrical, this representative must be the same for all r, so that it can depend only on the two eigenvalues labelled by a and b. We may therefore write it
$$\langle \alpha_r^a | U_r | \alpha_r^b \rangle = \langle \alpha^a | U | \alpha^b \rangle = \langle a | U | b \rangle \tag{23}$$

for brevity. We have
$$U_r | \alpha_1^{x_1} \alpha_2^{x_2} \alpha_3^{x_3} \cdots \rangle = \sum_a | \alpha_1^{x_1} \alpha_2^{x_2} \alpha_3^{x_3} \cdots \alpha_r^a \cdots \rangle \langle a | U | x_r \rangle. \tag{24}$$

Summing this equation for all values of r and applying the symmetrizing operator S to both sides, we get

$$SU_T|\alpha_1^{x_1}\alpha_2^{x_2}\alpha_3^{x_3}\cdots\rangle = \sum_r\sum_a S|\alpha_1^{x_1}\alpha_2^{x_2}\alpha_3^{x_3}\cdots\alpha_r^a\cdots\rangle\langle a|U|x_r\rangle. \quad (25)$$

Since U_T is symmetrical we can replace SU_T by $U_T S$ and can then substitute for the symmetrical kets in (25) their values given by (17). We get in this way

$$U_T\eta_{x_1}\eta_{x_2}\eta_{x_3}\cdots|0\rangle = \sum_a\sum_r \eta_a\eta_{x_r}^{-1}\eta_{x_1}\eta_{x_2}\eta_{x_3}\cdots|0\rangle\langle a|U|x_r\rangle$$

$$= \sum_{ab}\eta_a\sum_r \eta_{x_r}^{-1}\eta_{x_1}\eta_{x_2}\eta_{x_3}\cdots|0\rangle\delta_{bx_r}\langle a|U|b\rangle, \quad (26)$$

$\eta_{x_r}^{-1}$ meaning that the factor η_{x_r} must be cancelled out. Now from (15) and the commutation relations (11)

$$\bar{\eta}_b\eta_{x_1}\eta_{x_2}\eta_{x_3}\cdots|0\rangle = \sum_r \eta_{x_r}^{-1}\eta_{x_1}\eta_{x_2}\eta_{x_3}\cdots|0\rangle\delta_{bx_r} \quad (27)$$

(note that $\bar{\eta}_b$ is like the operator of partial differentiation $\partial/\partial\eta_b$), so (26) becomes

$$U_T\eta_{x_1}\eta_{x_2}\eta_{x_3}\cdots|0\rangle = \sum_{ab}\eta_a\bar{\eta}_b\eta_{x_1}\eta_{x_2}\eta_{x_3}\cdots|0\rangle\langle a|U|b\rangle. \quad (28)$$

The kets $\eta_{x_1}\eta_{x_2}\eta_{x_3}\cdots|0\rangle$ form a complete set, and hence we can infer from (28) the operator equation

$$U_T = \sum_{ab}\eta_a\langle a|U|b\rangle\bar{\eta}_b. \quad (29)$$

This gives us U_T in terms of the η and $\bar{\eta}$ variables and the matrix elements $\langle a|U|b\rangle$.

Now let us take a symmetrical function of the boson variables consisting of a sum of terms each referring to two bosons,

$$V_T = \sum_{r,s\neq r} V_{rs}. \quad (30)$$

We do not need to assume $V_{rs} = V_{sr}$. Corresponding to (23), V_{rs} has matrix elements

$$\langle \alpha_r^a\alpha_s^b|V_{rs}|\alpha_r^c\alpha_s^d\rangle = \langle ab|V|cd\rangle \quad (31)$$

for brevity. Proceeding as before we get, corresponding to (25),

$$SV_T|\alpha_1^{x_1}\alpha_2^{x_2}\alpha_3^{x_3}\cdots\rangle$$

242   X. THEORY OF RADIATION

$$= \sum_{r,s \neq r} \sum_{ab} S|\alpha_1^{x_1} \alpha_2^{x_2} \alpha_3^{x_3} \cdots \alpha_r^a \cdots \alpha_s^b \cdots\rangle \langle ab|V|x_r x_s\rangle \qquad (32)$$

and corresponding to (26)

$$V_T \eta_{x_1} \eta_{x_2} \eta_{x_3} \cdots |0\rangle$$
$$= \sum_{abcd} \eta_a \eta_b \sum_{r,s \neq r} \eta_{x_r}^{-1} \eta_{x_s}^{-1} \eta_{x_1} \eta_{x_2} \eta_{x_3} \cdots |0\rangle \delta_{cx_r} \delta_{dr_s} \langle ab|V|cd\rangle. \qquad (33)$$

We can deduce as an extension of (27)

$$\bar{\eta}_c \bar{\eta}_d \eta_{x_1} \eta_{x_2} \eta_{x_3} \cdots |0\rangle = \sum_{r,s \neq r} \eta_{x_r}^{-1} \eta_{x_s}^{-1} \eta_{x_1} \eta_{x_2} \eta_{x_3} \cdots |0\rangle \delta_{cx_r} \delta_{dx_s}, \qquad (34)$$

so that (33) becomes

$$V_T \eta_{x_1} \eta_{x_2} \eta_{x_3} \cdots |0\rangle = \sum_{abcd} \eta_a \eta_b \bar{\eta}_c \bar{\eta}_d \eta_{x_1} \eta_{x_2} \eta_{x_3} \cdots |0\rangle \langle ab|V|cd\rangle,$$

giving us the operator equation

$$V_T = \sum_{abcd} \eta_a \eta_b \langle ab|V|cd\rangle \bar{\eta}_c \bar{\eta}_d. \qquad (35)$$

The method can readily be extended to give any symmetrical function of the boson variables in terms of the η's and $\bar{\eta}$'s.

The foregoing theory can easily be generalized to apply to an assembly of bosons in interaction with some other dynamical system, which we shall call for definiteness the atom. We must introduce a set of basic kets, $|\zeta'\rangle$ say, for the atom alone. We can then get a set of basic kets for the whole system of atom and bosons together by multiplying each of the kets $|\zeta'\rangle$ into each of the kets (9). We may write these kets

$$|\zeta'\rangle, \quad |\zeta' \alpha^a\rangle, \quad S|\zeta' \alpha^a \alpha^b\rangle, \quad S|\zeta' \alpha^a \alpha^b \alpha^c\rangle, \quad \ldots. \qquad (36)$$

We may look upon the system as composed of the atom in interaction with a set of oscillators, so that it can be described in terms of the atom variables and the oscillator variables $\eta_a, \bar{\eta}_a$. Using again the standard ket $|0\rangle$ for the set of oscillators, we have

$$S|\zeta' \alpha^a \alpha^b \alpha^c \cdots\rangle = \eta_a \eta_b \eta_c \cdots |0\rangle |\zeta'\rangle, \qquad (37)$$

corresponding to (17), as the equation expressing the basic kets (36) in terms of the oscillator variables.

Any function of the atom variables and boson variables which is symmetrical between all the bosons is expressible as a function of the atom variables and the η's and $\bar{\eta}$'s. Consider first a function U_T of the form (22) with U_r a function only of the atom variables and the variables of the rth boson, so that it has a representative $\langle\zeta'\alpha_r^a|U_r|\zeta''\alpha_r^b\rangle$. This representative must be independent of r in order that U_T may be symmetrical between all the bosons, so we may write it $\langle\zeta'\alpha^a|U|\zeta''\alpha^b\rangle$. Now let us define $\langle a|U|b\rangle$ to be that function of the atom variables whose representative is $\langle\zeta'\alpha^a|U|\zeta''\alpha^b\rangle$, so that we have

$$\langle\zeta'\alpha_r^a|U_r|\zeta''\alpha_r^b\rangle = \langle\zeta'\alpha^a|U|\zeta''\alpha^b\rangle = \langle\zeta'|\langle a|U|b\rangle|\zeta''\rangle, \quad (38)$$

corresponding to (23). The equations (24)–(28) can now be taken over and applied to the present work if both sides of all these equations are multiplied by $|\zeta'\rangle$ on the right, with the result that formula (29) still holds. We can deal similarly with a symmetrical function V_T of the form (30) with V_{rs} a function only of the atom variables and the variables of the rth and sth bosons. Defining $\langle ab|V|cd\rangle$ to be that function of the atom variables whose representative is

$$\langle\zeta'\alpha_r^a\alpha_s^b|V_{rs}|\zeta''\alpha_r^c\alpha_s^d\rangle,$$

we find that formula (35) still holds.

61. Emission and Absorption of Bosons

Let us suppose that the oscillators of the preceding section are harmonic oscillators and there is no interaction between them. The energy of the ath oscillator is then, from (5) of § 34,

$$H_a = \hbar\omega_a\eta_a\bar{\eta}_a + \frac{1}{2}\hbar\omega_a.$$

We shall neglect the constant term $(1/2)\hbar\omega_a$, which is the energy of the oscillator in its lowest state—the so-called 'zero-point energy'. This neglect does not have any dynamical consequences, as explained at the beginning of § 30, and merely involves a redefinition of H_a. The total energy of all the oscillators is now

$$H_T = \sum_a H_a = \sum_a \hbar\omega_a\eta_a\bar{\eta}_a = \sum_a \hbar\omega_a n_a \quad (39)$$

with the help of (12). This is of the same form as (10), with $\hbar\omega_a$ for H^a. Thus *a set of harmonic oscillators is equivalent to an assembly of bosons in stationary states with no interaction between them. If an oscillator of the set is in its n'th quantum state, there are n' bosons in the associated boson state.*

In general the Hamiltonian for the set of oscillators will be a power series in the variables η_a, $\bar{\eta}_a$, say

$$H_T = H_P + \sum_a (U_a \eta_a + \bar{U}_a \bar{\eta}_a) + \sum_{ab}(U_{ab}\eta_a\bar{\eta}_b + V_{ab}\eta_a\eta_b + \bar{V}_{ab}\bar{\eta}_a\bar{\eta}_b) + \cdots, \tag{40}$$

where H_P, U_a, U_{ab}, V_{ab}, are numbers, H_P being real and $U_{ab} = \bar{U}_{ba}$. If the set of oscillators are in interaction with an atom, as we had at the end of the preceding section, the total Hamiltonian will still be of the form (40), with H_P, U_a, U_{ab}, V_{ab} functions of the atom variables, H_P in particular being the Hamiltonian for the atom by itself. A general treatment of this dynamical system would be rather complicated and for practical applications one assumes that the terms

$$H_P + \sum_a U_{aa}\eta_a\bar{\eta}_a \tag{41}$$

are large compared with the others and form by themselves an unperturbed system, the remaining terms being taken into account as a perturbation producing transitions in the unperturbed system, according to the theory of § 44. If, further, U_{aa} is independent of the atom variables, the unperturbed system with Hamiltonian (41) consists merely of an atom with Hamiltonian H_P and an assembly of bosons in stationary states with Hamiltonian of the form (39), with no interaction.

Let us consider what kinds of transitions are produced by the various perturbation terms in (40). Take a stationary state of the unperturbed system for which the atom is in a stationary state, ζ' say, and bosons are present in the stationary boson states, $a, b, c, \ldots$. This stationary state for the unperturbed system corresponds to the ket

$$\eta_a \eta_b \eta_c \cdots |0\rangle |\zeta'\rangle, \tag{42}$$

like (37). If the term $U_x \eta_x$ of (40) is multiplied into this ket, the result is a linear combination of kets like

$$\eta_x \eta_a \eta_b \eta_c \cdots |0\rangle |\zeta''\rangle, \tag{43}$$

ζ'' denoting any stationary state of the atom. The ket (43) refers to one more boson than the ket (42), the extra boson being in the state x. Thus the perturbation term $U_x \eta_x$ gives rise to transitions in which one boson is emitted into state x and the atom makes an arbitrary jump. If the term $\bar{U}_x \bar{\eta}_x$ of (40) is multiplied into (42), the result is zero unless (42) contains a factor η_x and is then a linear combination of kets like

$$\eta_x^{-1} \eta_a \eta_b \eta_c \cdots |0\rangle |\zeta''\rangle,$$

referring to one boson less in state x. Thus the perturbation term $\overline{U}_x \bar{\eta}_x$ gives rise to transitions in which one boson is absorbed from state x, the atom again making an arbitrary jump. Similarly, we find that a perturbation term $U_{xy}\eta_x\bar{\eta}_y$ ($x \neq y$) gives rise to processes in which a boson is absorbed from state y and one is emitted into state x, or, what is the same thing physically, one boson makes a transition from state y to state x. This kind of process would be produced by a term like the U_T of (22) and (29) in the perturbation energy, provided the diagonal elements $\langle a|U|a \rangle$ vanish. Again, the perturbation terms $V_{xy}\eta_x\eta_y$, $\overline{V}_{xy}\bar{\eta}_x\bar{\eta}_y$ give rise to processes in which two bosons are emitted or absorbed, and so on for more complicated terms. With any of these emission and absorption processes the atom can make an arbitrary jump.

Let us determine how the probability of occurrence of each of these transition processes depends on the numbers of bosons originally present in the various boson states. From §§ 44, 46 the transition probability is always proportional to the square of the modulus of the matrix element of the perturbation energy referring to the two states concerned. Thus the probability of a boson being emitted into state x with the atom making a jump from state ζ' to state ζ'' is proportional to

$$|\langle \zeta''|n_1' n_2' n_3' \cdots (n_x' + 1) \cdots |U_x \eta_x| n_1' n_2' n_3' \cdots n_x' \cdots \rangle \zeta' \rangle|^2, \quad (44)$$

the n''s being the numbers of bosons initially present in the various boson states. Now from (6) and (17), with reference to (4),

$$|n_1' n_2' n_3' \cdots \rangle = (n_1'! n_2'! n_3'! \cdots)^{-1/2} \eta_1^{n_1'} \eta_2^{n_2'} \eta_3^{n_3'} \cdots |0\rangle, \quad (45)$$

so that

$$\eta_x |n_1' n_2' n_3' \cdots n_x' \cdots \rangle = (n_x' + 1)^{1/2} |n_1' n_2' n_3' \cdots (n_x' + 1) \cdots \rangle. \quad (46)$$

Hence (44) is equal to

$$(n_x' + 1)|\langle \zeta''|U_x|\zeta' \rangle|^2, \quad (47)$$

showing that *the probability of a transition in which a boson is emitted into state x is proportional to the number of bosons originally in state x plus one.*

The probability of a boson being absorbed from state x with the atom making a jump from state ζ' to state ζ'' is proportional to

$$|\langle \zeta''|\langle n_1' n_2' n_3' \cdots (n_x' - 1) \cdots |\overline{U}_x \bar{\eta}_x| n_1' n_2' n_3' \cdots n_x' \cdots \rangle|\zeta' \rangle|^2, \quad (48)$$

the n''s again being the numbers of bosons initially present in the various boson states. Now from (45)

$$\bar{\eta}_x | n'_1 n'_2 n'_3 \cdots n'_x \cdots \rangle = n'^{1/2}_x | n'_1 n'_2 n'_3 \cdots (n'_x - 1) \cdots \rangle, \qquad (49)$$

so (48) is equal to

$$n'_x |\langle \zeta'' | \bar{U}_x | \zeta' \rangle|^2. \qquad (50)$$

Thus *the probability of a transition in which a boson is absorbed from state x is proportional to the number of bosons originally in state x.*

Similar methods may be applied to more complicated processes, and show that the probability of a process in which a boson makes a transition from state y to state x ($x \neq y$) is proportional to $n'_y(n'_x+1)$. More generally, the probability of a process in which bosons are absorbed from states $x, y, \ldots$ and emitted into states $a, b, \ldots$ is proportional

$$n'_x n'_y \cdots (n'_a + 1)(n'_b + 1) \cdots, \qquad (51)$$

the n''s being in each case the numbers of bosons originally present. These results hold both for direct transition processes and transition processes that take place through one or more intermediate states, in accordance with the interpretation given at the end of § 44.

62. Application to Photons

Since photons are bosons, the foregoing theory can be applied to them. A photon is in a stationary state when it is in an eigenstate of momentum. It then has two independent states of polarization, which may be taken to be two perpendicular states of linear polarization. The dynamical variables needed to describe the stationary states are then the momentum **p**, a vector, and a polarization variable **l**, consisting of a unit vector perpendicular to **p**. The variables **p** and **l** take the place of our previous α's. The eigenvalues of **p** consist of all numbers from $-\infty$ to ∞ for each of the three Cartesian components of **p**, while for each eigenvalue **p**$'$ of **p**, **l** has just two eigenvalues, namely two arbitrarily chosen vectors perpendicular to **p**$'$ and to one another. Owing to the eigenvalues of **p** forming a continuous range, there are a continuous range of stationary states, giving us the continuous basic kets $|\mathbf{p'l'}\rangle$. However, the foregoing theory was built up in terms of discrete basic kets $|\alpha'\rangle$ for a boson. There are two formalisms which one may use for getting over this discrepancy.

The first consists in replacing the continuous three-dimensional distribution of eigenvalues for **p** by a large number of discrete points lying

62. APPLICATION TO PHOTONS

very close together, forming a dust spread over the whole three-dimensional **p**-space. Let $s_{\mathbf{p}'}$ be the density of the dust (the number of points per unit volume) in the neighbourhood of any point $\mathbf{p}'$. Then $s_{\mathbf{p}'}$ must be large and positive, but is otherwise an arbitrary function of $\mathbf{p}'$. An integral over the **p**-space may be replaced by a sum over the dust of points, in accordance with the formula

$$\iiint f(\mathbf{p}')\,dp'_x dp'_y dp'_z = \sum f(\mathbf{p}')s_{\mathbf{p}'}^{-1}, \qquad (52)$$

which formula provides the basis of the passage from continuous $\mathbf{p}'$ values to discrete ones and vice versa. Any problem can be worked out in terms of the discrete $\mathbf{p}'$ values, for which the theory of §§ 59–61 can be used, and the results can be transformed back to refer to continuous $\mathbf{p}'$ values. The arbitrary density $s_{\mathbf{p}'}$ should then disappear from the results.

The second formalism consists in modifying the equations of the theory of §§ 59–61 so as to make them apply to the case of a continuous range of basic kets $|\alpha'\rangle$, by replacing sums by integrals and replacing the δ symbol in the commutation relations (11) by δ functions, so far as concerns the variables with continuous eigenvalues. Each of these formalisms has some advantages and some disadvantages. The first is usually more convenient for physical discussion, the second for mathematical development. Both will be developed here and one or other will be used according to which is more suitable at the moment.

The Hamiltonian describing an assembly of photons interacting with an atom will be of the general form (40), with the coefficients H_P, U_a, U_{ab}, V_{ab} involving the atom variables. This Hamiltonian may be written

$$H_T = H_P + H_Q + H_R, \qquad (53)$$

where H_P is the energy of the atom alone, H_R is the energy of the assembly of photons alone,

$$H_R = \sum_{\mathbf{p}'l'} n_{\mathbf{p}'l'} h\nu_{\mathbf{p}'}, \qquad (54)$$

$\nu_{\mathbf{p}'}$ being the frequency of a photon of momentum $\mathbf{p}'$, and H_Q is the interaction energy, which can be evaluated from analogy with the classical theory, as will be shown in the next section. The whole system can be treated by a perturbation method as discussed in the preceding section, H_P and H_R providing the energy (41) of the unperturbed system and H_Q being the perturbation energy, which gives rise to transition processes in which photons are emitted and absorbed and the atom jumps from one stationary state to another.

We saw in the preceding section that the probability of an absorption process is proportional to the number of bosons originally in the state from which a boson is absorbed. From this we can infer that the probability of a photon being absorbed from a beam of radiation incident on an atom is proportional to the intensity of the beam. We also saw that the probability of an emission process is proportional to the number of bosons originally in the state concerned plus one. To interpret this result we must make a careful study of the relations involved in replacing the continuous range of photon states by a discrete set.

Let us neglect for the present the polarization variable **l**. Let $|\mathbf{p}'D\rangle$ be the normalized ket corresponding to the discrete photon state $\mathbf{p}'$. Then from (22) of § 16

$$\sum_{\mathbf{p}'} = |\mathbf{p}'D\rangle\langle\mathbf{p}'D| = 1,$$

which gives from (52)

$$\int |\mathbf{p}'D\rangle\langle\mathbf{p}'D|s_{\mathbf{p}'}\, d^3p' = 1, \tag{55}$$

d^3p' being written for $dp'_x dp'_y dp'_z$, for brevity. Now if $|\mathbf{p}'\rangle$ is the basic ket corresponding to the continuous state $\mathbf{p}'$, we have according to (24) of § 16

$$\int |\mathbf{p}'\rangle\langle\mathbf{p}'|\, d^3p' = 1,$$

which shows, on comparison with (55), that

$$|\mathbf{p}'\rangle = |\mathbf{p}'D\rangle s_{\mathbf{p}'}^{1/2}. \tag{56}$$

The connexion between $|\mathbf{p}'\rangle$ and $|\mathbf{p}'D\rangle$ is like the connexion between the basic kets when one changes the weight function of the representation, as shown by (38) of § 16.

With $n'_{\mathbf{p}'}$ photons in each discrete photon state $\mathbf{p}'$, the Gibbs density ρ for the assembly of photons is, according to (68) of § 33,

$$\rho = \sum_{\mathbf{p}'} |\mathbf{p}'D\rangle n'_{\mathbf{p}'}\langle\mathbf{p}'D| = \int |\mathbf{p}'D\rangle n'_{\mathbf{p}'}\langle\mathbf{p}'D|s_{\mathbf{p}'}\, d^3p'$$

$$= \int |\mathbf{p}'\rangle n'_{\mathbf{p}'}\langle\mathbf{p}'|\, d^3p' \tag{57}$$

with the help of (56). The number of photons per unit volume in the neighbourhood of any point $\mathbf{x}'$ is then $\langle\mathbf{x}'|\rho|\mathbf{x}'\rangle$, according to (73) of § 33. From

62. APPLICATION TO PHOTONS

(57) this equals

$$\langle \mathbf{x'}|\rho|\mathbf{x'}\rangle = \int \langle \mathbf{x'}|\mathbf{p'}\rangle n'_{\mathbf{p'}} \langle \mathbf{p'}|\mathbf{x'}\rangle d^3 p'$$

$$= \int h^{-3} n'_{\mathbf{p'}} d^3 p' \tag{58}$$

if one puts in the value of the transformation function $\langle \mathbf{x'}|\mathbf{p'}\rangle$ given by (54) of § 23. Equation (58) expresses the number of photons per unit volume as an integral over the momentum space, so the integrand in (58) can be interpreted as the number of photons per unit of phase space. We obtain in this way the result that *the number of photons per unit of phase space is equal to h^{-3} times the number of photons per discrete state*, in other words, *a cell of volume h^3 in phase space is equivalent to a discrete state*. This result is a general one, holding for any kind of particle. If the polarization variable of the photons is not neglected, the result holds for each of the two independent states of polarization.

The momentum of a photon of frequency v is of magnitude hv/c, so the element of momentum space

$$dp_x dp_y dp_z = h^3 c^{-3} v^2 \, dv d\omega,$$

$d\omega$ being an element of solid angle for the direction of the vector $\mathbf{p}$. Thus a distribution of photons with $n'_\mathbf{p}$ per discrete state, which is equivalent to a distribution of $h^{-3} n'_\mathbf{p} \, d^3 p d^3 x$ photons in an element of volume $d^3 x$ and an element of momentum space $d^3 p$, equals a distribution of $n'_\mathbf{p} c^{-3} v^2 \, dv d\omega d^3 x$ photons in an element of volume $d^3 x$ and a frequency range dv and direction of motion $d\omega$. This corresponds to an energy density $n'_\mathbf{p} hc^{-3} v^3$ per unit solid angle per unit frequency range, or an intensity per unit frequency range (i.e. an energy crossing unit area per unit time per unit frequency range) of amount

$$I_v = \frac{n'_\mathbf{p} hv^3}{c^2}. \tag{59}$$

The result that the probability of a photon being emitted is proportional to $n'_{\text{pl}} + 1$, n'_{pl} being the number of photons initially present in the discrete state concerned, can now be interpreted as the probability being proportional to $I_{vl} + hv^3/c^2$, where I_{vl} is the intensity of the incident radiation per unit frequency range in the neighbourhood of the frequency of the emitted photon and having the same polarization $\mathbf{l}$ as the emitted photon. Thus with no incident radiation there is still a certain amount of emission, but the emission is increased or *stimulated* by incident radiation in the same direction and having the same frequency and polarization as the emitted radiation. The present theory of radiation thus completes the imperfect one of § 45 by giving both stimulated and spontaneous emission. The ratio it gives for the two kinds of emission, namely $I_{\nu\mathbf{l}} : h\nu^3/c^2$, is in agreement with that provided by Einstein's theory of statistical equilibrium mentioned in § 45.

The probability of a photon being scattered from the state $\mathbf{p}'\mathbf{l}'$ to the state $\mathbf{p}''\mathbf{l}''$ is proportional to $n_{\mathbf{p}'\mathbf{l}'}(n_{\mathbf{p}''\mathbf{l}''} + 1)$, the n's being the numbers of photons initially in the discrete states concerned. We can interpret this result as the probability being proportional to

$$I_{\nu'\mathbf{l}'}\left(I_{\nu''\mathbf{l}''} + \frac{h\nu''^3}{c^2}\right). \tag{60}$$

Similarly for a more general radiative process in which several photons are emitted and absorbed, the probability is proportional to a factor $I_{\nu\mathbf{l}}$ for each absorbed photon and a factor $I_{\nu\mathbf{l}} + h\nu^3/c^2$ for each emitted photon. Thus the process is stimulated by incident radiation in the same direction and with the same frequency and polarization as any of the emitted photons.

63. The Interaction Energy between Photons and an Atom

We shall now determine the interaction energy between an atom and an assembly of photons, i.e. the H_Q of equation (53), from analogy with the classical expression for the interaction energy between an atom and a field of radiation. For simplicity we shall suppose the atom to consist of a single electron moving in an electrostatic field of force. The field of radiation may be described by a scalar and a vector potential. These potentials are to a certain extent arbitrary and may be chosen so that the scalar potential vanishes. The field is then completely described by the vector potential A_x, A_y, A_z, or $\mathbf{A}$. The change that the field causes in the Hamiltonian describing the atom is now, as explained at the beginning of § 41,

$$H_Q = \frac{1}{2m}\left[\left(\mathbf{p} + \frac{e}{c}\mathbf{A}\right)^2 - \mathbf{p}^2\right] = \frac{e}{mc}(\mathbf{p}, \mathbf{A}) + \frac{e^2}{2mc^2}\mathbf{A}^2. \tag{61}$$

This is the classical interaction energy. The $\mathbf{A}$ that occurs here should be the value of the vector potential at the point where the electron is momentarily situated. It is, however, a good enough approximation if we take this $\mathbf{A}$ to be the vector potential at some fixed point in the atom, such as the nucleus, provided we are dealing with radiation whose wavelength is large compared with the dimensions of the atom.

Let us first consider the field of radiation classically and ignore its interaction with the atom. The vector potential **A** satisfies, according to Maxwell's theory, the equations

$$\Box \mathbf{A} = 0, \quad \text{div } \mathbf{A} = 0, \tag{62}$$

$\Box$ being short for $\partial^2/c^2\partial t^2 - \partial^2/\partial x^2 - \partial^2/\partial y^2 - \partial^2/\partial z^2$. The first of these equations shows that **A** can be resolved into Fourier components in the form

$$\mathbf{A} = \int [\mathbf{A_k} e^{-i(\mathbf{kx})+2\pi i v_\mathbf{k} t} + \overline{\mathbf{A}}_\mathbf{k} e^{i(\mathbf{kx})-2\pi i v_\mathbf{k} t}] d^3k, \tag{63}$$

each Fourier component representing a train of waves moving with the velocity of light, described by a vector **k** whose direction gives the direction of motion of the waves and whose magnitude $|\mathbf{k}|$ is connected with their frequency $v_\mathbf{k}$ by

$$2\pi v_\mathbf{k} = c|\mathbf{k}|. \tag{64}$$

The vector **k** is just the momentum of a photon which the quantum theory would associate with these waves, divided by $\hbar$. For each value of **k** we have an amplitude $\mathbf{A_k}$, which is in general a complex vector, and the integral in (63) extends over the whole of the three-dimensional **k**-space. The second of equations (62) gives

$$(\mathbf{k}, \mathbf{A_k}) = 0, \tag{65}$$

showing that, for each value of **k**, $\mathbf{A_k}$ is perpendicular to **k**. This expresses that the waves are transverse waves. $\mathbf{A_k}$ is determined by its two components in two directions perpendicular to each other and to **k**, these two components corresponding to two independent states of linear polarization.

The total energy of the radiation is given by the volume integral

$$H_R = (8\pi)^{-1} \int (\mathcal{E}^2 + \mathcal{H}^2) d^3x \tag{66}$$

taken over the whole of space, where the electric field $\mathcal{E}$ and the magnetic field $\mathcal{H}$ of the radiation are given by

$$\mathcal{E} = -\frac{1}{c}\frac{\partial \mathbf{A}}{\partial t}, \quad \mathcal{H} = \text{curl } \mathbf{A}. \tag{67}$$

Using standard formulas of vector analysis, we have

$$\text{div}[\mathbf{A} \times \mathcal{H}] = (\mathcal{H}, \text{curl } \mathbf{A}) - (\mathbf{A}, \text{curl } \mathcal{H}) = \mathcal{H}^2 - (\mathbf{A}, \text{curl curl } \mathbf{A})$$

$$= \mathcal{H}^2 + (\mathbf{A}, \nabla^2 \mathbf{A})$$

with the help of the second of equations (62). Thus (66) becomes, with neglect of a term which can be transformed to a surface integral at infinity,

$$H_R = [8\pi]^{-1} \int \left[\frac{1}{c^2} \left(\frac{\partial \mathbf{A}}{\partial t}, \frac{\partial \mathbf{A}}{\partial t} \right) - (\mathbf{A}, \nabla^2 \mathbf{A}) \right] d^3x. \tag{68}$$

By substituting for $\mathbf{A}$ here its value given by (63), we can get the energy of the radiation in terms of the Fourier amplitudes $\mathbf{A_k}$. The energy of the radiation is constant (since we are now ignoring the interaction of the radiation and the atom), so in this calculation we may take $t = 0$. This means taking

$$\mathbf{A} = \int (\mathbf{A_k} + \overline{\mathbf{A}}_{-\mathbf{k}}) e^{-i(\mathbf{kx})} d^3k, \tag{69}$$

$$\nabla^2 \mathbf{A} = -\int k^2 (\mathbf{A_k} + \overline{\mathbf{A}}_{-\mathbf{k}}) e^{-i(\mathbf{kx})} d^3k,$$

$$\frac{\partial \mathbf{A}}{\partial t} = ic \int |\mathbf{k}| (\mathbf{A_k} - \overline{\mathbf{A}}_{-\mathbf{k}}) e^{-i(\mathbf{kx})} d^3k. \tag{70}$$

Inserting these expressions in (68), we get

$$H_R = [8\pi]^{-1} \iiint [\mathbf{k}'^2 (\mathbf{A_k} + \overline{\mathbf{A}}_{-\mathbf{k}}, \mathbf{A_{k'}} + \overline{\mathbf{A}}_{-\mathbf{k'}})$$
$$- |\mathbf{k}||\mathbf{k}'| (\mathbf{A_k} - \overline{\mathbf{A}}_{-\mathbf{k}}, \mathbf{A_{k'}} - \overline{\mathbf{A}}_{-\mathbf{k'}})] e^{-i[\mathbf{kx}]} e^{-i[\mathbf{k'x}]} d^3k d^3k' d^3x$$
$$= \pi^2 \iint [\mathbf{k}'^2 (\mathbf{A_k} + \overline{\mathbf{A}}_{-\mathbf{k}}, \mathbf{A_{k'}} + \overline{\mathbf{A}}_{-\mathbf{k'}})$$
$$- |\mathbf{k}||\mathbf{k}'| (\mathbf{A_k} - \overline{\mathbf{A}}_{-\mathbf{k}}, \mathbf{A_{k'}} - \overline{\mathbf{A}}_{-\mathbf{k'}})] \delta[\mathbf{k} + \mathbf{k}'] d^3k d^3k',$$

with the help of formula (49) of § 23, $\delta(\mathbf{k} + \mathbf{k}')$ being the product of three factors, one for each component of $\mathbf{k}$. Hence

$$H_r = \pi^2 \int k^2 [(\mathbf{A_k} + \overline{\mathbf{A}}_{-\mathbf{k}}, \mathbf{A}_{-\mathbf{k}} + \overline{\mathbf{A}}_{\mathbf{k}}) - (\mathbf{A_k} - \overline{\mathbf{A}}_{-\mathbf{k}}, \mathbf{A}_{-\mathbf{k}} - \overline{\mathbf{A}}_{\mathbf{k}})] d^3k$$
$$= 2\pi^2 \int k^2 [(\mathbf{A_k}, \overline{\mathbf{A}}_{\mathbf{k}}) + (\mathbf{A}_{-\mathbf{k}}, \overline{\mathbf{A}}_{-\mathbf{k}})] d^3k$$
$$= 4\pi^2 \int k^2 (\mathbf{A_k}, \overline{\mathbf{A}}_{\mathbf{k}}) d^3k. \tag{71}$$

We can replace the continuous distribution of $\mathbf{k}$-values by a dust of discrete $\mathbf{k}$-values, like we did with the $\mathbf{p}$-values in the preceding section. The

63. THE INTERACTION ENERGY BETWEEN PHOTONS AND AN ATOM

integral (71) then goes over, according to formula (52), into the sum

$$H_R = 4\pi^2 \sum_{\mathbf{k}} \mathbf{k}^2 (\mathbf{A_k}, \overline{\mathbf{A}}_\mathbf{k}) s_\mathbf{k}^{-1},$$

$s_\mathbf{k}$ being the density of the discrete **k**-values. We may also write this as

$$H_R = 4\pi^2 \sum_{\mathbf{kl}} \mathbf{k}^2 A_{\mathbf{kl}} \overline{A}_{\mathbf{kl}} s_\mathbf{k}^{-1}, \tag{72}$$

$A_{\mathbf{kl}}$ being a component of $\mathbf{A_k}$ in a direction **l** perpendicular to **k** and the summation with respect to **l** referring to two directions **l** perpendicular to each other. Thus there is one term in (72) for each independent stationary state for a photon.

The field quantities $\mathcal{E}$ and $\mathcal{H}$ at any point **x** can be looked upon as dynamical variables. The quantities

$$A_{\mathbf{kl}t} = A_{\mathbf{kl}} e^{2\pi i v_\mathbf{k} t}, \quad \overline{A}_{\mathbf{kl}t} = \overline{A}_{\mathbf{kl}} e^{-2\pi i v_\mathbf{k} t}$$

are then dynamical variables at time t, since they are connected with $\mathcal{E}$ and $\mathcal{H}$ at various points **x** at time t by equations which do not involve t, as follows from (63) and (67). $A_{\mathbf{kl}}$ is constant, so $A_{\mathbf{kl}t}$ varies with t according to the simple harmonic law. Thus $A_{\mathbf{kl}t}$ is like the η_t of a harmonic oscillator, defined by (3) of § 34, the ω of the oscillator being $2\pi v_\mathbf{k}$. We may take each $A_{\mathbf{kl}t}$ to be proportional to the η_t of some harmonic oscillator and then the field of radiation becomes a set of harmonic oscillators.

Let us now pass over to the quantum theory and take the $A_{\mathbf{kl}t}$, $\overline{A}_{\mathbf{kl}t}$ to be dynamical variables in the Heisenberg picture. The expression (72) for the energy may be retained unchanged, the order in which the factors $A_{\mathbf{kl}}$, $\overline{A}_{\mathbf{kl}}$ there occur being the correct one to give no zero-point energy. The $A_{\mathbf{kl}t}$ then still vary with time according to the $e^{i\omega t}$ law and may still be taken to be proportional to the η_t's of harmonic oscillators. The factor of proportionality may be obtained by equating (72) to the expression (39) for the energy, with the label a replaced by the two labels **k** and **l** and with $hv_\mathbf{k}$ for $\hbar\omega_a$. This gives

$$4\pi^2 \sum_{\mathbf{kl}} \mathbf{k}^2 A_{\mathbf{kl}t} \overline{A}_{\mathbf{kl}t} s_\mathbf{k}^{-1} = \sum_{\mathbf{kl}} hv_\mathbf{k} \eta_{\mathbf{kl}t} \overline{\eta}_{\mathbf{kl}t},$$

the suffix t being inserted to show that we are dealing with Heisenberg dynamical variables (as we should when transferring equations of the classical theory to the quantum theory). Hence, using (64),

$$4\pi^2 A_{\mathbf{kl}t} = ch^{1/2} v_\mathbf{k}^{-1/2} \eta_{\mathbf{kl}t} s_\mathbf{k}^{1/2}, \tag{73}$$

with neglect of an unimportant arbitrary phase factor. In this way the Heisenberg dynamical variables $\eta_{\mathbf{k}lt}$, which describe the field of radiation as a set of oscillators, are introduced. The commutation relations between the $\eta_{\mathbf{k}lt}$ and $\bar{\eta}_{\mathbf{k}lt}$ are known, being given by (11), so equation (73) fixes the commutation relations between the $A_{\mathbf{k}lt}$ and $\bar{A}_{\mathbf{k}lt}$. It thus fixes the commutation relations between the potentials $\mathbf{A}$ and the field quantities $\mathcal{E}$ and $\mathcal{H}$ at various points $\mathbf{x}$ at the time t. (Incidentally, the commutation relations of the $A_{\mathbf{k}l}$, $\bar{A}_{\mathbf{k}l}$ are fixed, so the commutation relation of two potential or field quantities at two different times is also fixed.)

We can still use (73) when the interaction between the field of radiation and the atom is taken into account. This involves assuming that the interaction does not affect the commutation relations between the potentials and field quantities at a given time. The interaction causes the $\eta_{\mathbf{k}lt}$'s to cease to vary according to the simple harmonic law and the oscillators to cease to be harmonic. Thus it may affect the commutation relation between two potential or field quantities at two different times.

We can now take over the interaction energy (61) into the quantum theory, putting $\mathbf{p}_t$ for $\mathbf{p}$ to show it is a Heisenberg dynamical variable. Taking the atomic nucleus to be at the origin we get, by substituting (63) with $\mathbf{x} = 0$ into (61).

$$
\begin{aligned}
H_{Qt} &= \frac{e}{mc} \int (\mathbf{p}_t, \mathbf{A}_{\mathbf{k}t} + \bar{\mathbf{A}}_{\mathbf{k}t}) \, d^3k \\
&\quad + \frac{e^2}{2mc^2} \iint (\mathbf{A}_{\mathbf{k}t} + \bar{\mathbf{A}}_{\mathbf{k}t}, \mathbf{A}_{\mathbf{k}'t} + \bar{\mathbf{A}}_{\mathbf{k}'t}) \, d^3k \, d^3k' \\
&= \frac{e}{mc} \sum_{\mathbf{k}} (\mathbf{p}_t, \mathbf{A}_{\mathbf{k}t} + \bar{\mathbf{A}}_{\mathbf{k}t}) s_{\mathbf{k}}^{-1} + \frac{e^2}{2mc^2} \sum_{\mathbf{k}\mathbf{k}'} (\mathbf{A}_{\mathbf{k}t} + \bar{\mathbf{A}}_{\mathbf{k}t}, \mathbf{A}_{\mathbf{k}'t} + \bar{\mathbf{A}}_{\mathbf{k}'t}) s_{\mathbf{k}}^{-1} s_{\mathbf{k}'}^{-1}
\end{aligned}
$$

if we pass from continuous to discrete $\mathbf{k}$-values. Thus

$$
\begin{aligned}
H_{Qt} &= \frac{e}{mc} \sum_{\mathbf{k}l} p_{lt} (A_{\mathbf{k}lt} + \bar{A}_{\mathbf{k}lt}) s_{\mathbf{k}}^{-1} \\
&\quad + \frac{e^2}{2mc^2} \sum_{\mathbf{k}\mathbf{k}'ll'} (A_{\mathbf{k}lt} + \bar{A}_{\mathbf{k}lt})(A_{\mathbf{k}'l't} + \bar{A}_{\mathbf{k}'l't})(ll') s_{\mathbf{k}}^{-1} s_{\mathbf{k}'}^{-1},
\end{aligned}
$$

p_{lt} being the component of $\mathbf{p}_t$ in the direction $\mathbf{l}$. With the help of (73) we may express H_{Qt} in terms of the $\eta_{\mathbf{k}lt}$ and $\bar{\eta}_{\mathbf{k}lt}$, and we can then drop the suffix t (which means going over to Schrödinger dynamical variables), so that we obtain finally

$$H_Q = \frac{eh^{1/2}}{4\pi^2 m} \sum_{kl} p_l v_k^{-1/2} (\eta_{kl} + \bar{\eta}_{kl}) s_k^{-1/2}$$
$$+ \frac{e^2 h}{32\pi^4 m} \sum_{kk'll'} v_k^{-1/2} v_{k'}^{-1/2} (\eta_{kl} + \bar{\eta}_{kl})(\eta_{k'l'} + \bar{\eta}_{k'l'})(ll') s_k^{-1/2} s_{k'}^{-1/2}. \tag{74}$$

With the model of the atom we are using, the interaction energy appears as a linear plus a quadratic function in the η's and $\bar{\eta}$'s. The linear terms give rise to emission and absorption processes, the quadratic ones to scattering processes and processes in which two photons are absorbed or emitted simultaneously. The order of the factors η and $\bar{\eta}$ in the quadratic terms is not determined by the procedure of working from the classical theory, but this order is unimportant, since a change in it merely changes H_Q by a constant.

The matrix element of H_Q referring to the emission of a photon into the discrete state $\mathbf{k}l$, or into the discrete state $\mathbf{p}'l$, as it may also be labelled, with the atom jumping from state α^0 to state α', is

$$\langle \mathbf{p}'D l\alpha' | H_Q | \alpha^0 \rangle = \frac{eh^{1/2}}{4\pi^2 m v'^{1/2}} \langle \alpha' | p_l | \alpha^0 \rangle s_k^{-1/2}$$
$$= \frac{e}{mh(2\pi v')^{1/2}} \langle \alpha' | p_l | \alpha^0 \rangle s_\mathbf{p}^{-1/2}$$

since $s_\mathbf{k} = s_\mathbf{p} \hbar^3$. The p_l occurring here, referring to the momentum of the electron, is, of course, quite distinct from the other letters $\mathbf{p}$, referring to the momentum of the emitted photon. To avoid confusion we shall replace the electron momentum $\mathbf{p}$ by $m\dot{\mathbf{x}}$, these two dynamical variables being the same for the unperturbed atom. Passing over to continuous photon states by means of the conjugate imaginary of equation (56), we get

$$\langle \mathbf{p}'l\alpha' | H_Q | \alpha^0 \rangle = \frac{e}{h(2\pi v')^{1/2}} \langle \alpha' | \dot{x}_l | \alpha_0 \rangle. \tag{75}$$

Similarly, the matrix element of H_Q referring to the absorption of a photon from the continuous state $\mathbf{p}^0 l$ with the atom jumping from state α^0 to state α' is

$$\langle \alpha' | H_Q | \mathbf{p}^0 l\alpha^0 \rangle = \frac{e}{h(2\pi v^0)^{1/2}} \langle \alpha' | \dot{x}_l | \alpha^0 \rangle, \tag{76}$$

and the matrix element referring to the scattering of a photon from the continuous state $\mathbf{p}^0 l^0$ to the continuous state $\mathbf{p}'l'$ with the atom jumping from state α^0 to state α' is

$$\langle \mathbf{p'l'}\alpha'|H_Q|\mathbf{p}^0\mathbf{l}^0\alpha^0\rangle = \frac{e^2}{2\pi h^2 m(\nu^0)^{1/2}\nu'^{1/2}}(\mathbf{l'l}^0)\delta_{\alpha'\alpha^0}, \quad (77)$$

there being two terms in (74) which contribute to it. These matrix elements will be used in the next section. The matrix elements referring to the simultaneous absorption or emission of two photons may be written down in the same way, but they lead to physical effects too small to be of practical importance.

64. Emission, Absorption, and Scattering of Radiation

We can now determine directly the coefficients of emission, absorption, and scattering of radiation by substituting in the formulas of Chapter VIII the values for the matrix elements given by (75), (76), and (77).

For determining the emission probability we can use formula (56) of § 53. This shows that for an atom in a state α^0 the probability per unit time per unit solid angle of its spontaneously emitting a photon and dropping to a state α' of lower energy is

$$\frac{4\pi^2}{h}\frac{WP}{c^2}\left|\frac{e}{h}\frac{1}{(2\pi\nu)^{1/2}}\langle\alpha'|\dot{x}_1|\alpha^0\rangle\right|^2. \quad (78)$$

Now the energy and momentum of a photon of frequency ν are

$$W = h\nu, \quad P = \frac{h\nu}{c}.$$

Again, from the Heisenberg law (20) of § 29,

$$\langle\alpha'|\dot{x}_1|\alpha^0\rangle = -2\pi i\nu(\alpha^0\alpha')\langle\alpha'|x_1|\alpha^0\rangle,$$

$\nu(\alpha^0\alpha')$ being the frequency connected with transitions from state α^0 to state α', which in the present case is just the frequency ν of the emitted radiation. These results substituted in (78) make the emission coefficient reduce to

$$\frac{(2\pi\nu)^3}{hc^3}|\langle\alpha'|ex_1|\alpha^0\rangle|^2. \quad (79)$$

To obtain the rate of emission of energy per unit solid angle for a specified polarization, we must multiply this by $h\nu$. This gives for the total rate of emission of energy in all directions

$$\frac{4}{3}\frac{(2\pi\nu)^4}{c^3}|\langle\alpha'|e\mathbf{x}|\alpha^0\rangle|^2, \quad (80)$$

which is in agreement with expression (34) of § 45 and justifies Heisenberg's assumption for the interpretation of his matrix elements.

In the same way the absorption coefficient, given by formula (59) of § 53, becomes for photons

$$\frac{4\pi^2 h^2 W}{c^2 P}\left|\frac{e}{h}\frac{1}{(2\pi\nu)^{1/2}}\langle\alpha'|\dot{x}_1|\alpha^0\rangle\right|^2 = \frac{8\pi^3\nu}{c}|\langle\alpha'|ex_1|\alpha^0\rangle|^2.$$

This absorption coefficient refers to an incident beam of one photon crossing unit area per unit time per unit energy range. If we take one per unit frequency range instead of energy range, as is usual when dealing with radiation, the absorption coefficient becomes

$$\frac{8\pi^3\nu}{hc}|\langle\alpha'|ex_1|\alpha^0\rangle|^2.$$

This result is the same as (32) of § 45, if we substitute for the E_ν there the energy $h\nu$ of a single photon. Thus *the elementary theory of § 45, in which the radiation field is treated as an external perturbation, gives the correct value for the absorption coefficient.*

This agreement between the elementary theory and the present theory could be inferred from general arguments. The two theories differ only in that the field quantities all commute with one another in the elementary theory and satisfy definite commutation relations in the present theory, and this difference becomes unimportant for strong fields. Thus the two theories must give the same absorption and emission when strong fields are concerned. Since both theories give the rate of absorption proportional to the intensity of the incident beam, the agreement must hold also for weak fields in the case of absorption. In the same way the stimulated part of the emission in the present theory must agree with the emission in the elementary theory.

Let us now consider scattering. The direct scattering coefficient is given by formula (38) of § 50. Such scattering of photons will not be accompanied by any change of state of the atom on account of the factor $\delta_{\alpha'\alpha^0}$ in the expression for the matrix element (77). Thus the final energy W' of the photon will equal its initial energy W^0. The scattering coefficient now reduces to

$$\frac{e^4}{m^2 c^4}(\mathbf{l}'\mathbf{l}^0)^2.$$

This is the same as that given by classical mechanics for the scattering of radiation by a free electron. We thus see that the direct scattering of radiation by an electron in an atom is independent of the atom and is correctly given by the classical theory. This result, it should be remembered, holds only provided the wavelength of the radiation is large compared with the dimensions of the atom.

The direct scattering is a mathematical concept and cannot be separated out experimentally from the total scattering, given by formula (44) of § 51. Let us see what this total scattering is in the case of photons. We must be careful in our application of formula (44) of § 51. The summation $\sum_k$ in this formula may be considered as representing the contribution to the scattering of double transitions consisting of transitions firstly from the initial state to state k and secondly from state k to the final state. The first transition may be an absorption of the incident photon and the second an emission of the required scattered photon, but it is also possible for the first transition to be the emission and the second the absorption. It is clear from the general nature of the method used for deriving formula (44) of § 51 that both these kinds of double transitions must be included in the summation $\sum_k$ when this formula is applied to photons, although only the first of them appears in the actual derivation given in § 51, as the possibility of the particle being created or annihilated was not taken into account there.

We use zero, single prime, and double prime to refer to the initial, final, and intermediate states of the atom respectively, and zero and single prime to refer to the absorbed and emitted photons respectively. Then, for the double transition of absorption followed by emission, we must take for the matrix elements

$$\langle k|V|\mathbf{p}^0\alpha^0\rangle, \quad \langle \mathbf{p}'\alpha'|V|k\rangle$$

of the formula (44) of § 51

$$\langle k|V|\mathbf{p}^0\alpha^0\rangle = \langle \alpha''|H_Q|\mathbf{p}^0 1^0\alpha^0\rangle, \quad \langle \mathbf{p}'\alpha'|V|k\rangle = \langle \mathbf{p}'1'\alpha'|H_Q|\alpha''\rangle.$$

Also

$$E' - E_k = h\nu^0 + H_P(\alpha^0) - H_P(\alpha'') = h[\nu^0 - \nu(\alpha''\alpha^0)],$$

where

$$h\nu(\alpha''\alpha^0) = H_P(\alpha'') - H_P(\alpha^0).$$

Similarly, for the double transition of emission followed by absorption we must take

$$\langle k|V|\mathbf{p}^0\alpha^0\rangle = \langle \mathbf{p}'1'\alpha''|H_Q|\alpha^0\rangle, \quad \langle \mathbf{p}'\alpha'|V|k\rangle = \langle \alpha'|H_Q|\mathbf{p}^0 1^0\alpha''\rangle$$

and

$$E' - E_k = h\nu^0 + H_P(\alpha^0) - H_P(\alpha'') - h\nu^0 - h\nu' = -h[\nu' + \nu(\alpha''\alpha^0)],$$

there being now two photons, of frequencies ν^0 and ν', in existence for the intermediate state. Substituting in (44) of § 51 the values of the matrix elements given by (75), (76), and (77), we get for the scattering coefficients

$$\frac{e^4}{h^2c^4}\frac{v'}{v^0}\left|\frac{h}{m}(l'l^0)\delta_{\alpha'\alpha^0} + \sum_{\alpha'}\left[\frac{\langle\alpha'|\dot{x}_{l'}|\alpha''\rangle\langle\alpha''|\dot{x}_{l^0}|\alpha^0\rangle}{v^0 - v(\alpha''\alpha^0)}\right.\right.$$
$$\left.\left. - \frac{\langle\alpha'|\dot{x}_{l^0}|\alpha''\rangle\langle\alpha''|\dot{x}_{l'}|\alpha^0\rangle}{v' + v(\alpha''\alpha^0)}\right]\right|^2. \quad (81)$$

If we write (81) in terms of x instead of $\dot{x}$, we get

$$\frac{(2\pi e)^4}{h^2c^4}\frac{v'}{v^0}\left|\frac{\hbar}{2\pi m}(l'l^0)\delta_{\alpha'\alpha^0}\right.$$
$$\left. - \sum_{\alpha'} v(\alpha'\alpha'')v(\alpha''\alpha^0)\left[\frac{\langle\alpha'|x_{l'}|\alpha''\rangle\langle\alpha''|x_{l^0}|\alpha^0\rangle}{v^0 - v(\alpha''\alpha^0)}\right.\right.$$
$$\left.\left. - \frac{\langle\alpha'|x_{l^0}|\alpha''\rangle\langle\alpha''|x_{l'}|\alpha^0\rangle}{v' + v(\alpha''\alpha^0)}\right]\right|^2. \quad (82)$$

We can simplify (82) with the help of the quantum conditions. We have

$$x_{l'}x_{l^0} - x_{l^0}x_{l'} = 0,$$

which gives

$$\sum_{\alpha''}(\langle\alpha'|x_{l'}|\alpha''\rangle\langle\alpha''|x_{l^0}|\alpha^0\rangle - \langle\alpha'|x_{l^0}|\alpha''\rangle\langle\alpha''|x_{l'}|\alpha^0\rangle) = 0, \quad (83)$$

and also

$$x_{l'}\dot{x}_{l^0} - \dot{x}_{l^0}x_{l'} = \frac{1}{m}(x_{l'}p_{l^0} - p_{l^0}x_{l'}) = \frac{i\hbar}{m}(l'l^0),$$

which gives

$$\sum_{\alpha''}[\langle\alpha'|x_{l'}|\alpha''\rangle v(\alpha''\alpha^0)\langle\alpha''|x_{l^0}|\alpha^0\rangle - v(\alpha'\alpha'')\langle\alpha'|x_{l^0}|\alpha''\rangle\langle\alpha''|x_{l'}|\alpha^0\rangle]$$
$$= \frac{1}{2\pi i}\frac{i\hbar}{m}(l'l^0)\delta_{\alpha'\alpha^0} = \frac{\hbar}{2\pi m}(l'l^0)\delta_{\alpha'\alpha^0}. \quad (84)$$

Multiplying (83) by v' and adding to (84), we obtain

$$\sum_{\alpha''}\{\langle\alpha'|x_{l'}|\alpha''\rangle\langle\alpha''|x_{l^0}|\alpha^0\rangle[v' + v(\alpha''\alpha^0)]$$
$$- \langle\alpha'|x_{l^0}|\alpha''\rangle\langle\alpha''|x_{l'}|\alpha^0\rangle[v' + v(\alpha'\alpha'')]\} = \frac{\hbar}{2\pi m}(l'l^0)\delta_{\alpha'\alpha^0}.$$

If we substitute this expression for $\hbar/2\pi m \cdot (l'l^0)\delta_{\alpha'\alpha^0}$ in (82), we obtain, after a straightforward reduction making use of identical relations between the ν's,

$$\frac{(2\pi e)^4}{h^2 c^4} \nu^0 \nu'^3 \left| \sum_{\alpha''} \left[\frac{\langle \alpha'|x_{l'}|\alpha''\rangle\langle\alpha''|x_{l^0}|\alpha^0\rangle}{\nu^0 - \nu(\alpha''\alpha^0)} - \frac{\langle\alpha'|x_{l^0}|\alpha''\rangle\langle\alpha''|x_{l'}|\alpha^0\rangle}{\nu' + \nu(\alpha''\alpha^0)} \right] \right|^2. \tag{85}$$

This gives the scattering coefficient in the form of the effective area that a photon has to hit per unit solid angle of scattering. It is known as the *Kramers-Heisenberg dispersion formula*, having been first obtained by these authors from analogies with the classical theory of dispersion.

The fact that the various terms in (82) can be combined to give the result (85) justifies the assumption made in deriving formula (44) of § 51, that the matrix elements $\langle \mathbf{p}'\alpha'|V|\mathbf{p}''\alpha''\rangle$ of the interaction energy are of the second order of smallness compared with the $\langle \mathbf{p}'\alpha'|V|k\rangle$ ones, at any rate when the scattered particles are photons.

65. An Assembly of Fermions

An assembly of fermions can be treated by a method similar to that used in §§ 59 and 60 for bosons. With the kets (1) we may use the *antisymmetrizing operator* A defined by

$$A = u'!^{-1/2} \sum \pm P, \tag{2'}$$

summed over all permutations P, the $+$ or $-$ sign being taken according to whether P is even or odd. Applied to the ket (1) it gives

$$u'!^{-1/2} \sum \pm P |\alpha_1^a \alpha_2^b \alpha_3^c \cdots \alpha_{u'}^g\rangle = A|\alpha^a \alpha^b \alpha^c \cdots \alpha^g\rangle, \tag{3'}$$

a ket corresponding to a state for an assembly of u' fermions. The ket (3') is normalized provided the individual fermion kets $|\alpha^a\rangle, |\alpha^b\rangle, |\alpha^c\rangle, \ldots$ are all different, otherwise it is zero. In this respect the ket (3') is simpler than the ket (3). However, (3') is more complicated than (3) in that (3') depends on the order in which $\alpha^a, \alpha^b, \alpha^c, \ldots$ occur in it, being subject to a change of sign if an odd permutation is applied to this order.

We can, as before, introduce the numbers $n_1, n_2, n_3, \ldots$ of fermions in the states $\alpha^{(1)}, \alpha^{(2)}, \alpha^{(3)}, \ldots$ and treat them as dynamical variables or observables. They each have as eigenvalues only 0 and 1. They form a complete set of commuting observables for the assembly of fermions. The basic kets of a representation with the n's diagonal may be taken to be connected with the kets (3') by the equation

$$A|\alpha^a \alpha^b \alpha^c \cdots \alpha^g\rangle = \pm |n'_1 n'_2 n'_3 \cdots\rangle \tag{6'}$$

corresponding to (6), the n''s being connected with the variables $\alpha^a, \alpha^b, \alpha^c, \ldots$ by equation (4). The $\pm$ sign is needed in (6') since, for given n''s, the occupied states $\alpha^a, \alpha^b, \alpha^c, \ldots$ are fixed but not their order, so that the sign of the left-hand side of (6') is not fixed. To set up a rule which determines the sign in (6'), we must arrange all the states α for a fermion arbitrarily in some standard order. The α's occurring in the left-hand side of (6') form a certain selection from all the α's and the standard order for all the α's will give a standard order for this selection. We now make the rule that the $+$ sign should occur in (6') if the α's on the left-hand side can be brought into their standard order by an even permutation and the $-$ sign if an odd permutation is required. Owing to the complexity of this rule, the representation with the basic kets $|n'_1 n'_2 n'_3 \cdots\rangle$ is not a very useful one.

If the number of fermions in the assembly is variable, we can set up the complete set of kets

$$|\rangle, \quad |\alpha^a\rangle, \quad A|\alpha^a \alpha^b\rangle, \quad A|\alpha^a \alpha^b \alpha^c\rangle, \quad \ldots, \tag{9'}$$

corresponding to (9). A general ket is now expressible as a sum of the various kets (9').

To continue with the development we introduce a set of linear operators $\eta, \bar{\eta}$, one pair $\eta_a, \bar{\eta}_a$ corresponding to each fermion state α^a, satisfying the commutation relations

$$\begin{aligned} \eta_a \eta_b + \eta_b \eta_a &= 0, \\ \bar{\eta}_a \bar{\eta}_b + \bar{\eta}_b \bar{\eta}_a &= 0, \\ \bar{\eta}_a \eta_b + \eta_b \bar{\eta}_a &= \delta_{ab}. \end{aligned} \tag{11'}$$

These relations are like (11) with a $+$ sign instead of a $-$ on the left-hand side. They show that, for $a \neq b$, η_a and $\bar{\eta}_a$ anticommute with η_b and $\bar{\eta}_b$, while, putting $b = a$, they give

$$\eta_a^2 = 0, \quad \bar{\eta}_a^2 = 0, \quad \bar{\eta}_a \eta_a + \eta_a \bar{\eta}_a = 1. \tag{11''}$$

To verify that the relations (11') are consistent, we note that linear operators $\eta, \bar{\eta}$ satisfying the conditions (11') can be constructed in the following way. For each state α^a we take a set of linear operators $\sigma_{xa}, \sigma_{ya}, \sigma_{za}$ like the $\sigma_x, \sigma_y, \sigma_z$ introduced in § 37 to describe the spin of an electron and such that $\sigma_{xa}, \sigma_{ya}, \sigma_{za}$ commute with $\sigma_{xb}, \sigma_{yb}, \sigma_{zb}$ for $b \neq a$. We also take an independent set of linear operators ζ_a, one for each state α^a, which all anticommute with one another and have their squares unity, and commute with all the σ variables. Then, putting

$$\eta_a = \frac{1}{2}\zeta_a(\sigma_{xa} - i\sigma_{ya}), \quad \bar{\eta}_a = \frac{1}{2}\zeta_a(\sigma_{xa} + i\sigma_{ya}),$$

we have all the conditions (11′) satisfied.

From (11″)

$$(\eta_a\bar{\eta}_a)^2 = \eta_a\bar{\eta}_a\eta_a\bar{\eta}_a = \eta_a(1 - \eta_a\bar{\eta}_a)\bar{\eta}_a = \eta_a\bar{\eta}_a.$$

This is an algebraic equation for $\eta_a\bar{\eta}_a$, showing that $\eta_a\bar{\eta}_a$ is an observable with the eigenvalues 0 and 1. Also $\eta_a\bar{\eta}_a$ commutes with $\eta_b\bar{\eta}_b$ for $b \neq a$. These results allow us to put

$$\eta_a\bar{\eta}_a = n_a, \tag{12′}$$

the same as (12). From (11″) we get now

$$\bar{\eta}_a\eta_a = 1 - n_a, \tag{13′}$$

the equation corresponding to (13).

Let us write the normalized ket which is an eigenket of all the n's belonging to the eigenvalues zero as $|0\rangle$. Then

$$n_a|0\rangle = 0,$$

so from (12′)

$$\langle 0|\eta_a\bar{\eta}_a|0\rangle = 0.$$

Hence

$$\bar{\eta}_a|0\rangle = 0, \tag{15′}$$

like (15). Again

$$\langle 0|\bar{\eta}_a\eta_a|0\rangle = \langle 0|(1 - n_a)|0\rangle = \langle 0|0\rangle = 1,$$

showing that $\eta_a|0\rangle$ is normalized, and

$$n_a\eta_a|0\rangle = \eta_a\bar{\eta}_a\eta_a|0\rangle = \eta_a(1 - n_a)|0\rangle = \eta_a|0\rangle,$$

showing that $\eta_a|0\rangle$ is an eigenket of n_a belonging to the eigenvalue unity. It is an eigenket of the other n's belonging to the eigenvalues zero, since the other n's commute with η_a. By generalizing the argument we see that $\eta_a\eta_b\eta_c \cdots \eta_g|0\rangle$ is normalized and is a simultaneous eigenket of all the n's, belonging to the eigenvalues unity for $n_a, n_b, n_c, \ldots, n_g$ and zero for the other n's. This enables us to put

$$A|\alpha^a\alpha^b\alpha^c \cdots \alpha^g\rangle = \eta_a\eta_b\eta_c \cdots \eta_g|0\rangle, \tag{17′}$$

both sides being antisymmetrical in the labels $a, b, c, \ldots, g$. We have here the analogue of (17). The η's appear as creation operators for a fermion and the $\bar{\eta}$'s as annihilation operators.

65. AN ASSEMBLY OF FERMIONS

If we pass over to a different set of basic kets $|\beta^A\rangle$ for a fermion, we can introduce a new set of linear operators η_A corresponding to them. We then find, by the same argument as in the case of bosons, that the new η's are connected with the original ones by (21). This shows that there is a procedure of second quantization for fermions, similar to that for bosons, with the only difference that the commutation relations (11') must be employed for fermions to replace the commutation relations (11) for bosons.

A symmetrical linear operator U_T of the form (22) can be expressed in terms of the $\eta, \bar{\eta}$ variables by a method similar to that used for bosons. Equation (24) still holds, and so does (25) with S replaced by A. Instead of (26) we now have

$$
\begin{aligned}
U_T \eta_{x_1} \eta_{x_2} \eta_{x_3} &\cdots |0\rangle \\
&= \sum_a \sum_r (-)^{r-1} \eta_a \eta_{x_r}^{-1} \eta_{x_1} \eta_{x_2} \eta_{x_3} \cdots |0\rangle \langle a|U|x_r\rangle \\
&= \sum_{ab} \eta_a \sum_r (-1)^{r-1} \eta_{x_r}^{-1} \eta_{x_1} \eta_{x_2} \eta_{x_3} \cdots |0\rangle \delta_{bx_r} \langle a|U|b\rangle, \quad (26')
\end{aligned}
$$

$\eta_{x_r}^{-1}$ meaning that the factor η_{x_r} must be cancelled out, without its position among the other η_x's being changed before the cancellation. Instead of (27) we have

$$\bar{\eta}_b \eta_{x_1} \eta_{x_2} \eta_{x_3} \cdots |0\rangle = \sum_r (-)^{r-1} \eta_{x_r}^{-1} \eta_{x_1} \eta_{x_2} \eta_{x_3} \cdots |0\rangle \delta_{bx_r}, \quad (27')$$

so (28) holds unchanged and thus (29) holds unchanged. We have the same final form (29) for U_T in the fermion case as in the boson case. Similarly, a symmetrical linear operator V_T of the form (30) can be expressed as

$$V_T = \sum_{abcd} \eta_a \eta_b \langle ab|V|cd\rangle \bar{\eta}_d \bar{\eta}_c, \quad (35')$$

the same as one of the ways of writing (35).

The foregoing work shows that there is a deep-seated analogy between the theory of fermions and that of bosons, only slight changes having to be made in the general equations of the formalism when one passes from one to the other.

There is, however, a development of the theory of fermions that has no analogue for bosons. For fermions there are only the two alternatives of a state being occupied or unoccupied and *there is symmetry between these two alternatives*. One can demonstrate the symmetry mathematically by making a transformation which interchanges the concepts of 'occupied' and 'unoccupied', namely

$$\eta_a^* = \bar{\eta}_a, \quad \bar{\eta}_a^* = \eta_a,$$
$$n_a^* = \eta_a^* \bar{\eta}_a^* = 1 - n_a.$$

The creation operators of the unstarred variables are the annihilation operators of the starred variables, and vice versa. The starred variables are now seen to satisfy the same quantum conditions and to have all the same properties as the unstarred ones.

If there are only a few unoccupied states, a convenient standard ket to work with would be the one for which every state is occupied, namely $|0^*\rangle$ satisfying

$$n_a|0^*\rangle = |0^*\rangle.$$

It thus satisfies

$$n_a^*|0^*\rangle = 0,$$

or

$$\bar{\eta}_a^*|0^*\rangle = 0.$$

Other states for the assembly will now be represented by

$$\eta_a^* \eta_b^* \eta_c^* \cdots |0^*\rangle,$$

in which variables appear referring to the unoccupied fermion states $a, b, c, \ldots$. We may look upon these unoccupied fermion states as holes among the occupied ones and the η^* variables as the operators of creation of such holes. The holes are just as much physical things as the original particles and are also fermions.

CHAPTER XI

Relativistic Theory of the Electron

66. Relativistic Treatment of a Particle

The theory we have been building up so far is essentially a non-relativistic one. We have been working all the time with one particular Lorentz frame of reference and have set up the theory as an analogue of the classical non-relativistic dynamics. Let us now try to make the theory invariant under Lorentz transformations, so that it conforms to the special principle of relativity. This is necessary in order that the theory may apply to high-speed particles. There is no need to make the theory conform to general relativity, since general relativity is required only when one is dealing with gravitation, and gravitational forces are quite unimportant in atomic phenomena.

Let us see how the basic ideas of quantum theory can be adapted to the relativistic point of view that the four dimensions of space-time should be treated on the same footing. The general principle of superposition of states, as given in Chapter I, is a relativistic principle, since it applies to 'states' with the relativistic space-time meaning. However, the general concept of an observable does not fit in, since an observable may involve physical things at widely separated points at one instant of time. In consequence, if one works with a general representation referring to any complete set of commuting observables, the theory cannot display the symmetry between space and time required by relativity. In relativistic quantum mechanics one must be content with having one representation which displays this symmetry. One then has the freedom to transform to another representation referring to a special Lorentz frame of reference if it is useful for a particular calculation.

For the problem of a single particle, in order to display the symmetry between space and time we must use the Schrödinger representation. Let us put x_1, x_2, x_3 for x, y, z, and x_0 for ct. The time-dependent wave function then appears as $\psi(x_0 x_1 x_2 x_3)$ and provides us with a basis for treating the four x's on the same footing.

We shall use relativistic notation, writing the four x's as x_μ ($\mu = 0, 1, 2, 3$). Any space-time vector with four components which transform under Lorentz transformations like the four elements dx_μ will be written like a_μ with a lower Greek suffix. We may raise the suffix according to the rules

$$a^0 = a_0, \quad a^1 = -a_1, \quad a^2 = -a_2, \quad a^3 = -a_3. \tag{1}$$

The a_μ are called the contravariant components of the vector a, and the a^μ the covariant components. Two vectors a_μ and b_μ have a Lorentz-invariant scalar product

$$a_0 b_0 - a_1 b_1 - a_2 b_2 - a_3 b_3 = a^\mu b_\mu = a_\mu b^\mu,$$

a summation being implied over a repeated letter suffix. The fundamental tensor $g^{\mu\nu}$ is defined by

$$\begin{aligned} g^{00} &= 1, \quad g^{11} = g^{22} = g^{33} = -1, \\ g^{\mu\nu} &= 0 \quad \text{for } \mu \neq \nu. \end{aligned} \tag{2}$$

With its help the rules (1) connecting covariant and contravariant components may be written

$$a^\mu = g^{\mu\nu} a_\nu.$$

In the Schrödinger representation the momentum, whose components will now be written p_1, p_2, p_3 instead of p_x, p_y, p_z, is equal to the operator

$$p_r = -i\hbar \frac{\partial}{\partial x_r} \quad (r = 1, 2, 3). \tag{3}$$

Now the four operators $\partial/\partial x_\mu$ form the covariant components of a 4-vector whose contravariant components are written $\partial/\partial x^\mu$. So to bring (3) into a relativistic theory, we must first write it with its suffixes balanced,

$$p_r = i\hbar \frac{\partial}{\partial x^r},$$

and then extend it to the complete 4-vector equation

$$p_\mu = i\hbar \frac{\partial}{\partial x^\mu}. \tag{4}$$

We thus have to introduce a new dynamical variable p_0, equal to the operator $i\hbar \partial/\partial x_0$. Since it forms a 4-vector when combined with the momenta p_r, it must have the physical meaning of the energy of the particle divided by c. We can proceed to develop the theory treating the four p's on the same footing, like the four x's.

In the theory of the electron that will be developed here we shall have to introduce a further degree of freedom describing an internal motion of the electron. The wave function will thus have to involve a further variable besides the four x's.

67. The Wave Equation for the Electron

Let us consider first the case of the motion of an electron in the absence of an electromagnetic field, so that the problem is simply that of the free particle, as dealt with in § 30, with the possible addition of internal degrees of freedom. The relativistic Hamiltonian provided by classical mechanics for this system is given by equation (23) of § 30, and leads to the wave equation

$$[p_0 - (m^2c^2 + p_1^2 + p_2^2 + p_3^2)^{1/2}]\psi = 0, \tag{5}$$

where the p's are interpreted as operators in accordance with (4). Equation (5), although it takes into account the relation between energy and momentum required by relativity, is yet unsatisfactory from the point of view of relativistic theory, because it is very unsymmetrical between p_0 and the other p's, so much so that one cannot generalize it in a relativistic way to the case when there is a field present. We must therefore look for a new wave equation.

If we multiply the wave equation (5) on the left by the operator $[p_0 + (m^2c^2 + p_1^2 + p_2^2 + p_3^2)^{1/2}]$, we obtain the equation

$$(p_0^2 - m^2c^2 - p_1^2 - p_2^2 - p_3^2)\psi = 0, \tag{6}$$

which is of a relativistically invariant form and may therefore more conveniently be taken as the basis of a relativistic theory. Equation (6) is not completely equivalent to equation (5) since, although every solution of (5) is also a solution of (6), the converse is not true. Only those solutions of (6) belonging to positive values for p_0 are also solutions of (5).

The wave equation (6) is not of the form required by the general laws of the quantum theory on account of its being quadratic in p_0. In § 27 we deduced from quite general arguments that the wave equation must be linear in the operator $\partial/\partial t$ or p_0, like equation (7) of that section. We therefore seek a wave equation that is linear in p_0 and that is roughly equivalent to (6). In order that this wave equation shall transform in a simple way under a Lorentz transformation, we try to arrange that it shall be rational and linear in p_1, p_2, and p_3 as well as in p_0, and thus of the form

$$(p_0 - \alpha_1 p_1 - \alpha_2 p_2 - \alpha_3 p_3 - \beta)\psi = 0, \tag{7}$$

where the α's and β are independent of the p's. Since we are considering the case of no field, all points in space-time must be equivalent, so that the operator in the wave equation must not involve the x's. Thus the α's and β must also be independent of the x's, so that they must commute with the p's and the x's. They therefore describe some new degree of freedom, belonging to some internal motion in the electron. We shall see later that they bring in the spin of the electron.

Multiplying (7) by the operator $(p_0 + \alpha_1 p_1 + \alpha_2 p_2 + \alpha_3 p_3 + \beta)$ on the left, we obtain

$$\left\{ p_0^2 - \sum_{123}[\alpha_1^2 p_1^2 + (\alpha_1\alpha_2 + \alpha_2\alpha_1)p_1 p_2 + (\alpha_1\beta + \beta\alpha_1)p_1] - \beta^2 \right\}\psi = 0,$$

where $\sum_{123}$ refers to cyclic permutations of the suffixes $1, 2, 3$. This is the same as (6) if the α's and β satisfy the relations

$$\alpha_1^2 = 1, \qquad \alpha_1\alpha_2 + \alpha_2\alpha_1 = 0,$$
$$\beta^2 = m^2 c^2, \qquad \alpha_1\beta + \beta\alpha_1 = 0,$$

together with the relations obtained from these by permuting the suffixes $1, 2, 3$. If we write

$$\beta = \alpha_m mc,$$

these relations may be summed up in the single one,

$$\alpha_a\alpha_b + \alpha_b\alpha_a = 2\delta_{ab} \quad (a, b = 1, 2, 3, \text{ or } m). \tag{8}$$

The four α's all anticommute with one another and the square of each is unity.

Thus by giving suitable properties to the α's and β we can make the wave equation (7) equivalent to (6), in so far as the motion of the electron as a whole is concerned. We may now assume (7) is the correct relativistic wave equation for the motion of an electron in the absence of a field. This gives rise to one difficulty, however, owing to the fact that (7), like (6), is not exactly equivalent to (5), but allows solutions corresponding to negative as well as positive values of p_0. The former do not, of course, correspond to any actually observable motion of an electron. For the present we shall consider only the positive-energy solutions and shall leave the discussion of the negative-energy ones to § 73.

We can easily obtain a representation of the four α's. They have similar algebraic properties to the σ's introduced in § 37, which σ's can be represented by matrices with two rows and columns. So long as we keep to matrices with two rows and columns we cannot get a representation of more than three anticommuting quantities, and we have to go to four rows and columns to get a representation of the four anticommuting α's. It is convenient first to express the α's in terms of the σ's and also of a second similar set of three anticommuting variables whose squares are unity, ρ_1, ρ_2, ρ_3 say, that are independent of and commute with the σ's. We may take, amongst other possibilities,

$$\alpha_1 = \rho_1\sigma_1, \quad \alpha_2 = \rho_1\sigma_2, \quad \alpha_3 = \rho_1\sigma_3, \quad \alpha_m = \rho_3, \tag{9}$$

and the α's will then satisfy all the relations (8), as may easily be verified. If we now take a representation with ρ_3 and σ_3 diagonal, we shall get the following scheme of matrices:

$$\sigma_1 = \begin{pmatrix} 0 & 1 & 0 & 0 \\ 1 & 0 & 0 & 0 \\ 0 & 0 & 0 & 1 \\ 0 & 0 & 1 & 0 \end{pmatrix} \quad \sigma_2 = \begin{pmatrix} 0 & -i & 0 & 0 \\ i & 0 & 0 & 0 \\ 0 & 0 & 0 & -i \\ 0 & 0 & i & 0 \end{pmatrix} \quad \sigma_3 = \begin{pmatrix} 1 & 0 & 0 & 0 \\ 0 & -1 & 0 & 0 \\ 0 & 0 & 1 & 0 \\ 0 & 0 & 0 & -1 \end{pmatrix}$$

$$\rho_1 = \begin{pmatrix} 0 & 0 & 1 & 0 \\ 0 & 0 & 0 & 1 \\ 1 & 0 & 0 & 0 \\ 0 & 1 & 0 & 0 \end{pmatrix} \quad \rho_2 = \begin{pmatrix} 0 & 0 & -i & 0 \\ 0 & 0 & 0 & -i \\ i & 0 & 0 & 0 \\ 0 & i & 0 & 0 \end{pmatrix} \quad \sigma_3 = \begin{pmatrix} 1 & 0 & 0 & 0 \\ 0 & 1 & 0 & 0 \\ 0 & 0 & -1 & 0 \\ 0 & 0 & 0 & -1 \end{pmatrix}.$$

It should be noted that the ρ's and σ's are all Hermitian, which makes the α's also Hermitian.

Corresponding to the four rows and columns, the wave function ψ must contain a variable that takes on four values, in order that the matrices shall be capable of being multiplied into it. Alternatively, we may look upon the wave function as having four components, each a function only of the four x's. We saw in § 37 that the spin of the electron requires the wave function to have two components. The fact that our present theory gives four is due to our wave equation (7) having twice as many solutions as it ought to have, half of them corresponding to states of negative energy.

With the help of (9), the wave equation (7) may be written with three-dimensional vector notation

$$[p_0 - \rho_1(\boldsymbol{\sigma}, \mathbf{p}) - \rho_3 mc]\psi = 0. \tag{10}$$

To generalize this equation to the case when there is an electromagnetic field present, we follow the classical rule of replacing p_0 and $\mathbf{p}$ by $p_0 + e/c \cdot A_0$ and $\mathbf{p} + e/c \cdot \mathbf{A}$, A_0 and $\mathbf{A}$ being the scalar and vector potentials of the field at the place where the electron is. This gives us the equation

$$\left[p_0 + \frac{e}{c}A_0 - \rho_1\left(\boldsymbol{\sigma}, \mathbf{p} + \frac{e}{c}\mathbf{A}\right) - \rho_3 mc\right]\psi = 0, \tag{11}$$

which is the fundamental wave equation of the relativistic theory of the electron.

The four components of ψ in (10) or (11) should be pictured as written one below another, so as to form a single-column matrix. The square matrices ρ and σ then get multiplied into the single-column matrix ψ according to matrix multiplication, the product being in each case another single-column matrix. The conjugate imaginary wave function that represents a bra should be pictured as having its four components written one beside another, so as to form a single-row matrix, which can be multiplied from the right by a square matrix ρ or σ to give another single-row matrix. We denote this conjugate imaginary wave function pictured as a single-row matrix by $\overline{\psi}^\dagger$, using the symbol $\dagger$ to denote the transpose of any matrix, i.e. the result of interchanging the rows and columns. Then the conjugate imaginary of equation (11) reads

$$\overline{\psi}^\dagger \left[p_0 + \frac{e}{c} A_0 - \rho_1 \left(\sigma, \mathbf{p} + \frac{e}{c} \mathbf{A} \right) - \rho_3 mc \right] = 0, \qquad (12)$$

in which the operators p operate to the left. An operator of differentiation operating to the left must be interpreted according to (24) of § 22.

68. Invariance under a Lorentz Transformation

Before proceeding to discuss the physical consequences of the wave equation (11) or (12), we shall first verify that our theory really is invariant under a Lorentz transformation, or, stated more accurately, that the physical results the theory leads to are independent of the Lorentz frame of reference used. This is not by any means obvious from the form of the wave equation (11). We have to verify that, if we write down the wave equation in a different Lorentz frame, the solutions of the new wave equation may be put into one-one correspondence with those of the original one in such a way that corresponding solutions may be assumed to represent the same state. For either Lorentz frame, the square of the modulus of the wave function, summed over the four components, should give the probability per unit volume of the electron being at a certain place in that Lorentz frame. We may call this the *probability density*. Its values, calculated in different Lorentz frames for wave functions representing the same state, should be connected like the time components in these frames of some 4-vector. Further, the 4-dimensional divergence of this 4-vector should vanish, signifying conservation of the electron, or that the electron cannot appear or disappear in any volume without passing through the boundary.

For brevity it is convenient to introduce the symbol $\alpha_0 = 1$ and to suppose that the suffixes of the four α_μ ($\mu = 0, 1, 2, 3$) can be raised in

68. INVARIANCE UNDER A LORENTZ TRANSFORMATION

accordance with the rules (1), even though these four α's do not form the components of a 4-vector. We can now write the wave equation (11)

$$\left[\alpha^\mu \left(p_\mu + \frac{e}{c} A_\mu\right) - \alpha_m mc\right]\psi = 0. \tag{13}$$

The four α^μ satisfy

$$\alpha^\mu \alpha_m \alpha^\nu + \alpha^\nu \alpha_m \alpha^\mu = 2g^{\mu\nu}\alpha_m \tag{14}$$

with $g^{\mu\nu}$ defined by (2), as one can verify by taking separately the cases when μ and ν are both 0, when one of them is 0, and when neither of them is 0.

Let us apply an infinitesimal Lorentz transformation and distinguish quantities referring to the new frame of reference by a star. The components of the 4-vector p_μ will transform according to equations of the type

$$p_\mu^* = p_\mu + a_\mu{}^\nu p_\nu, \tag{15}$$

where the $a_\mu{}^\nu$ are small numbers of the first order. We shall neglect quantities that are quadratic in the a's and thus of the second order. The condition for a Lorentz transformation is that

$$p_\mu^* p^{\mu*} = p_\mu p^\mu,$$

which gives

$$a_\mu{}^\nu p_\nu p^\mu + p_\mu a^{\mu\nu} p_\nu = 0,$$

leading to

$$a^{\mu\nu} + a^{\nu\mu} = 0. \tag{16}$$

The components of A_μ will transform according to the same law, so we have

$$p_\mu + \frac{e}{c} A_\mu = p_\mu^* + \frac{e}{c} A_\mu^* - a_\mu{}^\nu \left(p_\nu^* + \frac{e}{c} A_\nu^*\right).$$

Thus the wave equation (13) becomes

$$\left[(\alpha^\mu - \alpha^\lambda a_\lambda{}^\mu)\left(p_\mu^* + \frac{e}{c} A_\mu^*\right) - \alpha_m mc\right]\psi = 0. \tag{17}$$

Define

$$M = \frac{1}{4} a_{\rho\sigma} \alpha^\rho \alpha_m \alpha^\sigma. \tag{18}$$

Then from (14)

$$\alpha^\mu \alpha_m M - M \alpha_m \alpha^\mu = \frac{1}{4} a_{\rho\sigma}[(\alpha^\mu \alpha_m \alpha^\rho + \alpha^\rho \alpha_m \alpha^\mu)\alpha_m \alpha^\sigma$$
$$- \alpha^\rho \alpha_m (\alpha^\mu \alpha_m \alpha^\sigma + \alpha^\sigma \alpha_m \alpha^\mu)]$$

XI. RELATIVISTIC THEORY OF THE ELECTRON

$$= \frac{1}{2}a_{\rho\sigma}(g^{\mu\rho}\alpha^\sigma - \alpha^\rho g^{\mu\sigma}) = -\alpha_\rho{}^\mu \alpha^\rho$$

with the help of (16), and hence

$$\alpha^\mu(1 + \alpha_m M) = (1 + M\alpha_m)(\alpha^\mu - a_\rho{}^\mu \alpha^\rho). \tag{19}$$

Thus, multiplying (17) by $(1 + M\alpha_m)$ on the left, we get

$$\left[\alpha^\mu(1 + \alpha_m M)\left(p_\mu^* + \frac{e}{c}A_\mu^*\right) - (\alpha_m + M)mc\right]\psi = 0.$$

So if we put

$$(1 + \alpha_m M)\psi = \psi^*, \tag{20}$$

we get

$$\left[\alpha^\mu\left(p_\mu^* + \frac{e}{c}A_\mu^*\right) - \alpha_m mc\right]\psi^* = 0. \tag{21}$$

This is of the same form as (13) with the starred variables p_μ^*, A_μ^*, ψ^*, and shows that (13) is invariant under an infinitesimal Lorentz transformation, provided ψ is subjected to the right transformation, given by (20). A finite Lorentz transformation can be built up from infinitesimal ones, so under a finite Lorentz transformation the wave equation (13) is also invariant. Note that the matrices α^μ do not get altered at all.

The invariance proved above means that the solutions ψ of the original wave equation (13) are in one-one correspondence with the solutions ψ^* of the new wave equation (21), corresponding solutions being connected by (20). We assume that corresponding solutions represent the same physical state. We must now verify that the physical interpretations of corresponding solutions, referred to their respective Lorentz frames of reference, are in agreement. This requires that $\overline{\psi}^\dagger \psi$ should give the probability density referred to the original frame and $\overline{\psi}^{*\dagger} \psi^*$ the probability density referred to the new frame. Let us examine the relationship between these quantities. $\overline{\psi}^\dagger \psi$ is the same as $\overline{\psi}^\dagger \alpha^0 \psi$ and forms one of the four quantities $\overline{\psi}^\dagger \alpha^\mu \psi$, which should be treated together.

Equations (18) and (16) show that M is pure imaginary. Thus the conjugate imaginary of equation (20) is

$$\overline{\psi}^{*\dagger} = \overline{\psi}^\dagger(1 - M\alpha_m).$$

Hence

$$\overline{\psi}^{*\dagger} \alpha^\mu \psi^* = \overline{\psi}^\dagger(1 - M\alpha_m)\alpha^\mu(1 + \alpha_m M)\psi$$
$$= \overline{\psi}^\dagger(1 - M\alpha_m)(1 + M\alpha_m)(\alpha^\mu - a_\nu{}^\mu \alpha^\nu)\psi$$

from (19). This reduces to

$$\overline{\psi}^{*\dagger}\alpha^\mu\psi^* = \overline{\psi}^\dagger(\alpha^\mu - a_\nu{}^\mu\alpha^\nu)\psi$$
$$= \overline{\psi}^\dagger\alpha^\mu\psi + a^\mu{}_\nu\overline{\psi}^\dagger\alpha^\nu\psi$$

with the help of (16). If we lower the suffix μ here, we get an equation of the same form as (15), which shows that the four quantities $\overline{\psi}^\dagger\alpha_\mu\psi$ transform like the contravariant components of a 4-vector. Thus $\overline{\psi}^\dagger\psi$ transforms like the time component of a 4-vector, which is the correct transformation law for a probability density. The space components of the 4-vector, namely $\overline{\psi}^\dagger\alpha_r\psi$, if multiplied by c, give the probability current, or the probability of the electron crossing unit area per unit time.

It should be noted that $\overline{\psi}^\dagger\alpha_m\psi$ is invariant, since

$$\overline{\psi}^{*\dagger}\alpha_m\psi^* = \overline{\psi}^\dagger(1 - M\alpha_m)\alpha_m(1 + \alpha_m M)\psi$$
$$= \overline{\psi}^\dagger\alpha_m\psi.$$

We must verify finally the conservation law, that the divergence

$$\frac{\partial}{\partial x_\mu}(\overline{\psi}^\dagger\alpha_\mu\psi) \tag{22}$$

vanishes. To prove this, multiply equation (13) by $\overline{\psi}^\dagger$ on the left. The result is

$$\overline{\psi}^\dagger\alpha^\mu\left(i\hbar\frac{\partial\psi}{\partial x^\mu} + \frac{e}{c}A_\mu\psi\right) - \overline{\psi}^\dagger\alpha_m mc\psi = 0.$$

The conjugate imaginary equation is

$$\left(-i\hbar\frac{\partial\overline{\psi}^\dagger}{\partial x^\mu} + \overline{\psi}^\dagger\frac{e}{c}A_\mu\right)\alpha^\mu\psi - \overline{\psi}^\dagger\alpha_m mc\psi = 0.$$

Subtracting and dividing $i\hbar$, we get

$$\overline{\psi}^\dagger\alpha^\mu\frac{\partial\psi}{\partial x^\mu} + \frac{\partial\overline{\psi}^\dagger}{\partial x^\mu}\alpha^\mu\psi = 0,$$

which just expresses the vanishing of (22). In this way we complete the proof that our theory gives consistent results in whichever frame of reference it is applied.

69. The Motion of a Free Electron

It is of interest to consider the motion of a free electron in the Heisenberg picture according to the above theory and to study the Heisenberg equations of motion. These equations of motion can be integrated exactly, as was first done by Schrödinger.[1] For brevity we shall omit the suffix t which the notation of § 28 requires to be inserted in dynamical variables that vary with time in the Heisenberg picture.

As Hamiltonian we must take the expression which we get as equal to cp_0 when we put the operator on ψ in (10) equal to zero, i.e.

$$H = c\rho_1(\boldsymbol{\sigma}, \mathbf{p}) + \rho_3 mc^2 = c(\boldsymbol{\alpha}, \mathbf{p}) + \rho_3 mc^2. \qquad (23)$$

We see at once that the momentum commutes with H and is thus a constant of the motion. Further, the x_1-component of the velocity is

$$\dot{x}_1 = [x_1, H] = c\alpha_1. \qquad (24)$$

This result is rather surprising, as it means an altogether different relation between velocity and momentum from what one has in classical mechanics. It is connected, however, with the expression $\overline{\psi}^\dagger c\alpha_1 \psi$ for a component of the probability current. The $\dot{x}_1$ given by (24) has as eigenvalues $\pm c$, corresponding to the eigenvalues ± 1 of α_1. As $\dot{x}_2$ and $\dot{x}_3$ are similar, we can conclude that *a measurement of a component of the velocity of a free electron is certain to lead to the result $\pm c$*. This conclusion is easily seen to hold also when there is a field present.

Since electrons are observed in practice to have velocities considerably less than that of light, it would seem that we have here a contradiction with experiment. The contradiction is not real, though, since the theoretical velocity in the above conclusion is the velocity at one instant of time while observed velocities are always average velocities through appreciable time intervals. We shall find upon further examination of the equations of motion that the velocity is not at all constant, but oscillates rapidly about a mean value which agrees with the observed value.

It may easily be verified that a measurement of a component of the velocity must lead to the result $\pm c$ in a relativistic theory, simply from an elementary application of the principle of uncertainty of § 24. To measure the velocity we must measure the position at two slightly different times and then divide the change of position by the time interval. (It will not do to measure the momentum and apply a formula, as the ordinary connexion between velocity and momentum is not valid.) In order that our measured velocity may approximate to the instantaneous velocity, the time interval between the two measurements of position must be very short and hence these measurements must be very accurate. The great accuracy with which the position of the electron is known during the time-interval must give rise, according to the principle of uncertainty,

---

[1] Schrödinger, *Sitzungsb. d. Berlin, Akad.*, 1930, 418.

69. THE MOTION OF A FREE ELECTRON

to an almost complete indeterminacy in its momentum. This means that almost all values of the momentum are equally probable, so that the momentum is almost certain to be infinite. An infinite value for a component of momentum corresponds to the value $\pm c$ for the corresponding component of velocity.

Let us now examine how the velocity of the electron varies with time. We have

$$i\hbar\dot{\alpha}_1 = \alpha_1 H - H\alpha_1.$$

Now since α_1 anticommutes with all the terms in H except $c\alpha_1 p_1$,

$$\alpha_1 H + H\alpha_1 = \alpha_1 c\alpha_1 p_1 + c\alpha_1 p_1 \alpha_1 = 2cp_1,$$

and hence

$$\begin{aligned} i\hbar\dot{\alpha}_1 &= 2\alpha_1 H - 2cp_1, \\ &= -2H\alpha_1 + 2cp_1. \end{aligned} \quad (25)$$

Since H and p_1 are constants, it follows from the first of equations (25) that

$$i\hbar\ddot{\alpha}_1 = 2\dot{\alpha}_1 H. \quad (26)$$

This differential equation in $\dot{\alpha}_1$ can be integrated immediately, the result being

$$\dot{\alpha}_1 = \dot{\alpha}_1^0 e^{-2iHt/\hbar}, \quad (27)$$

where $\dot{\alpha}_1^0$ is a constant, equal to the value of $\dot{\alpha}_1$ when $t = 0$. The factor $e^{-2iHt/\hbar}$ must be put to the right of the factor $\dot{\alpha}_1^0$ in (27) on account of the H occurring to the right of the $\dot{\alpha}_1$ in (26). The second of equations (25) leads in the same way to the result

$$\dot{\alpha}_1 = e^{2iHt/\hbar}\dot{\alpha}_1^0.$$

We can now easily complete the integration of the equation of motion for x_1. From (27) and the first of equations (25)

$$\alpha_1 = \frac{1}{2}i\hbar\dot{\alpha}_1^0 e^{-2iHt/\hbar} H^{-1} + cp_1 H^{-1}, \quad (28)$$

and hence the time-integral of equation (24) is

$$x_1 = -\frac{1}{4}c\hbar^2 \dot{\alpha}_1^0 e^{-2iHt/\hbar} H^{-2} + c^2 p_1 H^{-1} t + a_1, \quad (29)$$

a_1 being a constant.

From (28) we see that the x_1-component of velocity, $c\alpha_1$, consists of two parts, a constant part $c^2 p_1 H^{-1}$, connected with the momentum by the classical relativistic formula, and an oscillatory part

$$\frac{1}{2} ic\hbar \dot{\alpha}_1^0 e^{-2iHt/\hbar} H^{-1},$$

whose frequency is high, being $2H/h$, which is at least $2mc^2/h$. Only the constant part would be observed in a practical measurement of velocity, such a measurement giving the average velocity through a time-interval much larger than $h/2mc^2$. The oscillatory part secures that the instantaneous value of $\dot{x}_1$ shall have the eigenvalues $\pm c$. The oscillatory part of x_1 is small, being, according to (29),

$$-\frac{1}{4} c\hbar^2 \dot{\alpha}_1^0 e^{-2iHt/\hbar} H^{-2} = \frac{1}{2} ic\hbar(\alpha_1 - cp_1 H^{-1}) H^{-1},$$

which is of the order of magnitude $\hbar/mc$, since $(\alpha_1 - cp_1 H^{-1})$ is of the order of magnitude unity.

70. Existence of the Spin

In § 67 we saw that the correct wave equation for the electron in the absence of an electromagnetic field, namely equation (7) or (10), is equivalent to the wave equation (6) which is suggested from analogy with the classical theory. This equivalence no longer holds when there is a field. The wave equation to be expected from analogy with the classical theory in this case is

$$\left[\left(p_0 + \frac{e}{c} A_0 \right)^2 - \left(\mathbf{p} + \frac{e}{c} \mathbf{A} \right)^2 - m^2 c^2 \right] \psi = 0, \tag{30}$$

in which the operator is just the classical relativistic Hamiltonian. If we multiply (11) by some factor on the left to make it resemble (30) as closely as possible, namely the factor

$$p_0 + \frac{e}{c} A_0 + \rho_1 \left(\boldsymbol{\sigma}, \mathbf{p} + \frac{e}{c} \mathbf{A} \right) + \rho_3 mc,$$

we get

$$\left\{ \left(p_0 + \frac{e}{c} A_0 \right)^2 - \left(\boldsymbol{\sigma}, \mathbf{p} + \frac{e}{c} \mathbf{A} \right)^2 - m^2 c^2 - \rho_1 \left[\left(p_0 + \frac{e}{c} A_0 \right) \left(\boldsymbol{\sigma}, \mathbf{p} + \frac{e}{c} \mathbf{A} \right) - \left(\boldsymbol{\sigma}, \mathbf{p} + \frac{e}{c} \mathbf{A} \right) \left(p_0 + \frac{e}{c} A_0 \right) \right] \right\} \psi = 0. \tag{31}$$

70. EXISTENCE OF THE SPIN

We now use the general formula that, if **B** and **C** are any two three-dimensional vectors that commute with σ,

$$(\sigma, \mathbf{B})(\sigma, \mathbf{C}) = \sum_{123}(\sigma_1^2 B_1 C_1 + \sigma_1\sigma_2 B_1 C_2 + \sigma_2\sigma_1 B_2 C_1),$$

the summation referring to cyclic permutations of the suffixes 1, 2, 3, or

$$(\sigma, \mathbf{B})(\sigma, \mathbf{C}) = (\mathbf{B}, \mathbf{C}) + i\sum_{123}\sigma_3[B_1 C_2 - B_2 C_1]$$
$$= (\mathbf{B}, \mathbf{C}) + i(\sigma, \mathbf{B} \times \mathbf{C}). \tag{32}$$

Taking $\mathbf{B} = \mathbf{C} = \mathbf{p} + e/c \cdot \mathbf{A}$, we find, since

$$\left(\mathbf{p}+\frac{e}{c}\mathbf{A}\right) \times \left(\mathbf{p}+\frac{e}{c}\mathbf{A}\right) = \frac{e}{c}(\mathbf{p}\times\mathbf{A} + \mathbf{A}\times\mathbf{p})$$
$$= -i\frac{\hbar e}{c}\operatorname{curl}\mathbf{A} = -i\frac{\hbar e}{c}\mathcal{H},$$

where $\mathcal{H}$ is the magnetic field, that

$$\left(\sigma, \mathbf{p}+\frac{e}{c}\mathbf{A}\right)^2 = \left(\mathbf{p}+\frac{e}{c}\mathbf{A}\right)^2 + \frac{\hbar e}{c}(\sigma, \mathcal{H}). \tag{33}$$

Also we have

$$\left[p_0 + \frac{e}{c}A_0\right]\left(\sigma, \mathbf{p}+\frac{e}{c}\mathbf{A}\right) - \left(\sigma, \mathbf{p}+\frac{e}{c}\mathbf{A}\right)\left[p_0 + \frac{e}{c}A_0\right]$$
$$= \frac{e}{c}(\sigma, p_0\mathbf{A} - \mathbf{A}p_0 + A_0\mathbf{p} - \mathbf{p}A_0)$$
$$= \frac{i\hbar e}{c}\left(\sigma, \frac{1}{c}\frac{\partial \mathbf{A}}{\partial t} + \operatorname{grad} A_0\right) = -i\frac{\hbar e}{c}(\sigma, \mathcal{E}),$$

where $\mathcal{E}$ is the electric field. Thus (31) becomes

$$\left\{\left[p_0 + \frac{e}{c}A_0\right]^2 - \left[\mathbf{p}+\frac{e}{c}\mathbf{A}\right]^2 - m^2c^2 - \frac{\hbar e}{c}(\sigma, \mathcal{H}) + i\rho_1\frac{\hbar e}{c}(\sigma, \mathcal{E})\right\}\psi = 0. \tag{34}$$

This equation differs from (30) through having two extra terms in the operator. These extra terms involve some new physical effects, but since they are not real they do not lend themselves very directly to physical interpretation.

To get an understanding of the physical features involved in the difference between (34) and (30) it is better to work with the Heisenberg picture, this picture being always the more suitable one for comparisons between classical and quantum mechanics. The Heisenberg equations of motion are determined by the Hamiltonian

$$H = -eA_0 + c\rho_1\left(\sigma, \mathbf{p} + \frac{e}{c}\mathbf{A}\right) + \rho_3 mc^2, \tag{35}$$

the generalization of (23) to the case when there is a field. Equation (35) gives

$$\left(\frac{H}{c} + \frac{e}{c}A_0\right)^2 = \left[\rho_1\left(\sigma, \mathbf{p} + \frac{e}{c}\mathbf{A}\right) + \rho_3 mc\right]^2$$

$$= \left(\sigma, \mathbf{p} + \frac{e}{c}\mathbf{A}\right)^2 + m^2c^2$$

$$= \left(\mathbf{p} + \frac{e}{c}\mathbf{A}\right)^2 + m^2c^2 + \frac{\hbar e}{c}(\sigma, \mathcal{H}) \tag{36}$$

with the help of (33). We have here the real part of the extra terms in (34) appearing without the pure imaginary part. For an electron moving slowly (i.e. with small momentum), we may expect the Heisenberg equations of motion to be determined by a Hamiltonian of the form $mc^2 + H_1$, where H_1 is small compared with mc^2. Putting $mc^2 + H_1$ for H in (36) and neglecting H_1^2 and other terms involving c^{-2}, we get, on dividing by $2m$,

$$H_1 + eA_0 = \frac{1}{2m}\left(\mathbf{p} + \frac{e}{c}\mathbf{A}\right)^2 + \frac{\hbar e}{2mc}(\sigma, \mathcal{H}). \tag{37}$$

The Hamiltonian H_1 given by (37) is the same as the classical Hamiltonian for a slow electron, except for the last term

$$\frac{\hbar e}{2mc}(\sigma, \mathcal{H}).$$

This term may be considered as an additional potential energy which a slow electron has in the quantum theory and may be interpreted as arising from *the electron having a magnetic moment* $-\hbar e/2mc \cdot \sigma$. This magnetic moment is the one assumed in §§ 41 and 47 for dealing with the Zeeman effect and is in agreement with experiment.

The spin angular momentum does not give rise to any potential energy and therefore does not appear in the result of the preceding calculation. The simplest way of showing the existence of the spin angular momentum is to take the case of the motion of a free electron or an electron in a central field of force and determine the angular momentum integrals.

This means working with the Hamiltonian (23), or with the Hamiltonian (35) with $\mathbf{A} = 0$ and A_0 a function of the radius r, i.e.

$$H = -e\, A_0(r) + c\rho_1(\boldsymbol{\sigma}, \mathbf{p}) + \rho_3 mc^2, \tag{38}$$

and obtaining the Heisenberg equations of motion for the angular momentum. With either Hamiltonian we find for the rate of change of the x_1-component of orbital angular momentum, $m_1 = x_2 p_3 - x_3 p_2$, with the help of commutation relations proved in § 35,

$$i\hbar \dot{m}_1 = m_1 H - H m_1 = c\rho_1[m_1(\boldsymbol{\sigma}, \mathbf{p}) - (\boldsymbol{\sigma}, \mathbf{p})m_1]$$
$$= c\rho_1(\boldsymbol{\sigma}, m_1\mathbf{p} - \mathbf{p}m_1) = i\hbar c\rho_1(\sigma_2 p_3 - \sigma_3 p_2).$$

Thus $\dot{m}_1 \neq 0$ and the orbital angular momentum is not a constant of the motion. This result is to be expected from the integrated equation of motion (29), the oscillatory part of the motion here displayed giving rise to an oscillatory term in the angular momentum.

We have further

$$i\hbar \dot{\sigma}_1 = \sigma_1 H - H \sigma_1 = c\rho_1[\sigma_1(\boldsymbol{\sigma}, \mathbf{p}) - (\boldsymbol{\sigma}, \mathbf{p})\sigma_1]$$
$$= c\rho_1(\sigma_1 \boldsymbol{\sigma} - \boldsymbol{\sigma}\sigma_1, \mathbf{p}) = 2ic\rho_1(\sigma_3 p_2 - \sigma_2 p_3)$$

with the help of equations (51) of § 37. Hence

$$\dot{m}_1 + \frac{1}{2}\hbar\dot{\sigma}_1 = 0,$$

so that the vector $\mathbf{m} + (1/2)\hbar\boldsymbol{\sigma}$ is a constant of the motion. This result one can interpret by saying *the electron has a spin angular momentum* $(1/2)\hbar\boldsymbol{\sigma}$, which must be added to the orbital angular momentum $\mathbf{m}$ before one gets a constant of the motion. The spin angular momentum could alternatively be obtained from the rotation operators for states of spin in accordance with the general method of § 35.

The same vector $\boldsymbol{\sigma}$ fixes the directions of both the spin magnetic moment and the spin angular momentum. If an electron in a certain state of spin has a spin angular momentum of $(1/2)\hbar$ in a particular direction, it will have a magnetic moment $-e\hbar/2mc$ in the same direction.

We were led to the value $(1/2)\hbar$ for the spin of the electron by an argument depending simply on general principles of quantum theory and relativity. One could apply the same argument to other kinds of elementary particle and one would be led to the same conclusion, that the spin angular momentum is half a quantum. This would be satisfactory for the proton and the neutron, but there are some kinds of elementary particle (e.g. the photon and certain kinds of meson) whose spins are known experimentally to be different from $(1/2)\hbar$, so we have a discrepancy between our theory and experiment.

The answer is to be found in a hidden assumption in our work. Our argument is valid only provided the position of the particle is an observable. If this assumption holds, the particle must have a spin angular momentum of half a quantum. For those particles that have a different spin the assumption must be false and any dynamical variables x_1, x_2, x_3 that may be introduced to describe the position of the particle cannot be observables in accordance with our general theory. For such particles there is no true Schrödinger representation. One might be able to introduce a quasi wave function involving the dynamical variables x_1, x_2, x_3, but it would not have the correct physical interpretation of a wave function—that the square of its modulus gives the probability density. For such particles there is still a momentum representation, which is sufficient for practical purposes.

71. Transition to Polar Variables

For the further study of the motion of an electron in a central field of force with the Hamiltonian (38), it is convenient to make a transformation to polar coordinates, as was done in § 38 in the non-relativistic case. We can introduce r and p_r as before, but instead of k, the magnitude of the orbital angular momentum $\mathbf{m}$, which is no longer a constant of the motion, we must now use the magnitude of the total angular momentum $\mathbf{M} = \mathbf{m} + (1/2)\hbar\boldsymbol{\sigma}$. Let us put

$$j^2\hbar^2 = M_1^2 + M_2^2 + M_3^2 + \frac{1}{4}\hbar^2. \tag{39}$$

The eigenvalues of m_3 are integral multiples of $\hbar$, those of $(1/2)\hbar\sigma_3$ are $\pm(1/2)\hbar$, and hence those of M_3 must be half odd integral multiples of $\hbar$. It follows from the theory of § 36 that the eigenvalues of $|j|$ must be integers greater than zero.

If in formula (32) we take $\mathbf{B} = \mathbf{C} = \mathbf{m}$, we get

$$(\boldsymbol{\sigma}, \mathbf{m})^2 = \mathbf{m}^2 + i(\boldsymbol{\sigma}, \mathbf{m} \times \mathbf{m}) = \mathbf{m}^2 - \hbar(\boldsymbol{\sigma}, \mathbf{m})$$

$$= \left[\mathbf{m} + \frac{1}{2}\hbar\boldsymbol{\sigma}\right]^2 - 2\hbar(\boldsymbol{\sigma}, \mathbf{m}) - \frac{3}{4}\hbar^2.$$

Hence

$$[(\boldsymbol{\sigma}, \mathbf{m}) + \hbar]^2 = \mathbf{M}^2 + \frac{1}{4}\hbar^2.$$

71. TRANSITION TO POLAR VARIABLES

Thus $(\boldsymbol{\sigma}, \mathbf{m}) + \hbar$ is a quantity whose square is $\mathbf{M}^2 + (1/4)\hbar^2$ and we could, consistently with equation (39), define $j\hbar$ as $(\boldsymbol{\sigma}, \mathbf{m}) + \hbar$. This would not be the most convenient definition for j, however, since we would like to have j a constant of the motion and $(\boldsymbol{\sigma}, \mathbf{m}) + \hbar$ is not constant. We have, in fact, from applications of (32),

$$(\boldsymbol{\sigma}, \mathbf{m})(\boldsymbol{\sigma}, \mathbf{p}) = i(\boldsymbol{\sigma}, \mathbf{m} \times \mathbf{p})$$

and

$$(\boldsymbol{\sigma}, \mathbf{p})(\boldsymbol{\sigma}, \mathbf{m}) = i(\boldsymbol{\sigma}, \mathbf{p} \times \mathbf{m}),$$

so that

$$(\boldsymbol{\sigma}, \mathbf{m})(\boldsymbol{\sigma}, \mathbf{p}) + (\boldsymbol{\sigma}, \mathbf{p})(\boldsymbol{\sigma}, \mathbf{m}) = i \sum_{123} \sigma_1 (m_2 p_3 - m_3 p_2 + p_2 m_3 - p_3 m_2)$$

$$= i \sum_{123} \sigma_1 \cdot 2i\hbar p_1 = -2\hbar(\boldsymbol{\sigma}, \mathbf{p}),$$

or

$$[(\boldsymbol{\sigma}, \mathbf{m}) + \hbar](\boldsymbol{\sigma}, \mathbf{p}) + (\boldsymbol{\sigma}, \mathbf{p})[(\boldsymbol{\sigma}, \mathbf{m}) + \hbar] = 0.$$

Thus $(\boldsymbol{\sigma}, \mathbf{m}) + \hbar$ anticommutes with one of the terms in the expression (38) for H, namely the term $c\rho_1(\boldsymbol{\sigma}, \mathbf{p})$, and commutes with the other two. It follows that $\rho_3[(\boldsymbol{\sigma}, \mathbf{m}) + \hbar]$ commutes with all the three terms in H and is a constant of the motion. But the square of $\rho_3[(\boldsymbol{\sigma}, \mathbf{m}) + \hbar]$ is also $\mathbf{M}^2 + (1/4)\hbar^2$. We can therefore take

$$j\hbar = \rho_3[(\boldsymbol{\sigma}, \mathbf{m}) + \hbar], \tag{40}$$

which gives us a convenient rational definition for j which is consistent with (39) and makes j a constant of the motion. The eigenvalues of this j are all positive and negative integers, excluding zero.

By a further application of (32), we get

$$(\boldsymbol{\sigma}, \mathbf{x})(\boldsymbol{\sigma}, \mathbf{p}) = (\mathbf{x}, \mathbf{p}) + i(\boldsymbol{\sigma}, \mathbf{m})$$
$$= r p_r + i\rho_3 j\hbar - i\hbar, \tag{41}$$

with the help of (40) and also of equation (58) of § 38. We introduce the linear operator ϵ defined by

$$r\epsilon = \rho_1(\boldsymbol{\sigma}, \mathbf{x}). \tag{42}$$

Since r commutes with ρ_1 and with $(\boldsymbol{\sigma}, \mathbf{x})$, it must commute with ϵ. We thus have

$$r^2 \epsilon^2 = [\rho_1(\boldsymbol{\sigma}, \mathbf{x})]^2 = (\boldsymbol{\sigma}, \mathbf{x})^2 = \mathbf{x}^2 = r^2,$$

or

$$\epsilon^2 = 1.$$

Now $\rho_1(\sigma, \mathbf{p})$ commutes with j, and since there is symmetry between $\mathbf{x}$ and $\mathbf{p}$ so far as angular momentum is concerned, $\rho_1(\sigma, \mathbf{x})$ must also commute with j. Hence ϵ commutes with j. Further, ϵ must commute with p_r, since we have

$$(\sigma, \mathbf{x})(\mathbf{x}, \mathbf{p}) - (\mathbf{x}, \mathbf{p})(\sigma, \mathbf{x}) = (\sigma, \mathbf{x}(\mathbf{x}, \mathbf{p}) - (\mathbf{x}, \mathbf{p})\mathbf{x}) = i\hbar(\sigma, \mathbf{x}),$$

which gives

$$r\epsilon r p_r - r p_r r \epsilon = i\hbar r \epsilon,$$

or

$$r^2 \epsilon p_r - r^2 p_r \epsilon = 0.$$

From (41) and (42) we obtain

$$r\epsilon \rho_1(\sigma, \mathbf{p}) = r p_r + i\rho_3 j\hbar - i\hbar,$$

or

$$\rho_1(\sigma, \mathbf{p}) = \epsilon\left(p_r - \frac{i\hbar}{r}\right) + \frac{i\epsilon\rho_3 j\hbar}{r}.$$

Thus (38) becomes

$$\frac{H}{c} = -\frac{e}{c}A_0 + \epsilon\left(p_r - \frac{i\hbar}{r}\right) + \frac{i\epsilon\rho_3 j\hbar}{r} + \rho_3 mc.$$

This gives our Hamiltonian expressed in terms of polar variables. It should be noticed that ϵ and ρ_3 commute with all the other variables occurring in H and anticommute with one another. This means that we can take a representation with ρ_3 diagonal in which ϵ and ρ_3 are represented respectively by the matrices

$$\begin{pmatrix} 0 & -i \\ i & 0 \end{pmatrix}, \quad \begin{pmatrix} 1 & 0 \\ 0 & -1 \end{pmatrix}. \tag{43}$$

If r is also diagonal in the representation, the representative $\langle r' \rho_3' |\rangle$ of a ket will have two components, $\langle r', 1|\rangle = \psi_a(r')$ and $\langle r', -1|\rangle = \psi_b(r')$ say, referring to the two rows and columns of the matrices (43).

72. The Fine-structure of the Energy-levels of Hydrogen

We shall now take the case of the hydrogen atom, for which $A_0 = e/r$, and work out its energy-levels, given by the eigenvalues H' of H. The equation $(H' - H)|\rangle = 0$ which defines these eigenvalues, when written in terms of representatives in the representation discussed above with ϵ and ρ_3 represented by the matrices (43), gives the equations

$$\left(\frac{H'}{c} + \frac{e^2}{cr}\right)\psi_a + \hbar\left(\frac{\partial}{\partial r} + \frac{1}{r}\right)\psi_b + \frac{j\hbar}{r}\psi_b - mc\psi_a = 0,$$

$$\left(\frac{H'}{c} + \frac{e^2}{cr}\right)\psi_b - \hbar\left(\frac{\partial}{\partial r} + \frac{1}{r}\right)\psi_a + \frac{j\hbar}{r}\psi_a + mc\psi_b = 0.$$

If we put

$$\frac{\hbar}{mc - H'/c} = a_1, \quad \frac{\hbar}{mc + H'/c} = a_2, \tag{44}$$

these equations reduce to

$$\left(\frac{1}{a_1} - \frac{\alpha}{r}\right)\psi_a - \left(\frac{\partial}{\partial r} + \frac{j+1}{r}\right)\psi_b = 0,$$

$$\left(\frac{1}{a_2} + \frac{\alpha}{r}\right)\psi_b - \left(\frac{\partial}{\partial r} - \frac{j-1}{r}\right)\psi_a = 0, \tag{45}$$

where $\alpha = e^2/\hbar c$, which is a small number. We shall solve these equations by a similar method to that used for equation (73) in § 39.

Put

$$\psi_a = r^{-1} e^{-r/a} f, \quad \psi_b = r^{-1} e^{-r/a} g, \tag{46}$$

introducing two new functions, f and g, of r, where

$$a = (a_1 a_2)^{1/2} = \hbar\left(m^2 c^2 - \frac{H'^2}{c^2}\right)^{-1/2}. \tag{47}$$

Equations (45) become

$$\left(\frac{1}{a_1} - \frac{\alpha}{r}\right)f - \left(\frac{\partial}{\partial r} - \frac{1}{a} + \frac{j}{r}\right)g = 0,$$

$$\left(\frac{1}{a_2} + \frac{\alpha}{r}\right)g - \left(\frac{\partial}{\partial r} - \frac{1}{a} - \frac{j}{r}\right)f = 0. \tag{48}$$

We now try for a solution in which f and g are in the form of power series,

$$f = \sum_s c_s r^s, \quad g = \sum_s c'_s r^s, \tag{49}$$

in which consecutive values of s differ by unity though these values need not be integers. Substituting these expressions for f and g in (48) and picking out coefficients of r^{s-1}, we obtain

$$\frac{c_{s-1}}{a_1} - \alpha c_s - (s+j)c'_s + \frac{c'_{s-1}}{a} = 0,$$

$$\frac{c'_{s-1}}{a_2} + \alpha c'_s - (s-j)c_s + \frac{c_{s-1}}{a} = 0. \tag{50}$$

By multiplying the first of these equations by a and the second by a_2 and subtracting, we eliminate both c_{s-1} and c'_{s-1}, since from (47) $a/a_1 = a_2/a$. We are left with

$$[a\alpha - a_2(s-j)]c_s + [a_2\alpha + a(s+j)]c'_s = 0, \qquad (51)$$

a relation which shows the connexion between the primed and unprimed c's.

The boundary condition at $r = 0$ requires that $r\psi_a$ and $r\psi_b \to 0$ as $r \to 0$, so from (46) f and $g \to 0$ as $r \to 0$. Thus the series (49) must terminate on the side of small s. If s_0 is the minimum value of s for which c_s and c'_s do not both vanish, we obtain from (50), by putting $s = s_0$ and $c_{s_0-1} = c'_{s_0-1} = 0$,

$$\alpha c_{s_0} + (s_0 + j)c'_{s_0} = 0,$$
$$\alpha c'_{s_0} - (s_0 - j)c_{s_0} = 0, \qquad (52)$$

which give

$$\alpha^2 = -s_0^2 + j^2.$$

Since the boundary condition requires that the minimum value of s shall be greater than zero, we must take

$$s_0 = +\sqrt{j^2 - \alpha^2}.$$

To investigate the convergence of the series (49) we shall determine the ratio c_s/c_{s-1} for large s. Equation (51) and the second of equations (50) give approximately, when s is large,

$$a_2 c_s = ac'_s$$

and

$$sc_s = \frac{c_{s-1}}{a} + \frac{c'_{s-1}}{a_2}.$$

Hence

$$\frac{c_s}{c_{s-1}} = \frac{2}{as}.$$

The series (49) will therefore converge like

$$\sum_s \frac{1}{s!}\left(\frac{2r}{a}\right)^s,$$

or $e^{2r/a}$. This result is similar to that obtained in § 39 and allows us to infer, as in § 39, that all values of H' are permissible for which a is pure imaginary, i.e. from (47), for which $H' > mc^2$, while for $H' < mc^2$

we take a to be positive and then find that only those values of H' are permissible for which the series (49) terminate on the side of large s.

If the series (49) terminate with the terms c_s and c'_s, so that $c_{s+1} = c'_{s+1} = 0$, we obtain from (50) with $s+1$ substituted for s

$$\frac{c_s}{a_1} + \frac{c'_s}{a} = 0,$$
$$\frac{c'_s}{a_2} + \frac{c_s}{a} = 0.$$
(53)

These two equations are equivalent on account of (47). When combined with (51), they give

$$a_1[a\alpha - a_2(s-j)] = a[a_2\alpha + a(s+j)],$$

which reduces to

$$2a_1a_2 s = a(a_1 - a_2)\alpha,$$

or

$$\frac{s}{a} = \frac{1}{2}\left(\frac{1}{a_2} - \frac{1}{a_1}\right)\alpha = \frac{H'}{c\hbar}\alpha,$$

with the help of (44). Squaring and using (47), we obtain

$$s^2\left(m^2c^2 - \frac{H'^2}{c^2}\right) = \alpha^2 \frac{H'^2}{c^2}.$$

Hence

$$\frac{H'}{mc^2} = \left(1 + \frac{\alpha^2}{s^2}\right)^{-1/2}.$$

The s here, which specifies the last term in the series, must be greater than s_0 by some integer not less than zero. Calling this integer n, we have

$$s = n + \sqrt{j^2 - \alpha^2}$$

and thus

$$\frac{H'}{mc^2} = \left[1 + \frac{\alpha^2}{(n + \sqrt{j^2 - \alpha^2})^2}\right]^{-1/2}. \quad (54)$$

This formula gives the discrete energy-levels of the hydrogen spectrum and was first obtained by Sommerfeld working with Bohr's orbit theory. There are two quantum numbers n and j involved, but owing to α^2 being very small the energy depends almost entirely on $n + |j|$. Values of n and $|j|$ that give the same $n + |j|$ give rise to a set of energy-levels lying very close to one another, and to the energy-level given by the non-relativistic formula (80) of § 39 with $s = n + |j|$, apart from the constant term mc^2.

We used equations (53) by combining them with (51), but this does not make full use of (53) since the coefficients of c_s and c'_s in (51) may both vanish. In this case we get, multiplying the first coefficient by a_1 and the second by a and adding,

$$a(a_1 + a_2)\alpha + 2a_1 a_2 j = 0.$$

Thus j must be negative in this case. With the help of (44) and (47) we get further

$$-\frac{2j}{\alpha} = \frac{a}{a_2} + \frac{a}{a_1} = \frac{2mca}{\hbar} = \frac{2mc}{(m^2c^2 - H'^2/c^2)^{1/2}},$$

or

$$\frac{H'^2}{m^2 c^4} = 1 - \frac{\alpha^2}{j^2}.$$

Since H' must be positive, this leads to

$$\frac{H'}{mc^2} = \frac{\sqrt{j^2 - \alpha^2}}{|j|}, \tag{55}$$

which is the value of H' given by (54) when $n = 0$. The case $n = 0$ with j negative thus needs further investigation to see whether the conditions (53) are then fulfilled.

With $n = 0$, the maximum value of s is the same as the minimum, so equations (53) with s_0 substituted for s should agree with (52). Now (55) gives, from (44) and (47),

$$\frac{1}{a_1} = \frac{mc}{\hbar}\left(1 - \frac{\sqrt{j^2 - \alpha^2}}{|j|}\right), \quad \frac{1}{a} = \frac{mc}{\hbar}\frac{\alpha}{|j|},$$

so the first of equations (53) with s_0 substituted for s gives

$$c_{s_0}(|j| - \sqrt{j^2 - \alpha^2}) + c'_{s_0}\alpha = 0.$$

This agrees with the second of equations (52) only if j is positive. We can conclude that, for $n = 0$, j must be a positive integer, while for the other values of n all non-zero integral values of j are allowed.

73. Theory of the Positron

It has been mentioned in § 67 that the wave equation for the electron admits of twice as many solutions as it ought to, half of them referring to states with negative values for the kinetic energy $cp_0 + eA_0$. This difficulty was introduced as soon as we passed from equation (5) to equation (6) and is inherent in any relativistic theory. It occurs also in classical relativistic theory, but is not then serious since, owing to the continuity in the variation of all classical dynamical variables, if the kinetic energy $cp_0 + eA_0$ is initially positive (when it must be greater than or equal to mc^2), it cannot subsequently be negative (when it would have to be less than or equal to $-mc^2$). In the quantum theory, however, discontinuous transitions may take place, so that if the electron is initially in a state of positive kinetic energy it may make a transition to a state of negative kinetic energy. It is therefore no longer permissible simply to ignore the negative-energy states, as one can do in the classical theory.

Let us examine the negative-energy solutions of the equation

$$\left[\left(p_0 + \frac{e}{c}A_0\right) - \alpha_1\left(p_1 + \frac{e}{c}A_1\right)\right.$$
$$\left. - \alpha_2\left(p_2 + \frac{e}{c}A_2\right) - \alpha_3\left(p_3 + \frac{e}{c}A_3\right) - \alpha_m mc\right]\psi = 0 \quad (56)$$

a little more closely. For this purpose it is convenient to use a representation of the α's in which all the elements of the matrices representing α_1, α_2, and α_3 are real and all those of the matrix representing α_m are pure imaginary or zero. Such a representation may be obtained, for instance, from that of § 67 by interchanging the expressions for α_2 and α_m in (9). If equation (56) is expressed as a matrix equation in this representation and we put $-i$ for i all through it, we get, remembering the i in (4).

$$\left[\left(-p_0 + \frac{e}{c}A_0 - \alpha_1\left(-p_1 + \frac{e}{c}A_1\right)\right.\right.$$
$$\left.\left. - \alpha_2\left(-p_2 + \frac{e}{c}A_2\right) - \alpha_3\left(-p_3 + \frac{e}{c}A_3\right) + \alpha_m mc\right]\overline{\psi} = 0. \quad (57)$$

Thus each solution ψ of the wave equation (56) has for its conjugate complex $\overline{\psi}$ a solution of the wave equation (57). Further, if the solution ψ of (56) belongs to a negative value for $cp_0 + eA_0$, the corresponding solution $\overline{\psi}$ of (57) will belong to a positive value for $cp_0 - eA_0$. But the operator in (57) is just what one would get if one substituted $-e$ for e in the operator in (56). It follows that each negative-energy solution of (56) is the conjugate complex of a positive-energy solution of the wave equation obtained from (56) by substitution of $-e$ for e, which solution represents an electron of charge $+e$ (instead of $-e$, as we had up to the present) moving through the given electromagnetic field. Thus the unwanted solutions of (56) are connected with the motion of an electron with a charge $+e$. [It is not possible, of course, with an arbitrary electromagnetic field, to separate the solutions of (56) definitely into those referring to positive and those referring to negative values for $cp_0 + eA_0$, as such a separation would imply that transitions from one kind to the other do not occur. The preceding discussion is therefore only a rough one, applying to the case when such a separation is approximately possible.]

In this way we are led to infer that the negative-energy solutions of (56) refer to the motion of a new kind of particle having the mass of an electron and the opposite charge. Such particles have been observed experimentally and are called *positrons*. We cannot, however, simply assert that the negative-energy solutions represent positrons, as this would make the dynamical relations all wrong. For instance, it is certainly not true that a positron has a negative kinetic energy. We must therefore establish the theory of the positrons on a somewhat different footing. We assume that *nearly all the negative-energy states are occupied*, with one electron in each state in accordance with the exclusion principle of Pauli. An unoccupied negative-energy state will now appear as something with a positive energy, since to make it disappear, i.e. to fill it up, we should have to add to it an electron with negative energy. We assume that *these unoccupied negative-energy states are the positrons*.

These assumptions require there to be a distribution of electrons of infinite density everywhere in the world. A perfect vacuum is a region where all the states of positive energy are unoccupied and all those of negative energy are occupied. In a perfect vacuum Maxwell's equation

$$\text{div}\,\mathcal{E} = 0$$

must, of course, be valid. This means that the infinite distribution of negative-energy electrons does not contribute to the electric field. Only departures from the distribution in a vacuum will contribute to the electric density j_0 in Maxwell's equation

$$\text{div}\,\mathcal{E} = 4\pi j_0. \tag{58}$$

Thus there will be a contribution $-e$ for each occupied state of positive energy and a contribution $+e$ for each unoccupied state of negative energy.

The exclusion principle will operate to prevent a positive-energy electron ordinarily from making transitions to states of negative energy. It will still be possible, however, for such an electron to drop into an unoccupied state of negative energy. In this case we should have an electron and positron disappearing simultaneously, their energy being emitted in the form of radiation. The converse process would consist in the creation of an electron and a positron from electromagnetic radiation.

From the symmetry between occupied and unoccupied fermion states discussed at the end of § 65, the present theory is essentially symmetrical between the electrons and the positrons. We should have an equivalent theory if we supposed the positrons to be the basic particles, described by wave equations of the form (11) with $-e$ for e, and then supposed that nearly all the states of negative energy for the positrons are filled up, a hole in the distribution of negative-energy positrons being then interpreted as an ordinary electron. The theory could be developed consistently with the hypothesis that all the laws of physics are symmetrical between positive and negative electric charge.

CHAPTER XII

Quantum Electrodynamics

74. The Electromagnetic Field in the Absence of Matter

The theory of radiation that was set up in Chapter X involved some approximations in its handling of the interaction of the radiation with matter. The object of the present chapter is to remove these approximations and get, as far as possible, an accurate theory of the electromagnetic field interacting with matter, subject to the limitation that the matter consists only of electrons and positrons. Too little is known about other forms of matter, protons, neutrons, etc., for one to attempt at the present time to get an accurate theory of their interaction with the electromagnetic field. But there exists a precise theory of electrons and positrons, as given in the preceding chapter, which one can use for building up a precise theory of the interaction of the electromagnetic field with this form of matter. The theory must bring in the interaction of the electrons and positrons with one another, through their Coulomb forces, as well as their interaction with electromagnetic radiation, and it must, of course, conform to special relativity. For brevity in this chapter we shall take $c = 1$.

We must first consider the electromagnetic field without interaction with matter. Now in § 63 we set up first a treatment of the field of radiation without interaction of matter. Dynamical variables were there introduced to describe the field, commutation relations were established for them, and a Hamiltonian was found which made them vary correctly with the time. No approximations were made in this piece of work. The resulting theory would therefore be a satisfactory, exact theory of radiation without interaction with matter, were it not for one feature in it, namely our taking the scalar potential to be zero. This feature spoils the relativistic form of the theory and makes it unsuitable as a starting-point from which to develop a precise theory of the electromagnetic field in interaction with matter.

We must therefore extend the treatment of § 63 by leaving A_0 general and bringing it into the work along with the other potentials A_1, A_2, A_3. Thus we shall have the four A_μ and they will satisfy, as the generalization of (62) of § 63,

$$\Box A_\mu = 0, \quad \frac{\partial A_\mu}{\partial x_\mu} = 0. \qquad (1), (2)$$

For the present we shall ignore the second of these equations and work only from the first.

Equation (1) shows that each A_μ can be resolved into waves travelling with the velocity of light. Thus, corresponding to equation (63) of § 63,

$$A_\mu(x) = \int (A^c_{\mu\mathbf{k}} e^{ik\cdot x} + \overline{A}^c_{\mu\mathbf{k}} e^{-ik\cdot x}) \, d^3k, \qquad (3)$$

where $k \cdot x$ denotes the four-dimensional scalar product

$$k \cdot x = k_0 x_0 - (\mathbf{k}, \mathbf{x}),$$

k_ν being the 4-vector whose space components are the same as the components of the three-dimensional vector $\mathbf{k}$ of § 63 and whose time component $k_0 = |\mathbf{k}|$, and d^3k denotes $dk_1 dk_2 dk_3$, as in § 63. The index c in the coefficients $A^c_{\mu\mathbf{k}}$ indicates that they are constant in time. We shall later introduce some other Fourier coefficients $A_{\mu\mathbf{k}}$, not constant in time, which must be distinguished from the present ones.

The Fourier component $A^c_{\mu\mathbf{k}}$ has a part $A^c_{0\mathbf{k}}$ coming from $A_0(x)$ and a part $A^c_{r\mathbf{k}}$ ($r = 1, 2, 3$) which is a three-dimensional vector. The latter can be decomposed into two parts, a longitudinal part lying in the direction of $\mathbf{k}$, the direction of motion of the waves, and a transverse part perpendicular to $\mathbf{k}$. The longitudinal part is $k_r k_s / k_0^2 \cdot A^c_{s\mathbf{k}}$. The transverse part is

$$\left(\delta_{rs} - \frac{k_r k_s}{k_0^2}\right) A^c_{s\mathbf{k}} = \mathcal{A}^c_{r\mathbf{k}}, \qquad (4)$$

say. It satisfies

$$k_r \mathcal{A}^c_{r\mathbf{k}} = 0. \qquad (5)$$

It is known from the Maxwell theory of light that only the transverse part is effective for giving electromagnetic radiation. Chapter X dealt only with this transverse part, the $A_{r\mathbf{k}}$ of § 63 being the same as the present $\mathcal{A}^c_{r\mathbf{k}}$ and equation (65) of § 63 corresponding to the present equation (5). Nevertheless, the longitudinal part cannot be neglected in a complete theory of electrodynamics because of its connexion with the Coulomb forces, as will show up later.

We can now decompose the three-dimensional vector $A_r(x)$ into two parts, a transverse part and a longitudinal part. The former is

$$\mathcal{A}_r(x) = \int (A^c_{r\mathbf{k}} e^{ik \cdot x} + \overline{A}^c_{r\mathbf{k}} e^{-ik \cdot x}) \, d^3k$$

and satisfies

$$\frac{\partial \mathcal{A}_r(x)}{\partial x_r} = 0. \tag{6}$$

The longitudinal part may be expressed as the gradient $\partial V / \partial x_r$ of a scalar V given by

$$V = i \int \frac{k_s}{k_0^2} (A^c_{s\mathbf{k}} e^{ik \cdot x} - \overline{A}^c_{s\mathbf{k}} e^{-ik \cdot x}) \, d^3k. \tag{7}$$

Thus

$$A_r = \mathcal{A}_r + \frac{\partial V}{\partial x_r}. \tag{8}$$

The magnetic field is determined by the transverse part of A_r,

$$\mathcal{H} = \operatorname{curl} \mathbf{A} = \operatorname{curl} \mathcal{A}.$$

It is convenient to count $A_0(x)$ as longitudinal, so that the complete potentials $A_\mu(x)$ are separated into a transverse part $\mathcal{A}_r(x)$ and a longitudinal part A_0, $\partial V / \partial x_r$. This separation, of course, refers to a particular Lorentz frame of reference and must not be used when one wants to keep one's equations in a relativistic form.

Each Fourier coefficient $A^c_{\mu\mathbf{k}}$ occurs in (3) combined with the time factor $e^{ik_0 x_0}$. The product

$$A^c_{\mu\mathbf{k}} e^{ik_0 x_0} = A_{\mu\mathbf{k}} \tag{9}$$

say, forms a Hamiltonian dynamical variable in classical mechanics and a Heisenberg dynamical variable in quantum mechanics, like the $A_{\mathbf{k}lt}$ of § 63.

The work of § 63 gives us the P.B. relations for the transverse part of $A_{\mu\mathbf{k}}$. To connect up with it, we pass over to discrete $\mathbf{k}$-values in three-dimensional $\mathbf{k}$-space and take, for example, a particular discrete $\mathbf{k}$-value for which $k_1 = k_2 = 0$, $k_3 = k_0 > 0$. Then the polarization variable l can take on two values referring to the two directions 1 and 2 and equation (73) of § 63 gives, with the help of the commutation relations for the η's and $\overline{\eta}$'s, equations (11) of § 60,

$$[\overline{A}_{1\mathbf{k}}, A_{1\mathbf{k}}] = [\overline{A}_{2\mathbf{k}}, A_{2\mathbf{k}}] = -\frac{is_k}{4\pi^2 k_0}. \tag{10}$$

The work of § 63 gives us no information about $A_{3\mathbf{k}}$ and $A_{0\mathbf{k}}$.

However, we can now obtain the P.B. relations for $A_{3\mathbf{k}}$ and $A_{0\mathbf{k}}$ from the theory of relativity. Equations (10) have to be built up into a relativistic set and the only simple way of doing so is by adding to them the two further equations

$$[\overline{A}_{3\mathbf{k}}, A_{3\mathbf{k}}] = -[\overline{A}_{0\mathbf{k}}, A_{0\mathbf{k}}] = -\frac{is_{\mathbf{k}}}{4\pi^2 k_0}, \tag{11}$$

so that the four equations (10) and (11), together with the conditions that $\overline{A}_{\mu\mathbf{k}}$ and $A_{\nu\mathbf{k}}$ commute for $\mu \neq \nu$ (as they must do since they refer to different degrees of freedom), combine to form the single tensor equation

$$[\overline{A}_{\mu\mathbf{k}}, A_{\nu\mathbf{k}}] = \frac{ig_{\mu\nu}s_{\mathbf{k}}}{4\pi^2 k_0}. \tag{12}$$

We get in this way the P.B. relations for all the dynamical variables. Equation (12) can be extended to

$$[\overline{A}_{\mu\mathbf{k}}, A_{\nu\mathbf{k'}}] = \frac{ig_{\mu\nu}s_{\mathbf{k}}\delta_{\mathbf{kk'}}}{4\pi^2 k_0}. \tag{13}$$

Let us now return to continuous $\mathbf{k}$-values. To convert $\delta_{\mathbf{kk'}}$ to continuous $\mathbf{k}$-values we note that, for a general function $f(\mathbf{k})$ in three-dimensional $\mathbf{k}$-space,

$$\sum_{\mathbf{k}} f(\mathbf{k})\delta_{\mathbf{kk'}} = f(\mathbf{k'}) = \int f(\mathbf{k})\,\delta(\mathbf{k} - \mathbf{k'})\,d^3k, \tag{14}$$

where $\delta(\mathbf{k} - \mathbf{k'})$ is the three-dimensional δ function

$$\delta(\mathbf{k} - \mathbf{k'}) = \delta(k_1 - k'_1)\,\delta(k_2 - k'_2)\,\delta(k_3 - k'_3).$$

In order that (14) may conform to the standard formula connecting sums and integrals, equation (52) of § 62, we must have

$$s_{\mathbf{k}}\delta_{\mathbf{kk'}} = \delta(\mathbf{k} - \mathbf{k'}). \tag{15}$$

Thus (13) goes over to

$$[\overline{A}_{\mu\mathbf{k}}, A_{\nu\mathbf{k'}}] = \frac{ig_{\mu\nu}}{4\pi^2 k_0}\,\delta(\mathbf{k} - \mathbf{k'}). \tag{16}$$

This equation, together with the equations

$$[A_{\mu\mathbf{k}}, A_{\nu\mathbf{k'}}] = [\overline{A}_{\mu\mathbf{k}}, \overline{A}_{\nu\mathbf{k'}}] = 0, \tag{17}$$

provide the P.B. relations in the theory with continuous $\mathbf{k}$-values. It should be noted that these P.B. relations remain valid if we replace $A_{\mu\mathbf{k}}, \overline{A}_{\nu\mathbf{k}}$ by

$A^c_{\mu\mathbf{k}}$, $\overline{A}^c_{\nu\mathbf{k}}$. The same P.B. relations apply to the constant Fourier coefficients $\overline{A}^c_{\mu\mathbf{k}}$, $A^c_{\nu\mathbf{k}}$.

We must now obtain a Hamiltonian which makes each dynamical variable $A_{\mu\mathbf{k}}$ vary with the time $t = x_0$ in the Heisenberg picture according to the law (9) with $A^c_{\mu\mathbf{k}}$ constant. Calling this Hamiltonian H_F, we require

$$[A_{\mu\mathbf{k}}, H_F] = \frac{dA_{\mu\mathbf{k}}}{dx_0} = ik_0 A_{\mu\mathbf{k}}. \tag{18}$$

It is easily seen that this is satisfied by

$$H_F = -4\pi^2 \int k_0^2 A_{\mu\mathbf{k}} \overline{A}^{\mu}{}_{\mathbf{k}} \, d^3k. \tag{19}$$

We therefore take (19), with the possible addition of an arbitrary numerical term not involving any dynamical variables, as the Hamiltonian for the electromagnetic field in the absence of matter.

In § 63 we used our knowledge of the transverse part of the Hamiltonian to obtain the P.B.s of the transverse variables. We have now applied the reverse procedure to the longitudinal variables, using our knowledge of their P.B.s, obtained by a relativistic argument, to find the part of the Hamiltonian that refers to them so as to get agreement with (18).

If we write out the Hamiltonian (19) it appears as

$$H_F = 4\pi^2 \int k_0^2 (A_{1\mathbf{k}} \overline{A}_{1\mathbf{k}} + A_{2\mathbf{k}} \overline{A}_{2\mathbf{k}} + A_{3\mathbf{k}} \overline{A}_{3\mathbf{k}} - A_{0\mathbf{k}} \overline{A}_{0\mathbf{k}}) \, d^3k.$$

The first three terms of the integrand here have a transverse part which is just equal to the transverse energy given by (71) of § 63. The last term of the integrand, which is the part of H_F referring to the scalar potential A_0, appears with a minus sign. This minus sign is demanded by relativity and means that the dynamical system formed by the variables $A_{0\mathbf{k}}, \overline{A}_{0\mathbf{k}}$ is a *harmonic oscillator of negative energy*. It is rather surprising that such an unphysical idea as negative energy should appear in the theory in this way. We shall see in § 77 that the negative energy associated with the degrees of freedom connected with A_0 is always compensated by the positive energy associated with the other longitudinal degrees of freedom, so that it never shows up in practice.

75. Relativistic Form of the Quantum Conditions

The theory of the preceding section has relativistic field equations, namely equations (1). To establish that the theory is fully relativistic we must show further that the P.B. relations are relativistic. This is not at all evident from the form (16) in which they are written in terms of Fourier components. We shall obtain a relativistic form for the P.B.s by working out $[A_\mu(x), A_\nu(x')]$ with x and x' any two points in space-time. We must first, however, study a certain invariant singular function that exists in space-time.

The function $\delta(x_\mu x^\mu)$ is evidently Lorentz invariant. It vanishes everywhere except on the light-cone with the origin as vertex, i.e. the three-dimensional space $x_\mu x^\mu = 0$. This light-cone consists of two distinct parts, a *future part*, for which $x_0 > 0$, and a *past part*, for which $x_0 < 0$. The function which equals $\delta(x_\mu x^\mu)$ on the future part of the light-cone and $-\delta(x_\mu x^\mu)$ on the past part of the light-cone is also Lorentz invariant. This function, equal to $\delta(x_\mu x^\mu) x_0 / |x_0|$, plays an important role in the dynamical theory of fields, so we introduce a special notation for it. We define

$$\Delta(x) = \frac{2\delta(x_\mu x^\mu) x_0}{|x_0|}. \qquad (20)$$

This definition gives a meaning to the function Δ applied to any 4-vector. With the help of (9) of § 15, we can express $\delta(x_\mu x^\mu)$ in the form

$$\delta(x_\mu x^\mu) = \frac{1}{2} |\mathbf{x}|^{-1} [\delta(x_0 - |\mathbf{x}|) + \delta(x_0 + |\mathbf{x}|)], \qquad (21)$$

$|\mathbf{x}|$ being the length of the three-dimensional part of x_μ, and then $\Delta(x)$ takes the form

$$\Delta(x) = |\mathbf{x}|^{-1} [\delta(x_0 - |\mathbf{x}|) - \delta(x_0 + |\mathbf{x}|)]. \qquad (22)$$

$\Delta(x)$ is defined to have the value zero at the origin, and evidently $\Delta(-x) = -\Delta(x)$.

Let us make a Fourier analysis of $\Delta(x)$. Using d^4x to denote $dx_0 dx_1 dx_2 dx_3$ and d^3x to denote $dx_1 dx_2 dx_3$ we have, for any 4-vector k_μ,

$$\int \Delta(x) e^{ik\cdot x} d^4x = \int |\mathbf{x}|^{-1} [\delta(x_0 - |\mathbf{x}|) - \delta(x_0 + |\mathbf{x}|)] e^{i[k_0 x_0 - (\mathbf{kx})]} d^4x$$

$$= \int |\mathbf{x}|^{-1} (e^{ik_0|\mathbf{x}|} - e^{-ik_0|\mathbf{x}|}) e^{-i(\mathbf{kx})} d^3x.$$

By introducing polar coordinates $|x|, \theta, \phi$ in the three-dimensional $x_1 x_2 x_3$ space, with the direction of the three-dimensional part of k_μ as pole, we get

$$\int \Delta(x) e^{ik\cdot x} d^4x = \iiint (e^{ik_0|\mathbf{x}|} - e^{-ik_0|\mathbf{x}|}) e^{-i|\mathbf{k}||\mathbf{x}|\cos\theta} |\mathbf{x}| \sin\theta \, d\theta d\phi d|\mathbf{x}|$$

75. RELATIVISTIC FORM OF THE QUANTUM CONDITIONS

$$= 2\pi \int_0^\infty (e^{ik_0|\mathbf{x}|} - e^{-ik_0|\mathbf{x}|})\,d|\mathbf{x}| \int_0^\pi e^{-i|\mathbf{k}||\mathbf{x}|\cos\theta}|\mathbf{x}|\sin\theta\,d\theta$$

$$= 2\pi i|\mathbf{k}|^{-1} \int_0^\infty (e^{ik_0|\mathbf{x}|} - e^{-ik_0|\mathbf{x}|})\,d|\mathbf{x}|\,(e^{-i|\mathbf{k}||\mathbf{x}|} - e^{i|\mathbf{k}||\mathbf{x}|})$$

$$= 2\pi i|\mathbf{k}|^{-1} \int_{-\infty}^\infty (e^{i(k_0-|\mathbf{k}|)a} - e^{i(k_0+|\mathbf{k}|)a})\,da$$

$$= 4\pi^2 i|\mathbf{k}|^{-1}[\delta(k_0 - |\mathbf{k}|) - \delta(k_0 + |\mathbf{k}|)]$$

$$= 4\pi^2 i\,\Delta(k). \tag{23}$$

Thus the Fourier analysis gives the same function again, with the coefficient $4\pi^2 i$. Interchanging k and x in (23), we get

$$\Delta(x) = -\frac{i}{4\pi^2}\int \Delta(k)e^{ik\cdot x}\,d^4k. \tag{24}$$

Some of the important properties of $\Delta(x)$ can easily be deduced from its Fourier resolution. In the first place equation (24) shows that $\Delta(x)$ can be resolved into waves all travelling with the velocity of light. To get an equation for this result we apply the operator $\square$ to both sides of (24), thus

$$\square\Delta(x) = -\frac{i}{4\pi^2}\int \Delta(k)\square e^{ik\cdot x}\,d^4k = \frac{i}{4\pi^2}\int k_\mu k^\mu \Delta(k)e^{ik\cdot x}\,d^4k.$$

Now $k_\mu k^\mu \Delta(k) = 0$, and hence

$$\square\Delta(x) = 0. \tag{25}$$

This equation holds throughout space-time. We can give a meaning to $\square\Delta(x)$ at a point where $\Delta(x)$ is singular by taking the integral of $\square\Delta(x)$ over a small four-dimensional space surrounding the point and transforming it to a three-dimensional surface integral by Gauss's theorem. Equation (25) informs us that the three-dimensional surface integral always vanishes.

The function $\Delta(x)$ vanishes all over the three-dimensional surface $x_0 = 0$. Let us determine the value of $\partial\Delta(x)/\partial x_0$ on this surface. It evidently vanishes everywhere except at the point $x_1 = x_2 = x_3 = 0$, where it has a singularity which can be evaluated as follows. Differentiating both sides of (24) with respect to x_0, we get

$$\frac{\partial\Delta(x)}{\partial x_0} = \frac{1}{4\pi^2}\int k_0\,\Delta(k)e^{ik\cdot x}\,d^4k$$

$$= \frac{1}{4\pi^2}\int k_0|\mathbf{k}|^{-1}[\delta(k_0 - |\mathbf{k}|) - \delta(k_0 + |\mathbf{k}|)]e^{ik\cdot x}\,d^4k$$

$$= \frac{1}{4\pi^2} \int [\delta(k_0 - |\mathbf{k}|) + \delta(k_0 + |\mathbf{k}|)] e^{ik \cdot x} d^4k.$$

Putting $x_0 = 0$ on both sides here, we get

$$\left[\frac{\partial \Delta(x)}{\partial x_0}\right]_{x_0=0} = \frac{1}{4\pi^2} \int [\delta(k_0 - |\mathbf{k}|) + \delta(k_0 + |\mathbf{k}|)] e^{-i(\mathbf{k}x)} d^4k$$

$$= \frac{1}{2\pi^2} \int e^{-i(\mathbf{k}x)} d^3k$$

$$= 4\pi \, \delta(x_1) \delta(x_2) \delta(x_3) = 4\pi \, \delta(\mathbf{x}). \tag{26}$$

Thus the ordinary δ singularity, with the coefficient 4π, appears at the point $x_1 = x_2 = x_3 = 0$.

Let us now evaluate $[A_\mu(x), A_\nu(x')]$. We have from (3), (16), and (17)

$$[A_\mu(x), A_\nu(x')]$$

$$= \iint [A_{\mu\mathbf{k}} e^{ik \cdot x} + \overline{A}_{\mu\mathbf{k}} e^{-ik \cdot x}, A_{\nu\mathbf{k}'} e^{ik' \cdot x'} + \overline{A}_{\nu\mathbf{k}'} e^{-ik' \cdot x'}] d^3k \, d^3k'$$

$$= \frac{ig_{\mu\nu}}{4\pi^2} \iint k_0^{-1} (e^{-ik \cdot x} e^{ik' \cdot x'} - e^{ik \cdot x} e^{-ik' \cdot x'}) \delta(\mathbf{k} - \mathbf{k}') d^3k \, d^3k'$$

$$= \frac{ig_{\mu\nu}}{4\pi^2} \int k_0^{-1} [e^{-ik \cdot (x-x')} - e^{ik \cdot (x-x')}] d^3k. \tag{27}$$

The k_0 here is defined to be equal to $|\mathbf{k}|$ and is thus always positive. By putting $-\mathbf{k}$ for $\mathbf{k}$ in the second part of the integrand, one finds that (27) is equal to the four-dimensional integral.

$$\frac{ig_{\mu\nu}}{4\pi^2} \int |\mathbf{k}|^{-1} [\delta(k_0 - |\mathbf{k}|) - \delta(k_0 + |\mathbf{k}|)] e^{-ik \cdot (x-x')} d^4k$$

$$= \frac{ig_{\mu\nu}}{4\pi^2} \int \Delta(k) e^{-ik \cdot (x-x')} d^4k,$$

in which k_0 takes on all values, negative as well as positive. Evaluating this with the help of (24), we get finally

$$[A_\mu(x), A_\nu(x')] = g_{\mu\nu} \Delta(x - x'), \tag{28}$$

a result which shows that the P.B. relations are invariant under Lorentz transformations.

The formula (28) means that the potentials at two points in space-time always commute unless the line joining the two points is a null line, i.e. the track of a light-ray. The formula is consistent with the field equations $\Box A_\mu(x) = 0$, because $\Box$ applied to the right-hand side gives zero, from (25).

76. The Dynamical Variables at One Time

As a basis for a theory with interaction we must use the dynamical variables at one time. The relationships between the dynamical variables at one time (i.e. their P.B.s) are not affected by the introduction of interaction. On the other hand the relationships between the dynamical variables at different times (comprising the field equations as well as the P.B.s of variables at different times) are very much affected by the interaction. The dynamical variables at one time form a non-relativistic concept, but a very important concept in Hamiltonian theory.

For the case of the electromagnetic field the independent dynamical variables at one time are A_μ and $\partial A_\mu/\partial x_0$ for all values of x_1, x_2, x_3 for the given x_0. The higher time derivatives $\partial^2 A_\mu/\partial x_0^2, \ldots$, are not independent. Let us put

$$B_\mu = \frac{\partial A_\mu}{\partial x_0}. \tag{29}$$

Then we have $A_{\mu x}$, $B_{\mu x}$, with the suffix $\mathbf{x}$ denoting x_1, x_2, x_3, as the dynamical variables at one time.

The Fourier resolution of these variables is, from (3) and (9),

$$\begin{aligned} A_{\mu x} &= \int (A_{\mu k} + \overline{A}_{\mu -k}) e^{-i(\mathbf{k x})} d^3 k \\ B_{\mu x} &= i \int k_0 (A_{\mu k} - \overline{A}_{\mu -k}) e^{-i(\mathbf{k x})} d^3 k. \end{aligned} \tag{30}$$

We may reverse the Fourier transformation and express $A_{\mu k} + \overline{A}_{\mu -k}$ and $A_{\mu k} - \overline{A}_{\mu -k}$ in terms of $A_{\mu x}$ and $B_{\mu x}$ respectively. Thus $A_{\mu k}$ and $\overline{A}_{\mu k}$ are determined by $A_{\mu x}, B_{\mu x}$ for all $\mathbf{x}$ (at a given x_0). The equations connecting $A_{\mu k}, \overline{A}_{\mu k}$ with $A_{\mu x}, B_{\mu x}$ do not involve the time explicitly. Thus the $A_{\mu k}, \overline{A}_{\mu k}$ form an alternative set of one-time dynamical variables, on the same footing as the $A_{\mu x}, B_{\mu x}$.

When we work with the variables $A_{\mu x}, B_{\mu x}$ we shall need to know their P.B. relations. These may be obtained either from the Fourier expansions (30) together with (16) and (17) or from the general P.B. relation (28). The latter gives the required results more quickly. Putting $x_0' = x_0$ in (28), we get

$$[A_{\mu x}, A_{\nu x'}] = 0. \tag{31}$$

Differentiating (28) with respect to x_0 and then putting $x_0' = x_0$, we get, with the help of (26),

$$[B_{\mu x}, A_{\nu x'}] = 4\pi g_{\mu\nu} \delta(\mathbf{x} - \mathbf{x}'). \tag{32}$$

Differentiating (28) with respect to both x_0 and x_0' and then putting $x_0' = x_0$, we get

$$[B_{\mu x}, B_{\nu x'}] = 0, \tag{33}$$

since $\partial^2 \Delta(x)/\partial x_0^2 = 0$ for $x_0 = 0$. Equations (31), (32), and (33) give all the P.B. relations between the $A_{\mu x}$, $B_{\mu x}$ variables. They show that, apart from numerical coefficients, the $A_{\mu x}$ can be looked upon as a set of dynamical coordinates and the $B_{\mu x}$ as their conjugate momenta, there being a δ function on the right-hand side of (32) instead of a two-suffix δ symbol on account of the number of degrees of freedom being a continuous infinity.

We can decompose A_{rx} into a transverse and a longitudinal part, as shown by equations (8) and (6). We can do the same with B_{rx} and get

$$B_r = \mathcal{B}_r + \frac{\partial U}{\partial x_r} \tag{34}$$

with

$$\frac{\partial \mathcal{B}_r}{\partial x_r} = 0. \tag{35}$$

From (7) with $-\mathbf{k}$ substituted for $\mathbf{k}$ in the second term of the integrand,

$$V = i \int k_s k_0^{-2} (A_{sk} + \overline{A}_{s-\mathbf{k}}) e^{-i(kx)} \, d^3k. \tag{36}$$

The corresponding equation for U is, since $U = \partial V/\partial x_0$,

$$U = -\int k_s k_0^{-1} (A_{sk} - \overline{A}_{s-\mathbf{k}}) e^{-i(kx)} \, d^3k. \tag{37}$$

The electric field is given by

$$\mathcal{E}_r = -B_r - \frac{\partial A_0}{\partial x_r}$$

$$= -\mathcal{B}_r - \frac{\partial(A_0 + U)}{\partial x_r}. \tag{38}$$

Thus

$$\text{div}\, \mathcal{E} = -\frac{\partial \mathcal{B}_r}{\partial x_r} - \nabla^2 A_0$$

$$= -\nabla^2(A_0 + U). \tag{39}$$

It is evident that any longitudinal variable commutes with any transverse variable. Some useful P.B. relations will now be worked out. We shall use the notation—for any field function $f_\mathbf{x}$,

$$\frac{\partial f_\mathbf{x}}{\partial x_r} = f_\mathbf{x}^{\ r}, \quad \frac{\partial f_{\mathbf{x}'}}{\partial x'_r} = f_{\mathbf{x}'}^{\ r'}. \tag{40}$$

If in (32) we put $\mu = r$, $\nu = s$ and differentiate the equation with respect to x_r, we get

$$[B_{r\mathbf{x}}^{\ r}, A_{s\mathbf{x}'}] = 4\pi g_{rs}\, \delta^r(\mathbf{x} - \mathbf{x}') = -4\pi\, \delta^s(\mathbf{x} - \mathbf{x}'),$$

or, from (39),

$$[\text{div}\,\mathcal{E}_\mathbf{x}, A_{s\mathbf{x}'}] = 4\pi\, \delta^s(\mathbf{x} - \mathbf{x}'). \tag{41}$$

Now (39) shows that div $\mathcal{E}$ is a function only of the longitudinal variables, so (41) gives

$$[\text{div}\,\mathcal{E}_\mathbf{x}, V_{\mathbf{x}'}^{\ s'}] = 4\pi\, \delta^s(\mathbf{x} - \mathbf{x}') = -4\pi\, \delta^{s'}(\mathbf{x} - \mathbf{x}').$$

Integrating with respect to x'_s, we get

$$[\text{div}\,\mathcal{E}_\mathbf{x}, V_{\mathbf{x}'}] = -4\pi\, \delta(\mathbf{x} - \mathbf{x}'), \tag{42}$$

there being no constant of integration since the field functions $\mathcal{E}_\mathbf{x}$ and $V_\mathbf{x}$ are made up of waves of non-zero wave length. From (42) and (39)

$$\nabla^2 [U_\mathbf{x}, V_{\mathbf{x}'}] = 4\pi\, \delta(\mathbf{x} - \mathbf{x}').$$

Integrating with the help of formula (72) of § 38, we get

$$[U_\mathbf{x}, V_{\mathbf{x}'}] = -|\mathbf{x} - \mathbf{x}'|^{-1}, \tag{43}$$

there being no constant of integration or other terms not vanishing at infinity on the right-hand side, because $U_\mathbf{x}$ and $V_\mathbf{x}$ are made up of waves of non-zero wave length. We have from (38) and (43)

$$[\mathcal{E}_{r\mathbf{x}}, V_{\mathbf{x}'}] = -[U_\mathbf{x}^{\ r}, V_{\mathbf{x}'}] = -(x_r - x'_r)|\mathbf{x} - \mathbf{x}'|^{-3}. \tag{44}$$

We shall now obtain the Hamiltonian in terms of the $A_{\mu\mathbf{x}}$ and $B_{\mu\mathbf{x}}$ variables. We have from the second of equations (30)

$$\int B_{\mu\mathbf{x}} B^\mu_{\ \mathbf{x}}\, d^3x$$

$$= -\iiint k_0 k'_0 (A_{\mu\mathbf{k}} - \overline{A}_{\mu-\mathbf{k}})(A^\mu_{\ \mathbf{k}'} - \overline{A}^\mu_{\ -\mathbf{k}'}) e^{-i(kx)} e^{-i(k'x)}\, d^3k\, d^3k'\, d^3x$$

$$= -8\pi^3 \iint k_0 k'_0 (A_{\mu\mathbf{k}} - \overline{A}_{\mu-\mathbf{k}})(A^\mu_{\ \mathbf{k}'} - \overline{A}^\mu_{\ -\mathbf{k}'})\, \delta(\mathbf{k} + \mathbf{k}')\, d^3k\, d^3k'$$

$$= -8\pi^3 \int k_0^2 (A_{\mu\mathbf{k}} - \overline{A}_{\mu-\mathbf{k}})(A^\mu{}_{-\mathbf{k}} - \overline{A}^\mu{}_{\mathbf{k}}) d^3k.$$

Similarly, from the first of equations (30),

$$\int A_{\mu\mathbf{x}}{}^r A^\mu{}_\mathbf{x}{}^r d^3x$$

$$= -\iiint k_r k'_r (A_{\mu\mathbf{k}} + \overline{A}_{\mu-\mathbf{k}})(A^\mu{}_{\mathbf{k}'} + \overline{A}^\mu{}_{-\mathbf{k}'}) e^{-i(kx)} e^{-i(k'x)} d^3k \, d^3k' \, d^3x$$

$$= 8\pi^3 \int k_0^2 (A_{\mu\mathbf{k}} + \overline{A}_{\mu-\mathbf{k}})(A^\mu{}_{-\mathbf{k}} + \overline{A}^\mu{}_{\mathbf{k}}) d^3k.$$

Adding and dividing by -8π, we get

$$-(8\pi)^{-1} \int (B_\mu B^\mu + A_\mu{}^r A^{\mu r}) d^3x$$

$$= -2\pi^2 \int k_0^2 (A_{\mu\mathbf{k}} \overline{A}^\mu{}_\mathbf{k} + \overline{A}_{\mu-\mathbf{k}} A^\mu{}_{-\mathbf{k}}) d^3k.$$

This is equal to H_F given by (19), apart from an infinite numerical term. The formula (19) for H_F already involves an arbitrary numerical term, so we may take

$$H_F = -(8\pi)^{-1} \int (B_\mu B^\mu + A_\mu{}^r A^{\mu r}) d^3x \tag{45}$$

with an arbitrary numerical term, different from that of (19).

The Hamiltonian (45) can, of course, be used to give the Heisenberg equations of motion, and the arbitrary numerical term in it does not have any effect. One can easily check, using (31), (32) and (33), that

$$\frac{\partial A_\mu}{\partial x_0} = [A_\mu, H_F] = B_\mu, \tag{46}$$

and

$$\frac{\partial B_\mu}{\partial x_0} = [B_\mu, H_F] = \nabla^2 A_\mu, \tag{46}$$

agreeing with (29) and (1). It also gives the Schrödinger equation of motion

$$i\hbar \frac{d|P\rangle}{dx_0} = H_F |P\rangle$$

for a ket $|P\rangle$ representing a state in the Schrödinger picture. The arbitrary numerical term here has the effect of changing $|P\rangle$ by a phase factor, which is not of physical importance.

76. THE DYNAMICAL VARIABLES AT ONE TIME

We can decompose the expression (45) for H_F into a transverse part H_{FT} and a longitudinal part H_{FL}. We have from (34)

$$\int B_r B_r \, d^3x = \int (\mathcal{B}_r + U^r)(\mathcal{B}_r + U^r) \, d^3x$$
$$= \int \mathcal{B}_r \mathcal{B}_r \, d^3x + \int U^r U^r \, d^3x,$$

since the cross terms vanish on account of

$$\int U^r \mathcal{B}_r \, d^3x = -\int U \mathcal{B}_r^{\,r} \, d^3x = 0$$

from (35). Similarly we have from (8)

$$\int A_r^{\,s} A_r^{\,s} \, d^3x = \int \mathcal{A}_r^{\,s} \mathcal{A}_r^{\,s} \, d^3x + \int V^{rs} V^{rs} \, d^3x,$$

with the cross terms vanishing again. Thus (45) becomes

$$H_F = H_{FT} + H_{FL},$$

with

$$H_{FT} = (8\pi)^{-1} \int (\mathcal{B}_r \mathcal{B}_r + \mathcal{A}_r^{\,s} \mathcal{A}_r^{\,s}) \, d^3x \qquad (47)$$

and

$$H_{FL} = (8\pi)^{-1} \int (U^r U^r + V^{rs} V^{rs} - B_0 B_0 - A_0^{\,r} A_0^{\,r}) \, d^3x. \qquad (48)$$

It should be noted that the term

$$(8\pi)^{-1} \int \mathcal{A}_r^{\,s} \mathcal{A}_r^{\,s} \, d^3x$$

in H_{FT} can be transformed to

$$-(8\pi)^{-1} \int \mathcal{A}_r \mathcal{A}_{r,s}^{\,s} \, d^3x = -(8\pi)^{-1} \int \mathcal{A}_r (\mathcal{A}_{r,s}^{\,s} - \mathcal{A}_{s}^{rs}) \, d^3x$$
$$= (8\pi)^{-1} \int \mathcal{A}_r^{\,s} (\mathcal{A}_r^{\,s} - \mathcal{A}_s^{\,r}) \, d^3x$$
$$= (16\pi)^{-1} \int (\mathcal{A}_r^{\,s} - \mathcal{A}_s^{\,r})(\mathcal{A}_r^{\,s} - \mathcal{A}_s^{\,r}) \, d^3x$$
$$= (8\pi)^{-1} \int \mathcal{H}^2 \, d^3x,$$

so this term is just the magnetic energy. Some further partial integrations give

$$\int V^{rs} V^{rs} \, d^3x = \int V^{rr} V^{ss} \, d^3x,$$

so (48) may be written

$$H_{FL} = (8\pi)^{-1} \int [(U-A_0)^r(U+A_0)^r + (V^{rr}-B_0)(V^{ss}+B_0)] \, d^3x. \quad (49)$$

77. The Supplementary Conditions

We must now go back to the Maxwell equation (2), which we have ignored so far. We cannot take this equation over directly into the quantum theory without getting inconsistencies. The left-hand side of the equation does not commute with $A_\nu(x')$, according to the quantum conditions (28), so this left-hand side cannot vanish. The way out of the difficulty was shown by Fermi.[1] It consists in adopting a less stringent equation, namely the equation

$$\frac{\partial A_\mu}{\partial x_\mu}|P\rangle = 0, \quad (50)$$

and assuming it to hold for any $|P\rangle$ corresponding to a state that can actually occur in nature. There is one equation (50) for each point in space-time and these equations must all hold for any ket corresponding to a state that can actually occur.

We shall call a condition such as (50), which a ket has to satisfy to correspond to an actual state, a *supplementary condition*. The existence of supplementary conditions in the theory does not mean any departure from or modification in the general principles of quantum mechanics. The principle of superposition of states and the whole of the general theory of states, dynamical variables, and observables, as given in Chapter II, apply also when there are supplementary conditions, provided we impose a further requirement on a linear operator in order that it may represent an observable. We define a linear operator to be *physical* if it has the property that, when it operates on any ket satisfying the supplementary conditions, it produces another ket satisfying the supplementary conditions. In order that a linear operator may represent an observable it must evidently satisfy the requirement of being physical, in addition to the requirements of § 10.

We have already had an example of supplementary conditions in the theory of systems containing several similar particles. The condition that only symmetrical wave functions, or only antisymmetrical wave functions, represent states that can actually occur in nature, is precisely of the

---

[1] Fermi, *Reviews of Modern Physics*, **4** (1932), 125.

same type as condition (50) and is what we are now calling a supplementary condition. In this theory the requirement that a linear operator shall be physical is that it shall be symmetrical between the similar particles.

When we introduce supplementary conditions into our theory we must verify that they are consistent, i.e. not too restrictive to allow any ket at all to satisfy them. If we have more than one supplementary condition, we can deduce further supplementary conditions from them by taking P.B.s of the operators in them; thus if we have

$$U|P\rangle = 0, \quad V|P\rangle = 0, \tag{51}$$

we can deduce

$$[U, P]|P\rangle = 0, \quad [U, [U, V]]|P\rangle = 0, \tag{52}$$

and so on. To verify that our supplementary conditions are consistent we have to look into all the further supplementary conditions obtainable by this procedure to see that they can be satisfied, which we can usually do by showing that after a certain point the further supplementary conditions are all either identically satisfied or repetitions of the previous ones.

We must also verify that the supplementary conditions are in agreement with the equations of motion. In the Heisenberg picture, for which the ket $|P\rangle$ in (51) is fixed, we shall have different supplementary conditions referring to different times and they must all be consistent, in the way discussed above. In the Schrödinger picture, for which the ket $|P\rangle$ varies with the time in accordance with Schrödinger's equation, we require that if $|P\rangle$ satisfies the supplementary conditions initially it satisfies them always. This means that $d|P\rangle/dt$ must satisfy the supplementary conditions, or that $H|P\rangle$ must satisfy the supplementary conditions, or that H must be physical.

It is convenient when we have a supplementary condition $U|P\rangle = 0$ to write

$$U \approx 0 \tag{53}$$

and to call (53) a weak equation, in distinction to an ordinary or strong equation. A weak equation gives another weak equation if it is multiplied by any factor on the left, but does not in general give a valid equation if it is multiplied by a factor on the right. Thus a weak equation must not be used in working out P.B.s. With this way of speaking, the requirement (52) that the supplementary conditions are consistent becomes the requirement that the P.B.s of the operators in the supplementary conditions shall vanish weakly.

The condition for a dynamical variable ξ to be physical is that, for each supplementary condition $U|P\rangle = 0$, we have

$$U\xi|P\rangle = 0,$$

and hence

$$[U, \xi]|P\rangle = 0.$$

Thus the condition is that the P.B. of the dynamical variable with each of the operators of the supplementary conditions shall vanish weakly.

Let us now return to electrodynamics. We take equation (2) to be a weak equation, so it should be written

$$\frac{\partial A_\mu}{\partial x_\mu} \approx 0. \tag{54}$$

In the Heisenberg picture we have one of these equations for each point x. To check their consistency, we take two arbitrary points x and x' in space-time and form the P.B.

$$\left[\frac{\partial A_\mu(x)}{\partial x_\mu}, \frac{\partial A_\nu(x')}{\partial x'_\nu}\right] = \frac{\partial^2}{\partial x_\mu \partial x'_\nu}[A_\mu(x), A_\nu(x')].$$

Evaluating it with the help of (28), we get

$$g_{\mu\nu}\frac{\partial^2 \Delta(x-x')}{\partial x_\mu \partial x'_\nu} = -\Box \Delta(x-x') = 0$$

from (25), so the requirements for consistency are satisfied strongly. As we have verified that the supplementary conditions are consistent at all times in the Heisenberg picture, we have verified that they are in agreement with the equations of motion.

Since equation (54) is only a weak equation, any of its consequences in the ordinary Maxwell theory will be valid in the quantum theory only as weak equations. The equations

$$\operatorname{div} \mathcal{H} = 0, \quad \frac{\partial \mathcal{H}}{\partial t} = -\operatorname{curl} \mathcal{E}$$

follow simply from the definitions of $\mathcal{E}$ and $\mathcal{H}$ in terms of the potentials, so they are valid strongly in the quantum theory. The other Maxwell equations for empty space, namely

$$\operatorname{div} \mathcal{E} \approx 0, \quad \frac{\partial \mathcal{E}}{\partial t} \approx \operatorname{curl} \mathcal{H}, \tag{55}$$

are weak equations in the quantum theory, because one needs the help of (54) as well as (1) in deriving them.

77. THE SUPPLEMENTARY CONDITIONS

The field quantities $\mathcal{E}$ and $\mathcal{H}$ are components of the antisymmetric tensor $\partial A^\nu / \partial x_\mu - \partial A^\mu / \partial x_\nu$. The P.B. of the tensor with the operator of (54) at a general point x' is

$$\left[\frac{\partial A^\nu(x)}{\partial x_\mu} - \frac{\partial A^\mu(x)}{\partial x_\nu}, \frac{\partial A_\sigma(x')}{\partial x'_\sigma} \right] = g_\sigma^\nu \frac{\partial^2 \Delta(x-x')}{\partial x_\mu \partial x'_\sigma} - g_\sigma^\mu \frac{\partial^2 \Delta(x-x')}{\partial x_\nu \partial x'_\sigma} = 0.$$

It follows that $\mathcal{E}$ and $\mathcal{H}$ are physical. The potentials A_μ are not physical.

The supplementary conditions affecting the dynamical variables at a particular time are

$$\frac{\partial A_\mu}{\partial x_\mu} \approx 0, \quad \frac{\partial}{\partial x_0} \frac{\partial A_\mu}{\partial x_\mu} \approx 0. \tag{56}$$

Higher differentiations with respect to x_0 do not give independent equations, but equations which are consequences of these and the strong equation (1). Thus in terms of the Schrödinger variables of § 76, the supplementary conditions are

$$B_0 + A_r^{\ r} \approx 0 \tag{57}$$

and

$$(A_0^{\ r} + B_r)^r \approx 0. \tag{58}$$

Equation (58) is the same as the first of equations (55) and may also be written, from (39),

$$\nabla^2 (A_0 + U) \approx 0.$$

Since this holds throughout three-dimensional space, it leads to

$$A_0 + U \approx 0. \tag{59}$$

Noting that $A_r^{\ r} = V^{rr}$, we can now see from (49) that

$$H_{FL} \approx 0. \tag{60}$$

Thus *there is no longitudinal field energy for states that occur in nature*.

To set up a convenient representation, we introduce a standard ket $|0_F\rangle$ satisfying the supplementary conditions

$$(B_0 + A_r^{\ r})|0_F\rangle = 0, \quad (A_0 + U)|0_F\rangle = 0, \tag{61}$$

and also satisfying

$$\overline{\mathcal{A}}_{rk}|0_F\rangle = 0. \tag{62}$$

These conditions are consistent, because $\overline{\mathcal{A}}_{rk}$ commutes with the operators in (61), and they are sufficient to fix $|0_F\rangle$ completely, apart from a numerical factor, because the only independent dynamical variables that we have are $A_0, B_0, U, A_r^{\ r}, \mathcal{A}_{rk}, \overline{\mathcal{A}}_{rk}$, and of these $A_0 + U, B_0 + A_r^{\ r}, \overline{\mathcal{A}}_{rk}$

form a complete commuting set. With this standard ket we can express any ket as

$$\Psi(A_0, B_0, \mathcal{A}_{r\mathbf{k}})|0_F\rangle. \qquad (63)$$

Our representation is just the Fock representation so far as concerns the transverse dynamical variables $\mathcal{A}_{r\mathbf{k}}, \overline{\mathcal{A}}_{r\mathbf{k}}$, so Ψ must be a power series in the variables $\mathcal{A}_{r\mathbf{k}}$, with different terms in the series corresponding to the presence of different numbers of photons. The number of variables occurring in Ψ is a continuous infinity, so Ψ is what mathematicians call a 'functional'.

If the ket (63) satisfies the supplementary conditions, Ψ must be independent of A_0 and B_0, and thus a function only of the $\mathcal{A}_{r\mathbf{k}}$. So physical states are represented by kets of the form

$$\Psi(\mathcal{A}_{r\mathbf{k}})|0_F\rangle, \qquad (64)$$

with Ψ a power series in the variables $\mathcal{A}_{r\mathbf{k}}$. The standard ket $|0_F\rangle$ itself represents the physical state with no photons present, the perfect vacuum.

Our Hamiltonian H_F and its parts H_{FL}, H_{FT} have so far contained arbitrary numerical terms. It is convenient to choose these terms so that H_{FL}, H_{FT} are zero for the perfect vacuum. The result (60) shows that H_{FL} given by (48) or (49) has the numerical term in it correctly chosen to make H_{FL} have the value zero for the perfect vacuum, as well as for every other physical state. We must take H_{FT} to be

$$H_{FT} = 4\pi^2 \int k_0^2 \mathcal{A}_{r\mathbf{k}} \overline{\mathcal{A}}_{r\mathbf{k}} \, d^3k, \qquad (65)$$

the transverse part of (19), in order that the numerical term in it may be correctly chosen to give no zero-point energy for the photons. (47) differs from (65) by an infinite numerical term, consisting of a half-quantum of energy for each photon state.

78. Electrons and Positrons by Themselves

We now consider electrons and positrons in the absence of electromagnetic field. The state of an electron is described, as in Chapter XI, by a wave function ψ with four components ψ_a ($a = 1, 2, 3, 4$), satisfying the wave equation

$$i\hbar \frac{\partial \psi}{\partial x_0} = -i\hbar \alpha_r \frac{\partial \psi}{\partial x_r} + \alpha_m m \psi. \qquad (66)$$

78. ELECTRONS AND POSITRONS BY THEMSELVES

To get a many-electron theory we shall apply the method of second quantization of § 65, which involves changing the one-electron wave function into a set of operators satisfying certain anticommutation relations.

When we are dealing with ψ at various places at a given time we may write it $\psi_\mathbf{x}$, with $\mathbf{x}$ denoting x_1, x_2, x_3. Its components are then $\psi_{a\mathbf{x}}$. We pass to the momentum representation with the wave function $\psi_\mathbf{p}$ by a three-dimensional Fourier resolution

$$\psi_\mathbf{x} = h^{-3/2} \int e^{i(\mathbf{xp})/\hbar} \psi_\mathbf{p} \, d^3 p, \quad \psi_\mathbf{p} = h^{-3/2} \int e^{-i(\mathbf{xp})/\hbar} \psi_\mathbf{x} \, d^3 x. \quad (67)$$

$\psi_\mathbf{p}$ has four components $\psi_{a\mathbf{p}}$, corresponding to the four components of $\psi_\mathbf{x}$. In this representation the energy operator is

$$p_0 = \alpha_r p_r + \alpha_m m,$$

in which the momentum operators p_r are multiplying factors.

We can separate ψ into a positive-energy part ξ and a negative-energy part ζ,

$$\psi = \xi + \zeta,$$

ξ and ζ each having four components like ψ. In the momentum representation they are given by

$$\xi_\mathbf{p} = \frac{1}{2}\left[1 + \frac{\alpha_r p_r + \alpha_m m}{(\mathbf{p}^2 + m^2)^{1/2}}\right]\psi_\mathbf{p}, \quad \zeta_\mathbf{p} = \frac{1}{2}\left[1 - \frac{\alpha_r p_r + \alpha_m m}{(\mathbf{p}^2 + m^2)^{1/2}}\right]\psi_\mathbf{p}, \quad (68)$$

since these equations lead to

$$p_0 \xi_\mathbf{p} = (\alpha_r p_r + \alpha_m m)\xi_\mathbf{p} = \frac{1}{2}[\alpha_r p_r + \alpha_m m + (\mathbf{p}^2 + m^2)^{1/2}]\psi_\mathbf{p}$$
$$= (\mathbf{p}^2 + m^2)^{1/2} \xi_\mathbf{p},$$

and similarly

$$p_0 \zeta_\mathbf{p} = -(\mathbf{p}^2 + m^2)^{1/2} \zeta_\mathbf{p},$$

showing that $\xi_\mathbf{p}$ and $\zeta_\mathbf{p}$ are eigenfunctions of p_0 with the eigenvalues $(\mathbf{p}^2 + m^2)^{1/2}$ and $-(\mathbf{p}^2 + m^2)^{1/2}$ respectively. When one is working with the operators

$$\frac{1}{2}\left[1 + \frac{\alpha_r p_r + \alpha_m m}{(\mathbf{p}^2 + m^2)^{1/2}}\right], \quad \frac{1}{2}\left[1 - \frac{\alpha_r p_r + \alpha_m m}{(\mathbf{p}^2 + m^2)^{1/2}}\right],$$

one should note that their squares are equal to themselves and their product in either order is zero.

The second quantization makes the ψ's into operators like the $\bar{\eta}$'s of § 65, satisfying anticommutation relations like (11′) of § 65. Using the notation for the anticommutator

$$MN + NM = [M, N]_+, \tag{69}$$

we get

$$[\psi_{ax}, \psi_{bx'}]_+ = 0, \quad [\bar{\psi}_{ax}, \bar{\psi}_{bx'}]_+ = 0$$
$$[\psi_{ax}, \bar{\psi}_{bx'}]_+ = \delta_{ab}\,\delta(\mathbf{x} - \mathbf{x}'), \tag{70}$$

the function $\delta(\mathbf{x}-\mathbf{x}')$ appearing in the last equation owing to the $\mathbf{x}$'s taking on continuous ranges of values. On transforming to the $\mathbf{p}$-representation according to (67), we get

$$[\psi_{a\mathbf{p}}, \psi_{b\mathbf{p}'}]_+ = 0, \quad [\bar{\psi}_{a\mathbf{p}}, \bar{\psi}_{b\mathbf{p}'}]_+ = 0,$$
$$[\psi_{a\mathbf{p}}, \bar{\psi}_{b\mathbf{p}'}]_+ = \delta_{ab}\,\delta(\mathbf{p} - \mathbf{p}'). \tag{71}$$

With ξ and ζ defined again by (68), the last of equations (71) gives

$$[\xi_{a\mathbf{p}}, \bar{\xi}_{b\mathbf{p}'}]_+$$
$$= \frac{1}{2}\left\{1 + \frac{\alpha_r p_r + \alpha_m m}{(\mathbf{p}^2 + m^2)^{1/2}}\right\}_{ac} [\psi_{c\mathbf{p}}, \bar{\psi}_{d\mathbf{p}'}]_+ \frac{1}{2}\left\{1 + \frac{\alpha_s p'_s + \alpha_m m}{(\mathbf{p}'^2 + m^2)^{1/2}}\right\}_{db}$$
$$= \frac{1}{2}\left[1 + \frac{\alpha_r p_r + \alpha_m m}{(\mathbf{p}^2 + m^2)^{1/2}}\right]_{ab} \delta(\mathbf{p} - \mathbf{p}'), \tag{72}$$

and similarly

$$[\zeta_{a\mathbf{p}}, \bar{\zeta}_{b\mathbf{p}'}]_+ = \frac{1}{2}\left[1 - \frac{\alpha_r p_r + \alpha_m m}{(\mathbf{p}^2 + m^2)^{1/2}}\right]_{ab} \delta(\mathbf{p} - \mathbf{p}') \tag{73}$$

and

$$[\xi_{a\mathbf{p}}, \bar{\zeta}_{b\mathbf{p}'}]_+ = [\bar{\xi}_{a\mathbf{p}}, \zeta_{b\mathbf{p}'}]_+ = 0.$$

According to the interpretation of § 65, the operators $\psi_{a\mathbf{p}}$ are operators of annihilation of an electron of momentum $\mathbf{p}$ and the operators $\bar{\psi}_{a\mathbf{p}}$ are operators of creation of an electron of momentum $\mathbf{p}$. To avoid the unphysical notion of negative-energy electrons, we must pass over to a new interpretation based on the positron theory of § 73. The annihilation of a negative-energy electron is to be understood as the creation of a hole in the sea of negative-energy electrons, or the creation of a positron. So the operators $\zeta_{a\mathbf{p}}$ become operators of creation of a positron. The positron has the momentum $-\mathbf{p}$, because an amount $\mathbf{p}$ of momentum gets annihilated.

Similarly the $\bar{\zeta}_{ap}$ become operators of annihilation of a positron of momentum $-\mathbf{p}$. The ξ_{ap} and $\bar{\xi}_{ap}$ are operators of annihilation and creation respectively of an ordinary, positive-energy electron of momentum $\mathbf{p}$.

It should be noted that, although $\xi_\mathbf{p}$ has four components, only two of them are independent, because the four are connected by

$$\left[1 - \frac{\alpha_r p_r + \alpha_m m}{(\mathbf{p}^2 + m^2)^{1/2}}\right]\xi_\mathbf{p} = 0,$$

which involves two independent equations. The two independent components of $\xi_\mathbf{p}$ correspond to the annihilation of an electron in each of the two independent states of spin. Similarly $\zeta_\mathbf{p}$ has only two independent components, because of the equations

$$\left[1 + \frac{\alpha_r p_r + \alpha_m m}{(\mathbf{p}^2 + m^2)^{1/2}}\right]\zeta_\mathbf{p} = 0,$$

and they correspond to the creation of a positron in the two independent states of spin.

The vacuum state, for which there are no electrons or positrons present, is represented by the ket $|0_P\rangle$ satisfying

$$\xi_{ap}|0_P\rangle = 0, \quad \bar{\zeta}_{ap}|0_P\rangle = 0. \tag{74}$$

We can use this ket as the standard ket of a representation. We then have any ket expressed as

$$\Psi(\bar{\xi}_{ap}, \zeta_{ap})|0_P\rangle,$$

in which the function, or rather functional, Ψ is a power series in the variables $\bar{\xi}_{ap}, \zeta_{ap}$. Each term of Ψ is like (17′) of § 65. It must not contain any of its variables to a higher power than the first. It corresponds to the existence of certain (positive-energy) electrons and certain positrons, in states specified by the labels of the variables appearing in it.

From (12′) of § 65, the total number of electrons is $\int \bar{\psi}_{ap}\psi_{ap} d^3p$ summed over a. We may write it in the notation of equation (12) of § 67 as $\int \bar{\psi}_\mathbf{p}^\dagger \psi_\mathbf{p} d^3p$. Transforming it to the x-representation by (67), we get

$$h^{-3} \iiint e^{i(\mathbf{xp})/\hbar} e^{-i(\mathbf{x'p'})/\hbar} \bar{\psi}_\mathbf{x}^\dagger \psi_{\mathbf{x'}} d^3x d^3x' d^3p = \int \bar{\psi}_\mathbf{x}^\dagger \psi_\mathbf{x} d^3x,$$

showing that the density of the electrons is $\bar{\psi}_\mathbf{x}^\dagger \psi_\mathbf{x}$. This result includes an infinite constant representing the density of the sea of negative-energy electrons.

We get a quantity of more physical significance if we take the total charge Q, equal to the number of positive-energy electrons minus the number of holes or positrons, all multiplied by $-e$. Thus

$$Q = -e \int (\bar{\xi}_\mathbf{p}^\dagger \xi_\mathbf{p} - \zeta_\mathbf{p}^\dagger \bar{\zeta}_\mathbf{p}) \, d^3p. \tag{75}$$

We can evaluate this with the help of (68). Using the transpose of the second of these equations, namely

$$\zeta_\mathbf{p}^\dagger = \frac{1}{2} \psi_\mathbf{p}^\dagger \left[1 - \frac{\alpha_r^\dagger p_r + \alpha_m^\dagger m}{(\mathbf{p}^2 + m^2)^{1/2}} \right],$$

we get

$$Q = -e \int \left\{ \bar{\psi}_\mathbf{p}^\dagger \frac{1}{2} \left[1 + \frac{\alpha_r p_r + \alpha_m m}{(\mathbf{p}^2 + m^2)^{1/2}} \right] \psi_\mathbf{p} \right.$$

$$\left. - \psi_\mathbf{p}^\dagger \frac{1}{2} \left[1 - \frac{\alpha_r^\dagger p_r + \alpha_m^\dagger m}{(\mathbf{p}^2 + m^2)^{1/2}} \right] \bar{\psi}_\mathbf{p} \right\} d^3p.$$

Now for any matrix α whose diagonal sum is zero, the anticommutation relations (71) give

$$\bar{\psi}_\mathbf{p}^\dagger \alpha \psi_{\mathbf{p}'} + \psi_{\mathbf{p}'}^\dagger \alpha^\dagger \bar{\psi}_\mathbf{p} = \alpha_{ab}(\bar{\psi}_{a\mathbf{p}} \psi_{b\mathbf{p}'} + \psi_{b\mathbf{p}'} \bar{\psi}_{a\mathbf{p}}) = \alpha_{aa} \delta(\mathbf{p} - \mathbf{p}') = 0, \tag{76}$$

a result which we may assume still holds for $\mathbf{p}' = \mathbf{p}$. Then the expression for Q reduces to

$$Q = -e \int \frac{1}{2} (\bar{\psi}_\mathbf{p}^\dagger \psi_\mathbf{p} - \psi_\mathbf{p}^\dagger \bar{\psi}_\mathbf{p}) \, d^3p.$$

Transforming it to the $\mathbf{x}$-representation as before, we get

$$Q = -e \int \frac{1}{2} (\bar{\psi}_\mathbf{x}^\dagger \psi_\mathbf{x} - \psi_\mathbf{x}^\dagger \bar{\psi}_\mathbf{x}) \, d^3x,$$

showing that the charge density is

$$j_{0\mathbf{x}} = -\frac{1}{2} e (\bar{\psi}_\mathbf{x}^\dagger \psi_\mathbf{x} - \psi_\mathbf{x}^\dagger \bar{\psi}_\mathbf{x}). \tag{77}$$

The interpretation of the one-electron wave function in § 68 gives, besides the probability density $\bar{\psi}^\dagger \psi$, a probability current $\bar{\psi}^\dagger \alpha_r \psi$. With second quantization we shall have correspondingly a flow of electrons, given by the operator $\bar{\psi}_\mathbf{x}^\dagger \alpha_r \psi_\mathbf{x}$. The sea of negative-energy electrons produces no resultant flow of electrons, from symmetry, and so the electric current is

$$j_{r\mathbf{x}} = -e \bar{\psi}_\mathbf{x}^\dagger \alpha_r \psi_\mathbf{x}. \tag{78}$$

78. ELECTRONS AND POSITRONS BY THEMSELVES

The total energy of the electrons is, from formula (29) of § 60, which is valid also for fermions,

$$H_{P'} = \int \overline{\psi}_\mathbf{p}^\dagger p_0 \psi_\mathbf{p} \, d^3 p = \int \overline{\psi}_\mathbf{p}^\dagger (\alpha_r p_r + \alpha_m m) \psi_\mathbf{p} \, d^3 p. \tag{79}$$

It becomes, when transformed to the **x**-representation,

$$H_{P'} = \int \overline{\psi}_\mathbf{x}^\dagger (-i\hbar \alpha_r \psi_\mathbf{x}^r + \alpha_m m \psi_\mathbf{x}) \, d^3 x. \tag{80}$$

This total energy contains an infinite numerical term representing the energy of the sea of negative-energy electrons.

We get a quantity of more physical significance if we take the energy of all the electrons and positrons, reckoning the energy of the vacuum as zero. This quantity is

$$H_P = \int (\mathbf{p}^2 + m^2)^{1/2} (\overline{\xi}_\mathbf{p}^\dagger \xi_\mathbf{p} + \zeta_\mathbf{p}^\dagger \overline{\zeta}_\mathbf{p}) \, d^3 p \tag{81}$$

$$= \int (\mathbf{p}^2 + m^2)^{1/2} \left\{ \overline{\psi}_\mathbf{p}^\dagger \frac{1}{2} \left[1 + \frac{\alpha_r p_r + \alpha_m m}{(\mathbf{p}^2 + m^2)^{1/2}} \right] \psi_\mathbf{p} \right.$$

$$\left. + \psi_\mathbf{p}^\dagger \frac{1}{2} \left[1 - \frac{\alpha_r^\dagger p_r + \alpha_m^\dagger m}{(\mathbf{p}^2 + m^2)^{1/2}} \right] \overline{\psi}_\mathbf{p} \right\} d^3 p$$

$$- \int \frac{1}{2} [\overline{\psi}_\mathbf{p}^\dagger (\alpha_r p_r + \alpha_m m) \psi_\mathbf{p} - \psi_\mathbf{p}^\dagger (\alpha_r^\dagger p_r + \alpha_m^\dagger m) \overline{\psi}_\mathbf{p}] \, d^3 p$$

$$+ \int (\mathbf{p}^2 + m^2)^{1/2} \frac{1}{2} (\overline{\psi}_\mathbf{p}^\dagger \psi_\mathbf{p} + \psi_\mathbf{p}^\dagger \overline{\psi}_\mathbf{p}) \, d^3 p. \tag{82}$$

From (76), the first integral in (82) is the same as (79) and is just $H_{P'}$. The second integral is an infinite constant and is minus the energy of all the negative-energy electrons of the vacuum distribution.

We may take either H_P or $H_{P'}$ as the Hamiltonian. The Heisenberg equation of motion for $\psi_{a\mathbf{x}}$ is thus

$$\frac{\partial \psi_{a\mathbf{x}}}{\partial x_0} = [\psi_{a\mathbf{x}}, H_P] = [\psi_{a\mathbf{x}}, H_{P'}],$$

and if we work this out we just get back to the wave equation for ψ, namely (66).

We must now look into the question of whether the theory is relativistic. It is built up from operators ψ which satisfy the field equations (66). These equations are the same as the wave equation for the one-electron wave function and are known to be invariant under Lorentz transformations, provided ψ transforms according to the law (20) of Chapter XI. Our present theory goes beyond the one-electron theory in that anticommutation relations are introduced for the ψ's and $\overline{\psi}$'s, and it becomes necessary to verify that these anticommutation relations are Lorentz invariant.

We proceed by a method analogous to that of § 75. We take two general points x and x' in space-time and form the anticommutator

$$K_{ab}(x, x') = \psi_a(x)\overline{\psi}_b(x') + \overline{\psi}_b(x')\psi_a(x). \qquad (83)$$

We can evaluate it by working directly from the anticommutation relations (71) for the Fourier components of ψ and $\overline{\psi}$. A simpler way is to note certain properties that $K_{ab}(x, x')$ must have, namely

(i) it involves x_μ and x'_μ only through their difference $x_\mu - x'_\mu$;
(ii) it satisfies the wave equation

$$\left(i\hbar\frac{\partial}{\partial x_0} + i\hbar\alpha_r\frac{\partial}{\partial x_r} - \alpha_m m\right)_{ab} K_{bc}(x, x') = 0 \qquad (84)$$

on account of $\psi(x)$ satisfying (66);

(iii) for $x_0 = x'_0$ it has the value $\delta_{ab}\delta(\mathbf{x} - \mathbf{x}')$, as follows from the third of equations (70).

These properties are sufficient to fix $K_{ab}(x, x')$ completely, since (iii) fixes it for $x_0 = x'_0$, (ii) shows how it depends on x_0, and (i) then shows how it depends on x'_0. The solution is easily seen to be

$$K_{ab}(x, x') = h^{-3} \int \sum \frac{1}{2}\left(1 + \frac{\alpha_r p_r + \alpha_m m}{p_0}\right)_{ab} e^{-i(x-x')\cdot p/\hbar} d^3p, \qquad (85)$$

where the $\sum$ means a summation over the two values $\pm(\mathbf{p}^2 + m^2)^{1/2}$ for p_0 with particular values for p_1, p_2, p_3. It satisfies (ii) since the operator in (84) produces the factor $(p_0 - \alpha_r p_r - \alpha_m m)$ in the integrand of (85), which factor gives zero when multiplied on the left into the factor ( ). It satisfies (iii) since, with $x_0 = x'_0$, the summation over p_0 makes the second term in ( ) cancel out.

The law of transformation ψ and $\overline{\psi}$ given in § 68 has the effect of making the quantities $\overline{\psi}^\dagger(x')\alpha_\mu \psi(x)$ transform like the four components of a 4-vector and making $\overline{\psi}^\dagger(x')\alpha_m \psi(x)$ invariant. Thus

$$l^\mu \overline{\psi}^\dagger(x')\alpha_\mu \psi(x) + S \overline{\psi}^\dagger(x')\alpha_m \psi(x) \qquad (86)$$

is invariant with l^μ any 4-vector and S any scalar. The invariance of (86) must be sufficient to ensure the correct transformation law for ψ and $\overline{\psi}$, since it enables one to deduce the invariance of the wave equation for ψ, by taking $l^\mu = i\hbar\partial/\partial x_\mu$, $S = -m$.

The invariance of (86) leads to the invariance of

$$(l^\mu \alpha_\mu + S\alpha_m)_{ab}[\overline{\psi}_a(x')\psi_b(x) + \psi_b(x)\overline{\psi}_a(x')].$$

Thus

$$(l^\mu \alpha_\mu + S\alpha_m)_{ab} K_{ba}(x, x') \tag{87}$$

should be invariant with $K_{ab}(x, x')$ given by (85), and its invariance would be sufficient to ensure the invariance of the anticommutation relations. We get for (87)

$$h^{-3} \int \sum \frac{1}{2}(l^\mu \alpha_\mu + S\alpha_m)_{ab}(p_0 + \alpha_r p_r + \alpha_m m)_{ba} e^{-i(x-x')\cdot p/\hbar} p_0^{-1} d^3p$$

$$= h^{-3} \int \sum \frac{1}{2}[(l_0 - l_s \alpha_s + S\alpha_m)$$

$$(p_0 + \alpha_r p_r + \alpha_m m)]_{aa} e^{-i(x-x')\cdot p/\hbar} p_0^{-1} d^3p$$

$$= h^{-3} \int \sum 2(l_0 p_0 - l_r p_r + Sm) e^{-i(x-x')\cdot p/\hbar} p_0^{-1} d^3p. \tag{88}$$

This is Lorentz invariant because the differential element $p_0^{-1} d^3p$ is Lorentz invariant. Thus the relativistic invariance of the theory is proved.

79. The Interaction

The complete Hamiltonian for electrons and positrons interacting with the electromagnetic field is

$$H = H_F + H_P + H_Q, \tag{89}$$

where H_F is the Hamiltonian for the electromagnetic field alone, given by (19) or (45), H_P is the Hamiltonian for the electrons and positrons alone, given by (80) or (81), and H_Q is the interaction energy, involving the dynamical variables of the electrons and positrons as well as those of the electromagnetic field. We take

$$H_Q = \int A^\mu j_\mu d^3x, \tag{90}$$

with j_μ given by (77) and (78), as we shall see that this gives the correct equations of motion. Thus, with neglect of infinite numerical terms,

$$H = \int \left[\overline{\psi}^\dagger \alpha_r(-i\hbar \psi^r - eA^r \psi) + \overline{\psi}^\dagger \alpha_m m\psi - \frac{1}{2}eA^0(\overline{\psi}^\dagger \psi - \psi^\dagger \overline{\psi})\right] d^3x$$

$$- (8\pi)^{-1} \int (B_\mu B^\mu + A_\mu{}^r A^{\mu r}) d^3x. \tag{91}$$

Let us work out the Heisenberg equations of motion that follow from the Hamiltonian (91). We have

$$i\hbar \frac{\partial \psi_{ax}}{\partial x_0} = \psi_{ax} H - H \psi_{ax} = \psi_{ax}(H_P + H_Q) - (H_P + H_Q)\psi_{ax}$$

$$= \int [\psi_{ax}, \bar{\psi}_{bx'}]_+ \{\alpha_r(-i\hbar \psi_{x'}{}^{r'} - eA^r{}_{x'}\psi_{x'})$$

$$+ \alpha_m m \psi_{x'} - eA^0{}_{x'}\psi_{x'}\}_b \, d^3x'$$

$$= [\alpha_r(-i\hbar \psi_x{}^r - eA^r{}_x \psi_x) + \alpha_m m \psi_x - eA^0{}_x \psi_x]_a.$$

Thus

$$\left[\alpha_\mu \left(i\hbar \frac{\partial}{\partial x_\mu} + eA^\mu\right) - \alpha_m m\right]\psi = 0. \tag{92}$$

This agrees with the one-electron wave equation (11) of Chapter XI. Since H is real, the equation of motion for $\bar{\psi}$ will be the conjugate of the equation of motion for ψ and so will agree with (12) of Chapter XI. Thus the interaction (90) gives correctly the action of the field on the electrons and positrons, Further we have, making use of the P.B. relations in (46),

$$\frac{\partial A_\mu}{\partial x_0} = [A_\mu, H] = [A_\mu, H_F]$$

$$= B_\mu \tag{93}$$

and

$$\frac{\partial B_{\mu x}}{\partial x_0} = [B_{\mu x}, H] = [B_{\mu x}, H_F] + [B_{\mu x}, H_Q]$$

$$= \nabla^2 A_{\mu x} + \int [B_{\mu x}, A^\nu{}_{x'}] j_{\nu x'} \, d^3x'$$

$$= \nabla^2 A_{\mu x} + 4\pi j_{\mu x}. \tag{94}$$

(93) and (94) lead to

$$\Box A_\mu = 4\pi j_\mu, \tag{95}$$

which agrees with the Maxwell theory and shows that the interaction (90) gives correctly the action of the electrons and positrons on the field.

To complete the theory we must bring in the supplementary conditions (54). We must verify that they are in agreement with the equations of motion. The method used in § 77, which consisted in showing that the supplementary conditions at different times in the Heisenberg picture are consistent with one another, is no longer applicable, because the quantum conditions connecting dynamical variables at different times get altered by the interaction in a way that is too complicated to be worked out. So we shall obtain all the supplementary conditions affecting the dynamical variables at one instant of time and check whether they are consistent.

We have again equations (56). A further differentiation with respect to x_0 gives

$$\Box \frac{\partial A_\mu}{\partial x_\mu} \approx 0. \tag{96}$$

Now the equation of motion for ψ, namely (92), leads, as in § 68, to

$$\frac{\partial(\overline{\psi}^\dagger \alpha_\mu \psi)}{\partial x_\mu} = 0,$$

This is the same as

$$\frac{\partial j_\mu}{\partial x_\mu} = 0, \tag{97}$$

because the difference between $-e\overline{\psi}^\dagger \psi$ and j_0 is constant in time, even though it is infinite. From (95) we now see that (96) holds as a strong equation. Thus equations (56) are the only independent supplementary conditions affecting the dynamical variables at one instant of time. The first of them gives (57), as before, and the second now gives, with the help of (95) for $\mu = 0$,

$$(A_0{}^r + B_r)^r + 4\pi j_0 \approx 0. \tag{98}$$

This may be written

$$(A_0 + U)^{rr} + 4\pi j_0 \approx 0 \tag{99}$$

or, from (39),

$$\text{div } \mathcal{E} - 4\pi j_0 \approx 0, \tag{100}$$

and is just one of the Maxwell equations.

One can see without detailed calculation that, for any two points x and x' at the same time,

$$[j_{0\mathbf{x}}, j_{0\mathbf{x}'}] = 0,$$

since, from the form of (70), the P.B. must be a multiple of $\delta(\mathbf{x} - \mathbf{x}')$ and cannot contain derivatives of $\delta(\mathbf{x} - \mathbf{x}')$, while also it has to be antisymmetrical between $\mathbf{x}$ and $\mathbf{x}'$. Thus the extra terms $4\pi j_{0\mathbf{x}}$ in equations (98) for various values of $\mathbf{x}$, as compared with the corresponding equations (58), commute with one another as well as with all the other dynamical variables occurring in (58) and (57). It follows that these extra terms will not disturb the consistency of (58) and (57), and hence (98) and (57) are consistent.

Our method of introducing interaction into the theory was not relativistic, since the interaction energy (90) involves the dynamical variables at an instant of time in some Lorentz frame. It therefore becomes questionable whether the theory with interaction is a relativistic one. Our field equations, namely (92) and (95), are evidently relativistic and so are the supplementary conditions (54). It remains uncertain whether the quantum conditions are Lorentz invariant.

We know the quantum conditions connecting all our dynamical variables $A_{\mu\mathbf{x}}$, $B_{\mu\mathbf{x}}$, $\psi_{a\mathbf{x}}$, $\overline{\psi}_{a\mathbf{x}}$ at a given time x_0. We cannot, as mentioned above, work out the general quantum conditions connecting dynamical variables at any two points in space-time, because the interaction makes it too complicated. We shall therefore make an infinitesimal Lorentz transformation and work out the quantum conditions at a given time in the new frame of reference. If we can establish that the quantum conditions are invariant under infinitesimal Lorentz transformations, their invariance under finite Lorentz transformations will follow.

Let x_0^* be the time coordinate in the new frame of reference. It is connected with the original coordinates by

$$x_0^* = x_0 + \epsilon v_r x_r, \tag{101}$$

where ϵ is an infinitesimal number and v_r is a three-dimensional vector, ϵv_r being the relative velocity of the two frames. We shall neglect terms of order ϵ^2.

A field quantity κ at the place $\mathbf{x}$ at the time x_0^* in the new frame has the value

$$\kappa(\mathbf{x}, x_0^*) = \kappa(\mathbf{x}, x_0) + (x_0^* - x_0)\frac{\partial \kappa_\mathbf{x}}{\partial x_0} = \kappa(\mathbf{x}, x_0) + \epsilon v_r x_r [\kappa_\mathbf{x}, H]. \tag{102}$$

Its P.B. with another such field quantity $\lambda(\mathbf{x}', x_0^*)$ is

$$\begin{aligned}[\kappa(\mathbf{x}, x_0^*), \lambda(\mathbf{x}', x_0^*)] \\ = [\kappa(\mathbf{x}, x_0) + \epsilon v_r x_r [\kappa_\mathbf{x}, H], \lambda(\mathbf{x}', x_0) + \epsilon v_s x_s' [\lambda_{\mathbf{x}'}, H]] \\ = [\kappa(\mathbf{x}, x_0), \lambda(\mathbf{x}', x_0)] + \epsilon v_s x_s' [\kappa_\mathbf{x}, [\lambda_{\mathbf{x}'}, H]] + \epsilon v_r x_r [[\kappa_\mathbf{x}, H], \lambda_{\mathbf{x}'}] \\ = [\kappa(\mathbf{x}, x_0), \lambda(\mathbf{x}', x_0)] + \epsilon v_r (x_r' - x_r)[\kappa_\mathbf{x}, [\lambda_{\mathbf{x}'}, H]] \\ + \epsilon v_r x_r [[\kappa_\mathbf{x}, \lambda_{\mathbf{x}'}], H]. \end{aligned} \tag{103}$$

79. THE INTERACTION

If κ and λ are ψ or $\overline{\psi}$ variables, we should be interested in their anticommutator instead of their P.B. Using the notation (69) for the anticommutator, we have

$$[\kappa(\mathbf{x}, x_0^*), \lambda(\mathbf{x}', x_0^*)]_+$$
$$= [\kappa(\mathbf{x}, x_0), \lambda(\mathbf{x}', x_0)]_+ + \epsilon v_r x_r'[\kappa_\mathbf{x}, [\lambda_{\mathbf{x}'}, H]]_+ + \epsilon v_r x_r[[\kappa_\mathbf{x}, H], \lambda_{\mathbf{x}'}]_+$$
$$= [\kappa(\mathbf{x}, x_0), \lambda(\mathbf{x}', x_0)]_+ + \epsilon v_r(x_r' - x_r)[\kappa_\mathbf{x}, [\lambda_{\mathbf{x}'}, H]]_+$$
$$+ \epsilon v_r x_r[[\kappa_\mathbf{x}, \lambda_{\mathbf{x}'}]_+, H]. \tag{104}$$

With κ and λ any two of the basic variables A_μ, B_μ, ψ_a, $\overline{\psi}_a$, the P.B. $[\kappa_\mathbf{x}, \lambda_{\mathbf{x}'}]$ or anticommutator $[\kappa_\mathbf{x}, \lambda_{\mathbf{x}'}]_+$, as the case may be, is a number, and so the last term in (103) or (104) vanishes. We are left with

$$[\kappa(\mathbf{x}, x_0^*), \lambda(\mathbf{x}', x_0^*)]_\pm$$
$$= [\kappa(\mathbf{x}, x_0), \lambda(\mathbf{x}', x_0)]_\pm + \epsilon v_r(x_r' - x_r)[\kappa_\mathbf{x}, [\lambda_{\mathbf{x}'}, H_P + H_F]]_\pm$$
$$+ \epsilon v_r(x_r' - x_r)[\kappa_{\mathbf{x}'}, [\lambda_{\mathbf{x}'}, H_Q]]_\pm, \tag{105}$$

where $[\kappa, \lambda]_\pm$ denotes the P.B. or the anticommutator, as the case may be. From the form (90) for H_Q we see that $[\lambda_{\mathbf{x}'}, H_Q]$ can involve only the dynamical variables $A_{\mu\mathbf{x}'}$, $\psi_{a\mathbf{x}'}$, $\overline{\psi}_{a\mathbf{x}'}$ and cannot involve any derivatives of these variables. It follows that $[\kappa_\mathbf{x}, [\lambda_{\mathbf{x}'}, H_Q]]_\pm$, if it does not vanish, will be a multiple of $\delta(\mathbf{x} - \mathbf{x}')$ and will not contain terms with derivatives of $\delta(\mathbf{x} - \mathbf{x}')$. Hence the last term of (105) vanishes. We can conclude that $[\kappa(\mathbf{x}, x_0^*), \lambda(\mathbf{x}', x_0^*)]_\pm$ has the same value as when there is no interaction, and is thus Lorentz invariant from our earlier work.

A possible criticism of the above proof should be noted. At several places we worked out expressions in powers of ϵ and neglected ϵ^2. Such a procedure cannot be valid for calculating $[\kappa(x), \lambda(x')]_\pm$ with x and x' two general points in space-time lying close together, so that $x_\mu - x_\mu'$ is of order ϵ, because the result of the calculation should be a function of the $(x_\mu - x_\mu')$'s having a singularity when the 4-vector $x - x'$ lies on the light-cone and such a function, of course, cannot be expanded as a power series in the $(x_\mu - x_\mu')$'s.

To validate the argument we should reformulate it so as to avoid the use of the δ function. Instead of evaluating $[\kappa(\mathbf{x}, x_0^*), \lambda(\mathbf{x}', x_0^*)]_\pm$, we should evaluate

$$\left[\int a_\mathbf{x} \kappa(\mathbf{x}, x_0^*) d^3x, \int b_{\mathbf{x}'} \lambda(\mathbf{x}', x_0^*) d^3x'\right]_\pm, \tag{106}$$

where $a_\mathbf{x}$ and $b_\mathbf{x}$ are two arbitrary continuous functions of x_1, x_2, x_3. Then the quantities that we need to expand in powers of ϵ all vary continuously with a continuous change in the direction of the time-axis, and the expansions are justifiable. The equations that we now get are those of the previous argument multiplied by $a_x b_{x'} d^3x d^3x'$ and integrated. We are led to the same conclusion—that the P.B. or anticommutator has the same value as when there is no interaction.

It will be seen that the reason why the interaction does not disturb the quantum conditions is because it is so simple, involving only the basic dynamical variables and not their derivatives. The P.B.s and anticommutators have the same values as with no interaction provided they refer to variables at two points in space-time that are at the same time with respect to some observer. This means the two points must be outside each other's light-cones and may approach coincidence only along a path lying outside the light-cone.

80. The Physical Variables

A ket $|P\rangle$ that represents a physical state must satisfy the supplementary conditions

$$(B_0 + A_r{}^r)|P\rangle = 0, \quad (\text{div } \mathcal{E} - 4\pi j_0)|P\rangle = 0. \tag{107}$$

A dynamical variable is physical if, when multiplied into any ket satisfying these conditions, it gives another ket satisfying these conditions. This requires that it shall commute with the quantities

$$B_0 + A_r{}^r, \quad \text{div } \mathcal{E} - 4\pi j_0. \tag{108}$$

Let us see what simple dynamical variables have this property.

The transverse field variables $\mathcal{A}_r, \mathcal{B}_r$ evidently commute with the quantities (108) and are physical. The variable ψ_a commutes with the first of the quantities (108) but not the second and is thus not physical. We have

$$i\hbar[\psi_{ax}, \overline{\psi}_{bx'}\psi_{bx'}]$$
$$= (\psi_{ax}\overline{\psi}_{bx'} + \overline{\psi}_{bx'}\psi_{ax})\psi_{bx'} = \delta_{ab}\delta(\mathbf{x} - \mathbf{x}')\psi_{bx'} = \psi_{ax}\delta(\mathbf{x} - \mathbf{x}').$$

Thus

$$[\psi_{ax}, j_{0x'}] = \frac{ie}{\hbar}\psi_{ax}\delta(\mathbf{x} - \mathbf{x}'). \tag{109}$$

From (42)

$$[e^{ieV_x/\hbar}, \text{div } \mathcal{E}_{x'}] = \frac{4\pi ie}{\hbar}e^{ieV_x/\hbar}\delta(\mathbf{x} - \mathbf{x}').$$

Hence

$$[e^{ieV_x/\hbar}\psi_{ax}, \text{div } \mathcal{E}_{x'} - 4\pi j_{0x'}]$$

$$= [e^{ieV_x/\hbar}, \operatorname{div} \mathcal{E}_{x'}]\psi_{ax} - 4\pi e^{ieV_x/\hbar}[\psi_{ax}, j_{0x'}] = 0.$$

Thus if we put

$$\psi_{ax}^* = e^{ieV_x/\hbar}\psi_{ax}, \tag{110}$$

ψ_{ax}^* commutes with both expressions (108) and is physical. Similarly $\overline{\psi}_{ax}^*$ is physical. The variables $\mathcal{A}_r$, $\mathcal{B}_r$, ψ_a^*, $\overline{\psi}_a^*$ are the only independent physical variables, apart from the quantities (108) themselves.

We have

$$j_0 = -\frac{1}{2}e(\overline{\psi}^{*\dagger}\psi^* - \psi^{*\dagger}\overline{\psi}^*), \quad j_r = -e\overline{\psi}^{*\dagger}\alpha_r\psi^*. \tag{111}$$

Thus the charge density and current are physical. Also it is easily seen that $\mathcal{E}$ and $\mathcal{H}$ are physical, just as in the case when there are no electrons and positrons present. All those variables are physical that are unaffected by the arbitrariness that exists in the electromagnetic potentials in the Maxwell theory.

The operator ψ_{ax} represents the creation of a positron or the annihilation of an electron at the place $\mathbf{x}$. Let us see what is the physical significance of the operator ψ_{ax}^*. From (44)

$$i\hbar[e^{ieV_x/\hbar}, \mathcal{E}_{rx'}] = ee^{ieV_x/\hbar}(x_r - x_r')|\mathbf{x} - \mathbf{x}'|^{-3},$$

and hence

$$i\hbar[\psi_{ux}^*, \mathcal{E}_{rx'}] = e\psi_{ux}^*(x_r - x_r')|\mathbf{x} - \mathbf{x}'|^{-3}$$

or

$$\mathcal{E}_{rx'}\psi_{ax}^* = \psi_{ax}^*[\mathcal{E}_{rx'} + e(x_r' - x_r)|\mathbf{x}' - \mathbf{x}|^{-3}]. \tag{112}$$

Take a state $|P\rangle$ for which $\mathcal{E}_r$ at a certain point $\mathbf{x}'$ certainly has the numerical value c_r, so that

$$\mathcal{E}_{rx'}|P\rangle = c_r|P\rangle.$$

Then from (112)

$$\mathcal{E}_{rx'}\psi_{ax}^*|P\rangle = [c_r + e(x_r' - x_r)|\mathbf{x}' - \mathbf{x}|^{-3}]\psi_{ax}^*|P\rangle,$$

so for the state $\psi_{ax}^*|P\rangle$, $\mathcal{E}_r$ at the point $\mathbf{x}'$ certainly has the value

$$c_r + e(x_r' - x_r)|\mathbf{x}' - \mathbf{x}|^{-3}.$$

This means that the operator ψ_{ax}^*, besides creating a positron or annihilating an electron at the point $\mathbf{x}$, increases the electric field at the point $\mathbf{x}'$ by $e(x_r' - x_r)|\mathbf{x}' - \mathbf{x}|^{-3}$, which is just the classical Coulomb field at $\mathbf{x}'$ of a positron with charge e at the point $\mathbf{x}$. Thus the operator ψ_{ax}^* creates a positron at the point $\mathbf{x}$ *together with its Coulomb field*, or else annihilates an electron at $\mathbf{x}$ *together with its Coulomb field*.

XII. QUANTUM ELECTRODYNAMICS

For electrons and positrons interacting with the electromagnetic field it is the variables $\psi^*, \overline{\psi}^*$, rather than the variables $\psi, \overline{\psi}$, that correspond to the physical processes of creation and annihilation of electrons and positrons, since these processes must always be accompanied by the appropriate Coulomb change in the electric field around the point where the particle is created or annihilated. It is easily seen that the variables $\psi^*_{\alpha x}, \overline{\psi}^*_{\alpha x}$ satisfy the same anticommutation relations (70) as the unstarred variables. When we pass to the momentum representation the important quantities will be, not the unphysical variables ψ_p defined by (67), but the physical variables ψ^*_p defined by

$$\psi^*_x = h^{-3/2} \int e^{i(xp)/\hbar} \psi^*_p \, d^3p, \quad \psi^*_p = h^{-3/2} \int e^{-i(xp)/\hbar} \psi^*_x \, d^3x. \quad (113)$$

We must now replace (68) by

$$\xi^*_p = \frac{1}{2}\left[1 + \frac{\alpha_r p_r + \alpha_m m}{(p^2 + m^2)^{1/2}}\right]\psi^*_p, \quad \zeta^*_p = \frac{1}{2}\left[1 - \frac{\alpha_r p_r + \alpha_m m}{(p^2 + m^2)^{1/2}}\right]\psi^*_p,$$

and take ξ^*_p to represent the annihilation of an electron of momentum $\mathbf{p}$, $\overline{\xi}^*_p$ the creation of an electron of momentum $\mathbf{p}$, ζ^*_p the creation of a positron of momentum $-\mathbf{p}$ and $\overline{\zeta}^*_p$ the annihilation of a positron of momentum $-\mathbf{p}$. The variables $\psi^*_p, \overline{\psi}^*_p, \xi^*_p, \overline{\xi}^*_p, \zeta^*_p, \overline{\zeta}^*_p$ will all satisfy the same anticommutation relations as the corresponding unstarred variables.

We can express the Hamiltonian entirely in terms of physical variables. We have

$$\psi^{*r} = e^{ieV/\hbar}\left(\psi^r + \frac{ie}{\hbar}V^r\psi\right).$$

Thus

$$H_P + H_Q = \int \{\overline{\psi}^\dagger \alpha_r[-i\hbar\psi^r - e(A^r - V^r)\psi] + \overline{\psi}^\dagger \alpha_m m\psi + A^0 j_0\} \, d^3x$$

$$= \int [\overline{\psi}^{*\dagger}\alpha_r(-i\hbar\psi^{*r} - eA^r\psi^*) + \overline{\psi}^{*\dagger}\alpha_m m\psi^* + A^0 j_0] \, d^3x.$$

The last term in the integrand here should be combined with H_{FL}. From (49) and (57)

$$H_{FL} \approx -(8\pi)^{-1}\int (U - A_0)(U + A_0)^{rr} \, d^3x$$

$$\approx \frac{1}{2}\int (U - A_0) j_0 \, d^3x$$

80. THE PHYSICAL VARIABLES

with the help of (99). Thus

$$H_{FL} + \int A^0 j_0 \, d^3x \approx \frac{1}{2} \int (U + A_0) j_0 \, d^3x.$$

Integrating (99) with the help of formula (72) of § 38, we get

$$A_{0\mathbf{x}} + U_\mathbf{x} \approx \int \frac{j_{0\mathbf{x}'}}{|\mathbf{x} - \mathbf{x}'|} d^3x',$$

and hence

$$H_{FL} + \int A^0 j_0 \, d^3x \approx \frac{1}{2} \iint \frac{j_{0\mathbf{x}} j_{0\mathbf{x}'}}{|\mathbf{x} - \mathbf{x}'|} d^3x \, d^3x'.$$

Thus we get

$$H \approx H^*$$

with

$$H^* = \int [\overline{\psi}^{*\dagger} \alpha_r (-i\hbar \psi^{*r} - e\mathcal{A}^r \psi^*) + \overline{\psi}^{*\dagger} \alpha_m m \psi^*] \, d^3x$$

$$+ H_{FT} + \frac{1}{2} \iint \frac{j_{0\mathbf{x}} j_{0\mathbf{x}'}}{|\mathbf{x} - \mathbf{x}'|} d^3x \, d^3x'. \qquad (114)$$

We may use H^* instead of H as our Hamiltonian. It leads to the same Schrödinger equation for a physical ket, since if $|P\rangle$ is physical

$$H^*|P\rangle = H|P\rangle.$$

Also it leads to the same Heisenberg equations of motion for physical variables, since if ξ is a physical variable

$$[\xi, H^*] \approx [\xi, H].$$

Thus H^* and H are equivalent Hamiltonians for the physical quantities, and the others do not matter.

H^* involves only physical variables. The longitudinal field variables do not appear in it. Instead of them we have the last term of (114), which is just the Coulomb interaction energy of any charges that are present. The appearance of such a term in a relativistic theory is rather strange, as it is an energy associated with the instantaneous propagation of forces. It appears as a result of our having transformed the theory a long way from the Heisenberg form in which the relativistic invariance of the theory is manifest.

We could set up a representation by taking as standard ket the product of the standard ket $|0_F\rangle$ for the electromagnetic field alone, given by (61) and (62), with the standard ket $|0_P\rangle$ for the electrons and positrons alone, given by (74). This representation would not be a convenient one, however, because its standard ket does not satisfy the second of the supplementary conditions (107).

We get a more convenient representation if we take another standard ket $|Q\rangle$ satisfying

$$(B_0 + A_r{}^r)|Q\rangle = 0, \quad (\text{div } \mathcal{E} - 4\pi j_0)|Q\rangle = 0, \qquad (115)$$

$$\overline{\mathcal{A}}_{r\mathbf{k}}|Q\rangle = 0, \quad \xi^*_{a\mathbf{p}}|Q\rangle = 0, \quad \overline{\zeta}^*_{a\mathbf{p}}|Q\rangle = 0. \qquad (116)$$

These conditions are consistent, because the operators on $|Q\rangle$ in them all commute or anticommute with each other, and there are enough of them to fix $|Q\rangle$ completely, apart from a numerical factor, because there are as many of them as of the conditions for $|0_F\rangle|0_P\rangle$. The conditions (115) show that $|Q\rangle$ satisfies the supplementary conditions and so represents a physical state. The conditions (116) show that $|Q\rangle$ represents a state for which there are no photons, electrons, or positrons present.

Any ket $|P\rangle$ that satisfies the supplementary conditions (107) and so represents a physical state can be expressed as some physical variable multiplied into $|Q\rangle$. The only independent physical variables that give non-vanishing results when applied to $|Q\rangle$ are $\mathcal{A}_{r\mathbf{k}}, \overline{\xi}^*_{a\mathbf{p}}, \zeta^*_{a\mathbf{p}}$. Hence

$$|P\rangle = \Psi(\mathcal{A}_{r\mathbf{k}}, \overline{\xi}^*_{a\mathbf{p}}, \zeta^*_{a\mathbf{p}})|Q\rangle. \qquad (117)$$

Thus $|P\rangle$ is represented by a wave functional Ψ involving the variables $\mathcal{A}_{r\mathbf{k}}, \overline{\xi}^*_{a\mathbf{p}}, \zeta^*_{a\mathbf{p}}$. It is a power series in these variables, the various terms in it corresponding to the existence of various numbers of photons, electrons, and positrons, with the Coulomb fields around the electrons and positrons.

In using the representation (117) together with the Hamiltonian H^*, we have a form of the theory in which we can ignore the conditions (115), as they have no effect on the kets (117). We must retain the conditions (116). The longitudinal variables then no longer appear in the theory.

81. Interpretation

The foregoing work establishes the basic equations of quantum electrodynamics. There are two forms of the theory, involving the Hamiltonians H and H^* respectively. We must now consider the interpretation and application of the theory. We shall take the H^* form for definiteness. The argument would be essentially the same with the H form.

The ket $|Q\rangle$ represents a state for which there are no photons, electrons, or positrons present. One would be inclined to suppose this state to be the perfect vacuum, but it cannot be, because it is not stationary. For it to be stationary we should need to have

$$H^*|Q\rangle = C|Q\rangle$$

with C a number. Now H^* contains the terms

$$-e\int \overline{\psi}^{*\dagger}\alpha_r \mathcal{A}^r \psi^* d^3x + \frac{1}{2}\iint \frac{j_{0x}j_{0x'}}{|\mathbf{x}-\mathbf{x}'|}d^3x d^3x', \qquad (118)$$

which do not give numerical factors when applied to $|Q\rangle$ and which therefore spoil the stationary character of $|Q\rangle$.

Let us call the state Q represented by $|Q\rangle$ the no-particle state at a certain time. If we start with the no-particle state it does not remain the no-particle state. Particles get created where none previously existed, their energy coming from the interaction part of the Hamiltonian.

To study this spontaneous creation of particles, we take the ket $|Q\rangle$ as initial ket in the Schrödinger picture and treat the terms (118) as a perturbation giving rise to a probability of the state Q jumping into another state, in accordance with the theory of § 44. The first of them, resolved into its Fourier components, contains a part

$$-e(\alpha_r)_{ab} \iint \mathcal{A}^r{}_\mathbf{k} \overline{\xi}^*_{a\mathbf{p}} \zeta^*_{b\mathbf{p}+\mathbf{k}\hbar} d^3k d^3p, \qquad (119)$$

which causes transitions in which a photon is emitted and simultaneously an electron-positron pair is created. After a short time the transition probability is proportional to the squared length of the ket formed by multiplying (119) into the initial ket $|Q\rangle$, which is

$$e^2(\overline{\alpha}_r)_{ab}(\alpha_s)_{cd}$$
$$\iiiint \langle Q|\overline{\zeta}^*_{a\mathbf{p}+\mathbf{k}\hbar}\xi^*_{b\mathbf{p}}\overline{\mathcal{A}^r}_\mathbf{k}\mathcal{A}^s{}_{\mathbf{k}'}\overline{\xi}^*_{c\mathbf{p}'}\zeta^*_{d\mathbf{p}'+\mathbf{k}'\hbar}|Q\rangle d^3k d^3p d^3k' d^3p'$$
$$= e^2(\overline{\alpha}_r)_{ab}(\alpha_s)_{cd}\iiiint \langle Q|i\hbar[\overline{\mathcal{A}^r}_\mathbf{k},\mathcal{A}^s{}_{\mathbf{k}'}][\xi^*_{b\mathbf{p}},\overline{\xi}^*_{c\mathbf{p}'}]_+$$
$$[\overline{\zeta}^*_{a\mathbf{p}+\mathbf{k}\hbar},\zeta^*_{d\mathbf{p}'+\mathbf{k}'\hbar}]_+|Q\rangle d^3k d^3p d^3k' d^3p'.$$

Using the values of the P.B. and anticommutators given by (4), (16), (72), (73), we get an integrand which depends on the $\mathbf{k}$, $\mathbf{k}'$ variables according to the law $|\mathbf{k}|^{-1}\delta(\mathbf{k}-\mathbf{k}')$ for large values of $\mathbf{k}$ and $\mathbf{k}'$. This gives an integral that diverges, so the transition probability is infinite.

The second term of (118), resolved into its Fourier components, contains terms like $\overline{\xi}^*_\mathbf{p}\overline{\xi}^*_{\mathbf{p}'}\zeta^*_{\mathbf{p}''}\zeta^*_{\mathbf{p}+\mathbf{p}'-\mathbf{p}''}$, which cause transitions in which two electron-positron pairs are created simultaneously. One can calculate the transition probability as before, and one finds again that it is infinite. From these calculations one can conclude that the state Q is not even approximately stationary.

A theory which gives rise to infinite transition probabilities of course cannot be correct. We can infer that there is something wrong with quantum electrodynamics. This result need not surprise us, because quantum electrodynamics does not provide a complete description of nature. We know from experiment that there exist other kinds of particles, which can get created when large amounts of energy are available. All that we can expect from a theory of quantum electrodynamics is that it shall be valid for processes in which there is not enough energy available for these other particles to be created to an appreciable extent, say for energies up to a few hundred MeV. Thus the high-energy part of the interaction energy (118) is quite unreliable, and it is this high-energy part that is responsible for the infinities.

It appears that we must modify the high-energy part of the interaction. At present there does not exist any detailed theory of the other particles and so it is not possible to say how it ought to be modified. The best we can do is to cut it out from the theory altogether, and so remove the infinities. The precise form of the cut-off and the energy where it is applied will be left unspecified. Of course, the cut-off spoils the relativistic invariance of the theory. This is a blemish which cannot be avoided in our present state of ignorance of high-energy processes.

Even with a cut-off the no-particle state Q is not approximately stationary. It therefore differs very much from the vacuum state. The vacuum state must contain many particles, which may be pictured as in a state of transient existence with violent fluctuations.

Let us introduce the ket $|V\rangle$ to represent the vacuum state. It is the eigenket of H^* belonging to the lowest eigenvalue. Here and subsequently H^* denotes the expression (114) modified by the cut-off. One might try to calculate $|V\rangle$ as a perturbation of the ket $|Q\rangle$, but such a method would be of doubtful validity, because the difference between $|V\rangle$ and $|Q\rangle$ is not small. No satisfactory way of calculating $|V\rangle$ is known. In any case the result would depend strongly on the cut-off, and since the cut-off is unspecified the result would not be a definite one.

It follows that we must develop the theory without knowing $|V\rangle$. This is not a great hardship, because we are not mainly interested in the vacuum state. We are mainly interested in states which differ from the vacuum through having a few particles present in addition to those associated with the vacuum fluctuations, and we want to know how these extra paricles behave. For this purpose we focus our attention on an operator K representing the creation of the extra particles, so that the state we are interested in appears as $K|V\rangle$.

We do not know how the ket $|V\rangle$ varies with the time in the Schrödinger picture, since we do not know the lowest eigenvalue of H^*. To avoid this difficulty we work in the Heisenberg picture in which $|V\rangle$ is constant. We then require $K|V\rangle$ to represent another state in the Heisenberg picture and thus to be another constant ket. This leads to

$$\frac{dK}{dt} = 0. \tag{120}$$

Usually K will involve the time explicitly as well as Heisenberg dynamical variables, so (120) gives

$$i\hbar\frac{\partial K}{\partial t} + KH^* - H^*K = 0. \tag{121}$$

We now have each physical state determined by a solution K of (120) or (121). We obtained this result without knowing the vacuum ket $|V\rangle$, and we can proceed to study K without knowing $|V\rangle$. The only further information about K that we would have if we did know $|V\rangle$ would be that two K's, say K_1 and K_2, would correspond to the same state if we had $(K_1 - K_2)|V\rangle = 0$. But we can get on without this further information and count all different K's satisfying (121) as corresponding to different states.

We are thus led to a drastic alteration of one of the basic ideas of quantum mechanics, namely *to represent a state by a linear operator and not a ket vector*. This alteration is brought about by the complexities of applying quantum mechanics to a field and by our ignorance of high-energy processes.

A trivial solution of (120) or (121) is $K = 1$. This evidently corresponds to the vacuum state.

A general solution may be put in the form of an explicit function of t and of the dynamical variables at time t. Let us use the symbol η_t to denote collectively the emission operators at time t. Thus η_t equals one of the variables $A_{r\mathbf{k}}, \bar{\xi}^*_{a\mathbf{p}}, \zeta^*_{a\mathbf{p}}$ at the time t in the Heisenberg picture. The absorption operators are then $\bar{\eta}_t$. A solution of (121) then appears as

$$K = f(t, \eta_t, \bar{\eta}_t). \tag{122}$$

We require some physical interpretation for the state represented by this K, as the usual physical interpretation of quantum mechanics, requiring a state to be represented by a ket, is no longer applicable. We shall need to make some new assumptions.

Keeping to the Heisenberg picture, we introduce at each time t the ket $|Q_t\rangle$ satisfying the conditions (116) with respect to the Heisenberg dynamical variables at time t. These conditions may now be written

$$\bar{\eta}_t |Q_t\rangle = 0.$$

The ket $|Q_t\rangle$ corresponds to no particles existing at the time t and it provides a reference ket for the discussion of general states at time t.

For any state fixed by a solution K of (121) we form $K|Q_t\rangle$ and assume that *this ket determines what can be observed at the time t and is to be interpreted according to the standard rules*. We obtain K in the form (122) and then arrange it so that in each term all the absorption operators $\bar{\eta}_t$ are to the right of all the emission operators η_t. It is then said to be in the normal order. Any term in K containing an absorption operator then contributes nothing to $K|Q_t\rangle$. The surviving terms in $K|Q_t\rangle$ will contain only emission operators, like (117). Each surviving term is associated with certain particles in particular states, and the square of the modulus of its coefficient (with the appropriate factors $n!$ when there is more than one boson in the same state) is assumed to be, after normalization, the probability of these particles existing in these particular states at the time t.

We now have a general method of physical interpretation which is rather similar to the usual one, but there are important differences. A term in K with an absorption operator on the right will not contribute to $K|Q_t\rangle$ and so will not contribute anything observable at time t. We may call it a latent term at the time t. Such a term cannot be discarded as non-existent, because it will contribute observable effects at other times. These latent terms are a new feature of the theory and are to be understood as an incompleteness in the description of a state in terms merely of the particles which can be observed to be present at a certain time.

As a consequence of the occurrence of latent terms, if $K|Q_t\rangle$ is normalized at one time, it will usually not be normalized at other times. We thus have to carry out a separate normalization for each time in order to derive the probabilities.

82. Applications

There are two important applications of the foregoing theory in which effects are calculated that cannot be obtained from a more primitive theory. These applications are concerned with a single electron in a static

82. APPLICATIONS

electric or magnetic field. As a consequence of the interaction of the electron with electromagnetic waves, the energy levels are shifted somewhat from their values given by the elementary theory. The important cases are:

(i) An electron in the Coulomb field of a proton. The theory here leads to a shift in the energy levels of the hydrogen atom. It is named the *Lamb shift*, after its discoverer.

(ii) An electron in a uniform magnetic field. The extra energy is here interpreted as arising from an extra magnetic moment of the electron, called the anomalous magnetic moment.

To take a static field into account one merely has to introduce potentials to describe it and add them on to the potentials in the Hamiltonian. The potentials of the static field are functions of x_1, x_2, x_3 only, and are numbers for each x_1, x_2, x_3, not dynamical variables, so their introduction does not increase the number of degrees of freedom.

The calculations of the Lamb shift and anomalous magnetic moment are rather complicated. They are given in detail, working from the Hamiltonian H, in the author's book *Lectures on Quantum Field Theory* (Academic Press, 1966). The results are in good agreement with experiment and provide a confirmation of the theory.

These calculations were made in terms of the Heisenberg picture throughout. One may tackle quantum electrodynamics on the Schrödinger picture, looking for a solution of the Schrödinger equation by taking the no-particle ket, or a ket corresponding to just a few particles present, as the initial ket of a perturbation procedure and applying the standard perturbation technique. One finds that the later terms are large and depend strongly on the cut-off, or are infinite if there is no cut-off. The perturbation procedure is not logically valid under these conditions.

Nevertheless people have developed this method a long way and have devised working rules for discarding infinities (in a theory without cut-off) in a systematic manner, so that finite residual effects remain. The procedure is described in many books, e.g. Heitler's *Quantum Theory of Radiation* (Clarendon Press, 1954). The original calculations of the Lamb shift and anomalous magnetic moment were carried out on these lines, long before the corresponding calculations in the Heisenberg picture. The results are the same by both methods.

I do not see how these calculations based on the Schrödinger picture, supplemented by some working rules, can be presented as a logical development of the standard principles of quantum mechanics. The

Schrödinger picture is unsuited for dealing with quantum electrodynamics, because the vacuum fluctuations play such a dominant role in it. These fluctuations present great mathematical difficulties, and also they are not of physical importance. They get bypassed when one uses the Heisenberg picture, and one is then able to concentrate on quantities that are of physical importance.

Quantum mechanics may be defined as the application of equations of motion to atomic particles. It was first shown that atomic particles are subject to equations of motion when Bohr set up his theory of the hydrogen atom. The big development was made when Heisenberg discovered the need for non-commutative multiplication. The domain of applicability of the theory is mainly the treatment of electrons and other charged particles interacting with the electromagnetic field—a domain which includes most of low-energy physics and chemistry.

Now there are other kinds of interactions, which are revealed in high-energy physics and are important for the description of atomic nuclei. These interactions are not at present sufficiently well understood to be incorporated into a system of equations of motion. Theories of them have been set up and much developed and useful results obtained from them. But in the absence of equations of motion these theories cannot be presented as a logical development of the principles set up in this book. We are effectively in the *pre-Bohr era* with regard to these other interactions.

It is to be hoped that with increasing knowledge a way will eventually be found for adapting the high-energy theories into a scheme based on equations of motion, and so unifying them with those of low-energy physics.

Index

action, 131
adjoint, 26
angular momentum, 143
anticommutator, 310
anticommute, 154
antilinear, 20
antisymmetrical
   ket, 219
   state, 219
antisymmetrizing operator, 260

bar notation, 19
basic
   bras, 53
   kets, 57
belonging to an eigenvalue, 30
Bohr's frequency condition, 120, 183
Bose statistics, 220
boson, 220
boundary condition, 160
bra, 18

canonical coordinates and momenta, 86
casuality, 4
central field, 156
character (of a group), 225
class of permutations, 222
closed state, 161
combination law, 2, 120
combination of angular momenta, 151
commutation relation, 85
commute, 24
compatible observations, 52

complete set of
   bras, 53
   commuting observables, 57
   states, 36
conjugate
   complex, 19
   complex linear operator, 27
   imaginary, 19
conservation laws, 118, 147
constant of the motion, 118
contact transformation, 107
contravariant, 266
Coulomb interaction energy, 323
covariant, 266
cut-off, 326

de Broglie waves, 123
degenerate sysetem, 177
dependent, 15, 16
diagonal
   element, 68
   in a representation, 74
   matrix, 68, 70, 71
   with respect to an observable, 78
displacement operator, 103
dual vector, 17
δ_{rs}, 62
δ function, 58
Δ function, 296

e, 161
eigen, 30
eigenfunction, 120
Einstein's photo-electric law, 7

element of a matrix, 68
even permutation, 218
exclusion principle, 221
exclusive set of states, 225

Fermi statistics, 220
fermion, 220
Fock's representation, 142, 238
functional, 308

Gibbs ensemble, 134
Green's theorem, 199
group velocity, 124

h, $\hbar$, 88
half-width of absorption line, 212
Hamiltonian, 116, 117
Hamilton-Jacobi equation, 125
Heisenberg
   dynamical variable, 116
   picture, 115
   representation, 120
Hermitian matrix, 68, 70
Hilbert space, 40
holes, 264

identical permutation, 221
improper function, 58
independent, 15, 16
intermediate state, 182

ket, 15
Kramers-Heisenberg dispersion
   formula, 260

Lagrangian, 131
Landé's formula, 191
length of a
   bra, 20
   ket, 20
linear operator, 23
longitudinal
   energy, 303, 307
   field, 293

magnetic
   anomaly of the spin, 171
   moment of electron, 170, 278
magnitude of angular momentum, 150

matrix, 68, 69
Maxwell's equations, 306, 317
momentum representation, 98
multiplet, 189, 233

non-degenerate system, 177
no-partical state, 326
normal order, 328
normal state, 143
normalization, 21

observable, 37, 304
   having a value, 46
   having an average value, 46
odd permutation, 218
orbital
   angular momentum, 146, 153
   variable, 230
orthogonal
   bras, 20
   kets, 20
   representation, 54
   states, 20, 36
orthogonality theorem, 32
oscillator, 139, 237

P.B., 86
Pauli's exclusion principle, 221
permutation, 218, 221
phase
   factor, 21
   space, 134
physical variable, 304
Planck's constant, 88
Poisson bracket, 86
positive square root, 45
positron, 288
probability
   amplitude, 74
   coefficient, 186
   current, 273
   density, 270
   of observable having a value, 48
proper-energy, 185

quantum condition, 85

radial momentum, 157

real linear operator, 27
reciprocal
    of an observable, 44
    permutation, 222
reciprocity theorem, 77
relative probability amplitude, 74
representation, 53
representative, 53, 67
rotation operator, 145

scatterer, 193
Schrödinger
    dynamical variable, 116
    picture, 114
Schrödinger's
    representation, 95
    wave equation, 114
second quantization, 240, 263
selection rule, 164
self-adjoint, 27
similar permutations, 222
simultaneous eigenstate, 49
Sommerfeld's formula, 285
spherical
    harmonic, 159
    symmetry, 147
spin
    angular momentum, 146, 279
    of electron, 153, 279
square root of an observable, 45
standard ket, 80
state, 11
    of absorption, 195
    of motion, 11
    of polarization, 4
stationary state, 119
stimulated emission, 183, 250
strong equation, 305
superposition of states, 11
supplementary condition, 304
symmetrical
    ket, 218
    representation, 218
    state, 219
symmetrizing operator, 235

time-dependent wave function, 114

transformation function, 76
translational state, 7
transverse
    energy, 303, 308
    field, 293

uncertainty principle, 100
unit matrix, 68, 70
unitary, 106

wave
    equation, 114
    function, 81
    mechanics, 13
    packet, 99, 124
weak equation, 305
weight function, 67
well-ordered function, 133